Amsco's Preparing for Qualifying Examinations in **Mathematics**

Richard J. Andres, Ph.D.
Mathematics Teacher
Jericho High School
Jericho, New York

Joyce Bernstein, Ed.D.
Director of Mathematics
Bethpage Union Free School District
Bethpage, New York

AMSCO SCHOOL PUBLICATIONS, INC.
315 Hudson Street, New York, N.Y. 10013

Dedication

In remembrance of Edward P. Keenan, our Amsco mentor.

Special thanks to Peg and Marc for their encouragement, patience, and understanding.

Reviewers

Rosalie David
  Assistant Principal—Mathematics
  A. Philip Randolf Campus High School, NY, NY

Rhonda Eisenberg
  Assistant Principal—Mathematics
  South Shore High School, Brooklyn, NY

Joel Friedberg
  Assistant Principal—Mathematics
  Lehman High School, Bronx, NY

Dr. Stevan R. Peters
  Assistant Principal—Mathematics
  Ralph McKee Vocational and Technical High School, Staten Island, NY

Ruby M. Sylvester
  Assistant Principal—Mathematics
  East New York High School of Transit Technology, Brooklyn, NY

Bruce Waldner
  District Mathematics Chairperson
  Malverne UFSD, Malverne, NY

Susan P. Willner
  Assistant Principal—Mathematics
  Samuel Gompers Vocational & Technical High School, Bronx, NY

Brief portions of this book were adapted from the following Amsco publications:

Amsco's Preparing for the ACT Mathematics and Science Reasoning
Amsco's Preparing for the SAT I Mathematics
Florida: Preparing for FCAT Mathematics Grade 10
Michigan: Preparing for the MEAP/HST in Mathematics
Preparing for the Georgia High School Graduation Test: Mathematics

Please visit our Web site at:
  www.amscopub.com

When ordering this book, please specify either **R 712 W** or
PREPARING FOR QUALIFYING EXAMINATIONS IN MATHEMATICS

ISBN 1-56765-539-4

Copyright © 2001 by Amsco School Publications, Inc.
No part of this book may be reproduced in any form without written permission from the publisher.

Printed in the United States of America

3 4 5 6 7 8 9 10   04 03 02

Contents

Getting Started — 1
About This Book — 1
Test-Taking Strategies — 1
   General Strategies — 1
   Specific Strategies — 2
Problem Solving — 3
   Steps in Problem Solving — 3
   Strategies for Problem Solving — 3
Problem-Solving Review — 13

Chapter 1: Numeration — 14
Sets of Real Numbers — 14
Factors and Multiples — 17
The Order of Real Numbers — 19
Converting Between Percents, Fractions, and Decimals — 22
Exponents and Scientific Notation — 25
Chapter Review — 28

Chapter 2: Operations With Real Numbers — 29
Fundamental Operations — 29
   Properties of Operations — 29
   Closure — 32
   Order of Operations — 33
Operations With Signed Numbers — 35
   Absolute Value — 35
   Adding and Subtracting Signed Numbers — 36
   Multiplying and Dividing Signed Numbers — 38

Operations With Fractions	40
Adding and Subtracting Fractions	40
Multiplying and Dividing Fractions	43
Operations With Decimals	47
Operations With Exponents	49
Radical Notation and Operations	52
Simplifying Square Roots	52
Operations With Square Roots	54
Ratios, Rates, and Proportions	56
Ratios and Rates	56
Using Proportions	59
Applications of Percent	62
Computing With Percents	62
Percent of Increase and Decrease	64
Chapter Review	66

Chapter 3: Algebraic Operations and Reasoning   68

The Language of Algebra	68
Evaluating Algebraic Expressions	72
Adding and Subtracting Algebraic Expressions	74
Multiplying Algebraic Expressions	76
Special Products of Binomials	79
Dividing Algebraic Expressions by a Monomial	81
Operations with Rational Expressions	83
Literal Equations and Formulas	85
Solving Linear Equations	88
Solving Equations of the Form $x + a = b$	89
Solving Equations of the Form $ax = b$	90
Solving Equations of the Form $ax + b = c$	92
Solving Equations With Like Terms	94
Solving Equations With Parentheses or Rational Expressions	96
Solving Literal Equations	100
Linear Inequalities in One Variable	102
Solving Systems of Linear Equations	106
Chapter Review	111

Chapter 4: Factoring and Quadratic Equations   112

Factoring	112
Common Monomial Factors	112
Factoring the Difference of Two Squares	114
Factoring Trinomials	116

Factoring Completely	119
A Special Case: Factoring Out 1	121
Solving Factorable Quadratic Equations	123
Standard Form	123
Quadratic Trinomial Equations	124
Incomplete Quadratic Equations	130
Fractional Quadratic Equations	134
Solving Quadratic-Linear Pairs Algebraically	137
Verbal Problems Involving Quadratics	140
Number Problems	140
Real-World Applications	142
Chapter Review	145

Chapter 5: Functions and the Coordinate Plane — 146

Relations and Functions	147
Lines on the Coordinate Plane	150
Graphing a Linear Equation	150
Horizontal and Vertical Lines	152
Slope of a Line	154
Direct Variation	157
Slope-Intercept and Point-Slope Forms	161
Graphing a Line From Partial Information	166
Quadratic Equations on the Coordinate Plane	168
Graphing a Parabola	168
Solving Quadratic Equations Graphically	172
Graphing Systems of Equations and Inequalities	175
Systems of Linear Equations	175
Linear Inequalities	178
Systems of Linear Inequalities	181
Quadratic-Linear Pairs	184
Graphic Solutions of Real Situations	188
Chapter Review	193

Chapter 6: Foundations of Geometry — 196

Basic Elements	196
Points, Lines, and Planes	196
Angles	198
Line and Angle Relationships	201
Plane Figures	205
Circles	206
Polygons: General Properties	208

Triangles	213
Quadrilaterals	219
Solid Figures	224
Congruence and Similarity	228
Congruent Figures	228
Similar Figures	232
Chapter Review	238

Chapter 7: Mathematical Reasoning   240

Statements, Truth Values, and Negations	240
Open and Closed Sentences	240
Negations	242
Compound Statements	243
Conjunctions	243
Disjunctions	244
Conditionals	246
Converses, Inverses, and Contrapositives	249
Biconditionals	251
Chapter Review	254

Chapter 8: Measurement and Coordinate Geometry   255

Measurement and Dimensional Analysis	255
Systems of Measurement and Conversions	255
Perimeter and Circumference	260
Area	264
Surface Area and Volume	269
Measures of Similar Figures	275
Error in Measurement	279
Triangles and Measurement	282
Proportions in a Right Triangle	282
The Pythagorean Theorem	285
Special Right Triangles	287
Trigonometric Ratios	290
Indirect Measure	293
Coordinate Geometry	298
Distance Formula	298
Midpoint Formula	302
Finding Areas in Coordinate Geometry	305
Coordinate Geometry Proofs	308
Chapter Review	313

Chapter 9: Transformation, Locus, and Construction — 315

- Duplicating an Image — 315
  - Translations — 315
  - Constructing Duplicate or Congruent Figures — 318
- Reflections and Symmetry — 320
  - Line Reflection and Symmetry — 320
  - Point Reflection and Symmetry — 325
  - Constructing a Perpendicular Line — 327
- Dilations — 329
  - Dilations in the Coordinate Plane — 329
  - Constructing Similar Figures — 331
- Rotations — 334
- Properties Under Transformations — 337
- Locus — 340
  - 1. Distance From a Point: The Circle — 340
  - 2. Equidistant From Two Points: The Perpendicular Bisector — 344
  - 3. Distance From a Line: Two Parallel Lines — 347
  - 4. Equidistant From Two Parallel Lines: One Parallel Line — 348
  - 5. Equidistant From Intersecting Lines: The Angle Bisector — 351
- *Chapter Review* — 353

Chapter 10: Statistics — 355

- Sampling and Collecting Data — 355
- Organizing Data — 356
  - Stem-and-Leaf Plots — 356
  - Tally and Frequency Tables — 358
  - Frequency Histograms — 359
  - Bar Graphs — 360
  - Broken-Line Graphs — 362
  - Scatter Plots — 363
  - Circle Graphs — 364
- Analyzing Data — 367
  - Measures of Central Tendency: Mean, Median, and Mode — 367
  - Range, Quartiles, and Box-and-Whisker Plots — 370
  - Percentiles and Cumulative Frequency Histograms — 373
- *Chapter Review* — 377

Chapter 11: Probability — 380

Principles of Probability — 380
    Theoretical and Empirical Probability — 380
    Rules of Probability — 380
    Random Variables and Probability Distributions — 382
Calculating Sample Spaces — 385
    Tree Diagrams — 386
    The Counting Principle — 386
    Sets of Ordered Pairs — 388
    Permutations — 389
    Combinations — 391
Probability Problems Involving Permutations and Combinations — 394
Chapter Review — 397

Appendix: Using Your Calculator — 398

by Kim E. Genzer and David Vintinner

Assessment: Cumulative Reviews and Practice Examinations — 406

Ten Cumulative Reviews — 406
Practice Examinations — 431

Index — 461

Getting Started

About This Book

Preparing for Qualifying Examinations in Mathematics is intended to help you review the knowledge and skills needed by high school students as described in NCTM's *Principles and Standards for School Mathematics*. This book contains Model Problems, Practice, Chapter Reviews, and Cumulative Reviews to help you study for your qualifying examination. These practice problems incorporate both content and problem-solving situations that are similar to those indicated in state guidelines and frameworks. The book also contains Practice Examinations you can use before taking the real thing.

Test-Taking Strategies

General Strategies

- <u>Become familiar with the directions and format of the test ahead of time.</u> The exams in this book have four parts with a total of 35 questions. You must answer *all* the questions. There are 20 multiple-choice questions and 15 questions where you must justify your answer. For the 5 questions that appear in Parts II, III, and IV, you must show the steps you used to solve the problem, including formulas, diagrams, graphs, charts, and so on, where appropriate.
- <u>Pace yourself.</u> Do not race to answer every question immediately. On the other hand do not linger over any problem too long. The 20 multiple-choice questions are each worth 2 points. The remaining 15 questions are worth 2 points each in Part II, 3 points each in Part III, and 4 points each in Part IV. Keep in mind that you will need more time to complete the questions in Parts II, III, and IV.
- <u>Speed comes from practice.</u> The more you practice, the faster you will become and the more comfortable you will be with the material. Practice as often as you can.

- Keep track of your place on the answer sheet if there is one. This way you can mark each successive answer easily. If you find yourself bogged down, skip the problem and move on to the next. Return to the problem later if you have time. Make a note in the margin of the test booklet so that you can locate the skipped problem easily. Be careful when you skip a problem. Be sure to leave blank the answer line that corresponds to the question you skipped.
- Be cautious about making wild guesses. If you find that a question is difficult, skip the problem and return to it later.

Specific Strategies

- Always scan the answer choices before beginning to work on a multiple-choice question. This will help you to focus on the kind of answer that is required. Are you looking for fractions, decimals, percents, integers, squares, cubes, and so on? Eliminate choices that clearly do not answer the question asked.
- Do not assume that your answer is correct just because it appears among the choices. The wrong choices are usually there because they represent common student errors. After you find an answer, always reread the problem to make sure you have chosen the answer to the question that is asked, not the question you have in your mind.
- Sub-in. To sub-in means to substitute. You can sub-in friendly numbers for the variables to find a pattern and determine the solution to the problem.
- Backfill. If a problem is simple enough and you want to avoid doing the more complex algebra, or if a problem presents a phrase such as $x = ?$, then just fill in the answer choices that are given in the problem until you find the one that works.
- Fill in what is known to find what is unknown. In geometry problems where the figure is drawn to scale, do the work on the given figure. The picture itself provides visual information useful in solving the problem. Feel free to draw in lines and angles, and extend lines if needed. When you mark given measurements on the diagram, you clarify the problem. Be aware that sometimes the picture is not drawn to scale. The obvious answer choice or obvious answer based on the visual setup may be incorrect, and you should probably redraw the figure accurately. This new drawing should help you to see the real relationships.
- Do the math. This is the ultimate strategy. Don't go wild searching in your mind for tricks, gimmicks, or math magic to solve every problem. Most of the time the best way to get the right answer is to do the math and solve the problem.

Problem Solving

Steps in Problem Solving

A **problem** is defined as a situation that can be resolved if certain questions are answered. **Problem solving** is the process used to answer these questions. Problem solving often takes place in four steps: (1) Read and think about the problem. (2) Explore it. (3) Solve it. (4) Look back at it.

Step 1: Read and Think About the Problem

- To begin, look for the question or questions that must be answered to solve the problem.
- Identify the given facts.
- Be sure that you understand the vocabulary involved in the problem.
- Then check the given information. Is enough information given to solve the problem? Is more information given than you actually need?
- Think about the form of the answer. For example, will it be a measurement? If so, what will the unit of measure be? Make an estimate of the answer. This will help you recognize the answer when you see it.

Step 2: Explore the Problem

- Use tables, charts, and diagrams to organize the given information.
- Identify any questions and answers that will lead you to the solution.
- Decide on a strategy for reaching the solution. (Several strategies are described in this Introduction.)
- Choose the methods you need for your strategy.

Step 3: Solve the Problem

- Choose the operations you need for your strategy.
- Make an estimate, if appropriate.
- If a precise calculation is needed, try to choose the best method of finding the answer: mental computation, paper and pencil, or a calculator.

Step 4: Look Back at the Problem

- Is your answer reasonable? Does it answer the question asked? If not, work backward to find your mistake.
- Is your answer in the appropriate form? If the answer is a measurement, does the unit of measure make sense?
- Is your answer close to your estimate?
- If there is a mathematical check you can use, do the check.

Strategies for Problem Solving

In solving many problems, it is important to have a **strategy**—a plan or method for attacking the problem and finding the answer. You can also think of a strategy as a useful tool. Practice in using a variety of problem-solving tools will help you decide which strategy or strategies will be appropriate for solving a particular problem. Practice will also help you discover which strategies are easiest for you to understand and apply.

Drawing a Diagram   A diagram is a visual display, and its purpose is to organize information. This can be very helpful in problem solving.

MODEL PROBLEMS

1. Eight baseball teams are competing for a league championship. A team is eliminated when it loses a game. How many games must be played to determine the winner?

SOLUTION

Draw a diagram.

The standard diagram for an elimination tournament looks like this:

Answer: You find the answer by counting the number of blank lines, which represent games to be played. The diagram shows that 7 games are needed to determine the winner.

2. There are 32 students in a science class. Twelve of these students belong to the Math Club, 20 students belong to the Chemistry Club, and 9 students belong to both. How many students in the science class belong to neither the Math Club nor the Chemistry Club?

SOLUTION

Draw a diagram. A problem like this can be solved using a Venn diagram. A **Venn diagram** usually consists of circles within a rectangle. The rectangle represents the "universe"—the set of all the things being considered. The circles represent subsets of the universe. Intersections of the circles represent elements that are members of more than one subset. A Venn diagram can be a useful tool for counting the number of members of a set when sets overlap, that is, when members are in two or more sets at the same time.

In this case, use overlapping circles and begin with the number of students that belong to both clubs. Then use this number to determine how many belong to only the Math Club and how many belong to only the Chemistry Club.

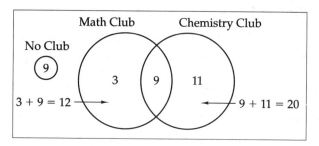

Add the numbers in the diagram (3 + 9 + 11 = 23). Thus there are 23 students in both clubs. The difference (32 − 23 = 9) means that there are 9 students who do not belong to either club.

3. The distance between two cities, M and N, is 300 miles. The distance between cities N and O is 500 miles. Which could be the distance between cities O and M?
   (1) 100
   (2) 500
   (3) 850
   (4) 1,000

SOLUTION

Draw a picture.

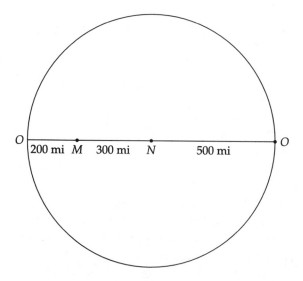

Point O can lie anywhere on the circle with N as its center. The distance from N to O, or the radius of the circle, is 500 miles, so the distance from M to O could be at most 800 miles. The smallest possible distance from M to O is 200 miles. The only answer choice between 200 miles and 800 miles is 500 miles.

Answer: (2) 500 miles

Draw a diagram to solve each problem. Explain your reasoning.

1. Lupe started a pep club. On Monday, she got 2 friends to join. On Tuesday, each of these friends got 2 more people to join. On Wednesday, each person who joined on Tuesday got 2 more people to join. If this pattern continues, how many people (including Lupe) are in the club by the end of Saturday?

2. The numbers 1, 4, 9, and 16 are called squares. Find the tenth square and its pattern.

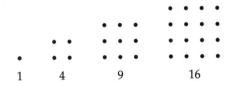

3. A string ensemble has 15 members. If 4 members play the cello and 9 members play the violin, but 4 members play neither of those instruments, how many members play both the violin and the cello?

4. Bus routes connect these cities: Ashton and Denby, Denby and Beacon, Elkton and Crestwood, Beacon and Crestwood, and Denby and Elkton. How many ways are there to travel from Elkton to Beacon? What are they?

5. In a middle school with 600 students, 100 are in the band, 70 are in the orchestra, and 12 are in both the band and the orchestra. How many students are not in either the band or the orchestra?

Using Tables and Lists Like a diagram, a table or a list is a visual display. In a table, data are arranged in rows and columns.

MODEL PROBLEMS

1. The problem above about the 8-team tournament can also be solved by using a table. If 8 teams are competing for a championship, and a team is eliminated when it loses a game, how many games are needed to determine the winner?

SOLUTION

Set up a table to show the following information: 8 teams play 4 games, resulting in 4 winners and 4 losers. The 4 winners then play 2 games, resulting in 2 winners and 2 losers. These 2 winners play one game, which determines the final winner.

Teams	Games	Winners	Losers
8	4	4	4
4	2	2	2
2	1	1	1

The answer is found by adding up the numbers in the column headed "Games." The table shows that 7 games are needed.

2. Ann, Bryan, Carla, Don, and Ernesto have applied for three job openings at the library. What are all the possible groups of three that could be hired?

SOLUTION

Make a list of groups of 3 students using their initials. Note that in this situation the group ABC is the same as ACB, BCA, CAB, and so on.

ABC	ACD	BCD	CDE
ABD	ACE	BCE	
ABE	ADE	BDE	

Answer: There are 10 possible groups of 3 students.

Problem Solving **5**

Use a table or a list to solve each problem. Explain your reasoning.

1. How many triangles are in the figure?

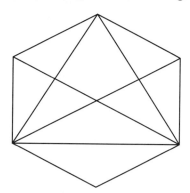

2. How many different ways can you make change for a quarter using pennies, nickels, and dimes?

3. Mr. Jenkins has 50 feet of fencing to enclose a rectangular pen. What will the length and the width have to be to give the pen the largest possible area if only whole feet are used?

4. Hot dogs come in packages of 8. Hot-dog rolls come in packages of 6. What is the least number of packages of hot dogs and hot-dog rolls needed to have the same number of hot dogs and rolls?

Working Backward In **working backward**, we start with a given result and then go back, step by step, to the beginning.

Often, working backward involves thinking what operation (or operations) we would have used to get the given result, and then using the *opposite* operation (or operations) to find the original number.

MODEL PROBLEMS

1. If we multiply a number by 6, subtract 357, and then divide by 9, the result is 237. Find the original number.

SOLUTION

To solve a problem like this, we can work backward. We start with the result (237) and reverse our steps to get to the original number.

$237 \times 9 = 2{,}133$	Multiply, the opposite of dividing.
$2{,}133 + 357 = 2{,}490$	Add, the opposite of subtracting.
$2{,}490 \div 6 = 415$	Divide, the opposite of multiplying.

Answer: 415

ALTERNATIVE SOLUTION

This problem can also be solved algebraically. Let the original number be n. Then solve the equation:

$$\frac{6n - 357}{9} = 237$$

2. Helen, Ira, Jan, and Ken shared a box of pens. Helen took half. Ira took half of what was left. Jan took a third of what was left. Ken took the remaining 6 pens. How many pens were in the box?

SOLUTION

Work backward.

Ken took 6 pens. This was two-thirds of n, if n is what was left after Ira took his pens. $6 = \frac{2}{3}n$, so $n = 9$.

If there were 9 pens left after Ira took half, there were 18 before he picked. Similarly, if there were 18 pens after Helen took half, there were 36 pens in the box to start.

6 Getting Started

3 Mr. James spent $32,500 for a sports car. This amount was 30 percent more than the sticker price. The manufacturer sets its sticker price at 150 percent of the production cost. How much did it cost to produce Mr. James's car?

SOLUTION

- Start at the end. Mr. James buys the car for $32,500, which is 30 percent more than the sticker price.
- To find the sticker price, go back one step. Let S be the sticker price. Think: "30 percent more than S" means 100 percent of S plus 30 percent of S, or 130 percent. Change this to its decimal equivalent, 1.30. Then:

$1.30S = \$32{,}500$

$S = \$32{,}500 \div 1.30 = \$25{,}000$

- To find the production cost, go back another step. Let P be the production cost. Sticker price S is 150 percent of P. Change 150 percent to its decimal equivalent, 1.50. Then:

$1.50P = \$25{,}000$

$P = \$25{,}000 \div 1.50 = \$16{,}666.66\ldots$, or approximately $16,667

Answer: It cost $16,667 to produce the car.

Practice

1 Marcia's school starts at 8:30 A.M. Marcia must make three stops before she gets to school. It takes 15 minutes to get to her first stop from her house, 30 minutes to get to the second stop from the first stop, 30 minutes to get to the third stop from the second stop, and 30 minutes to get to school from the third stop. She spends 2 minutes at each stop. What is the latest time Maria can leave home to get to school on time? Solve by working backward. Explain your answer.

2 The price tag below shows that the price of a CD player is reduced for the second time. The current price, $39.20, represents an additional 20 percent discount after a 30 percent discount. What was the original price?

3 Three years ago, Mrs. Canada bought a new car. The first year the car lost 20 percent of its value. For each of the next 2 years it lost 10 percent of its remaining value. Its current value is $25,920. How much did Mrs. Canada pay for the car when it was new?

4 A football team gained 7 yards, lost 5 yards, gained 6 yards, then lost 3 yards. The team is now at its 30-yard line. From what line did the team start?

Problem Solving

Guessing and Checking An effective procedure for solving a problem is to **guess and check**. When you use this strategy, you guess an answer and then check to see if it is correct. If it is not correct, you guess again, choosing a larger number or a smaller number. Then you check your second guess. Repeat this process until you find the right answer. Try to make each guess *better* than the preceding guess. In other words, each guess should bring you closer to the right answer.

MODEL PROBLEMS

1 If $5x + 6 = 61$, find the value of x.

SOLUTION

Use guessing and checking. Guess a value of x. Substitute that value in the given equation and then evaluate the equation. If your guess gives a result greater than 61, try a lower value for x. If your guess gives a result smaller than 61, try a higher value for x. Keep guessing, substituting, and evaluating until you get the correct value.

Guesses may vary. Here is one example.

Guess	Check: $5x + 6 = 61$	Result
8	$(5 \cdot 8) + 6 = 40 + 6 = 46$	46 is too low. Try a greater value for x.
10	$(5 \cdot 10) + 6 = 50 + 6 = 56$	56 is still too low. Try a greater value for x.
11	$(5 \cdot 11) + 6 = 55 + 6 = 61$ ✓	Therefore the value of x is 11.

Answer: $x = 11$

2 There are 31 students in a math class. There are 5 more girls than boys. How many boys are there?

SOLUTION

Use guessing and checking. Guess the number of girls and the number of boys. (Remember that the number of boys must be 5 less than the number of girls.)

Add these numbers. If the sum is 31, your guesses were correct. If the sum is less than 31, guess again, using greater numbers, and then add again. If the sum is more than 31, use smaller numbers on your next guess. Keep guessing and checking until you get the right answer.

Guesses may vary. Here is one example.

Guess	Check	Result
15 girls, 10 boys	$15 + 10 = 25$	25 is too low. Try greater numbers of boys and girls.
20 girls, 15 boys	$20 + 15 = 35$	35 is too high. Try numbers less than 20 and 15 but more than 15 and 10 (the first guess).
18 girls, 13 boys	$18 + 13 = 31$ ✓	Right.

Answer: There are 18 girls and 13 boys.

8 Getting Started

Solve by guessing and checking.

1. What two consecutive numbers add up to 39?

2. An Internet provider is hiring 1,500 new workers. Of the new hires, there will be 250 more computer programmers than salespeople. How many computer programmers will be hired?

3. Chandler Used Car Company sells more recreational vehicles than regular automobiles. Last week Chandler sold 60 units. If the number of recreational vehicles sold is multiplied by the number of regular automobiles sold, the product is 779. How many recreational vehicles were sold?

4. Arrange the numbers $-4, -5, 3, 2, 6,$ and 7 so that the sum of the numbers on each side of the triangle is 4.

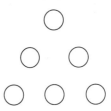

5. Arrange the numbers from 1 to 9 in the **magic square** so that the sum of each row, each column, and each diagonal is 15.

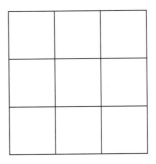

Finding a Pattern   To solve a problem, we sometimes need to **find a pattern** in a given sequence of numbers. Each number in the sequence is called a **term**. Each term has a position in the sequence: first, second, third, and so on. Most patterns can be found by comparing a few terms to determine their relationship to one another. This gives us a rule for the relationship between consecutive terms so that the successive terms in the sequence can be found.

One way to find a pattern is to ask the following four questions, in order:

- Is the second term greater or less than the first term?
- What is the difference between the first term and the second?
- Is the difference the same between each term and the next term? If not, what is the pattern of the differences?
- What operation or operations can be performed on the first term to get the second term? This operation or combination of operations is the rule for the pattern.

MODEL PROBLEMS

1 What is the next number in the sequence 3, 8, 13, 18, 23, . . . ? Give a rule for finding other terms.

SOLUTION

To find a pattern, answer the four questions.

- The second term is greater than the first term.
- The difference between the first term and the second term is $8 - 3 = 5$.
- The difference, 5, is the same between consecutive terms:

    $13 - 8 = 5 \quad 18 - 13 = 5 \quad 23 - 18 = 5$

- To get from the first term to the second term, the operation is addition: add 5.

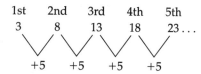

Answer: To get the next term after 23, add 5: $23 + 5 = 28$. The next term is 28.

The rule is: 5 added to each term gives the next term.

2 What is the next number in the sequence 1, 4, 9, 16, 25, 36, . . . ? Give a rule for finding other terms.

SOLUTION

Find a pattern.

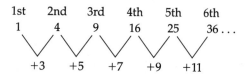

As shown above, the numbers added to each term are odd and increase by 2.

Therefore, the next term can be found by adding $11 + 2$, or 13, to 36.

Answer: To get the next term, add: $36 + 13 = 49$. The next term is 49.

The rule is: To find successive terms of the sequence, add positive odd numbers that increase by 2 each time.

ALTERNATIVE SOLUTION

The rule for this type of pattern can also be shown using multiplication.

1st	2nd	3rd	4th	5th	6th
1	4	9	16	25	36 . . .
1×1	2×2	3×3	4×4	5×5	6×6
1^2	2^2	3^2	4^2	5^2	6^2

Answer: The sequence is the perfect squares, in increasing order. Therefore the next term is the next perfect square: $7 \times 7 = 7^2 = 49$.

The rule is: Find the sequence of perfect squares in increasing order.

3 A store clerk is making a pyramid display of soup cans. The bottom row consists of 10 cans. Then, going upward, the next row has 9 cans, the next 8, and so on. To complete the pyramid, how many cases of soup should the clerk bring from the warehouse if each case contains 12 cans?

SOLUTION

Find a pattern.

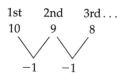

The rule for finding the next term is: subtract 1. There will be 10 rows. Apply the rule to find the numbers in the pattern.

$10 + 9 + 8 + 7 + 6 + 5 + 4 + 3 + 2 + 1 = 55$

To calculate the number of cases needed, divide: $55 \div 12 = 4\frac{7}{12}$. Assume that the clerk can bring only whole cases from the warehouse.

Answer: The clerk must bring 5 cases.

The rule is: To find the next term, subtract 1.

10 Getting Started

1. For each sequence state the rule and give the next term.

   a. 9, 13, 17, 21, …

   b. 1, 5, 8, 1, 5, …

   c. 12, 17, 22, 27, …

   d. 26, 24, 22, 20, …

   e. 2, 5, 11, 23, …

   f. 3, 6, 3, 6, …

Solve each problem by finding a pattern. Explain your reasoning.

2. John is job hunting. He is 25 years old, and his goal is to be earning $40,000 a year by the time he is 35. He sees a job advertised that pays $24,000 per year now, with a raise of $1,000 each year. If John gets this job, will he reach his goal?

3. Paula received $1 for her first birthday, $2 for her second birthday, $4 for her third birthday, and $8 for her fourth birthday. At this rate, how much money will she have received in all after she celebrates her ninth birthday?

4. Mary bought a used car 2 years ago for $30,000. This model **depreciates** (loses value) at the following rate: $1,000 the first year; $2,000 the second year; $3,000 the third year, and so on. How much will the car be worth 5 years from now?

5. When a ball is dropped from a height of 64 inches, it bounces back half the original height. If it continues to bounce in this manner, what is the total distance that it will have traveled when it hits the ground for the fifth time?

Solving a Simpler Problem We can often solve a difficult problem more easily if we first solve a simpler problem. One way to do this is to use easier numbers. Using simpler numbers helps us to focus on the operations needed to solve the problem. Another way is to break a long problem into smaller parts.

MODEL PROBLEMS

1. Find the sum of the series
   2 + 4 + 6 + 8 + … + 996 + 998 + 1,000

   SOLUTION

   Solve a simpler problem, the sum of the series 2 + 4 + 6 + 8 + 10 + 12:

   2 + 4 + 6 + 8 + 10 + 12

   with pairs summing to 14, 14, 14

   The sum of each *pair* is 14. There are 6 members of the series, so there are 3 pairs.

   Therefore, the sum is 3 × 14 = 42.

Use this technique to solve the given problem:

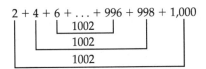

Now the sum of each pair is 1,002. There are 500 even numbers from 2 to 1,000, so there are 250 pairs. To find the sum, multiply:

250 × 1,002 = 250,500

Problem Solving **11**

2 Ballpoint pens are on sale at 3 for $1.29. Mr. Miller needs 100 pens for his department. How much will the pens cost?

SOLUTION

Break this problem into two simpler parts. First, find the cost of 1 pen, called the unit price. To do this, divide:

$$\$1.29 \div 3 = \$0.43$$

Next, multiply the unit price by 100:

$$\$0.43 \times 100 = \$43$$

Answer: The pens will cost $43.

3 A cookbook recommends roasting a chicken $\frac{1}{2}$ hour per pound at 375°. How much time is needed to roast a $4\frac{1}{2}$-pound chicken?

SOLUTION

Solve a simpler problem. Suppose that the recommendation was 1 hour per pound. Then the solution is easy: $1 \cdot 4\frac{1}{2} = 4\frac{1}{2}$ hours. The operation used is multiplication. Since the given recommendation is $\frac{1}{2}$ hour, you need to multiply $4\frac{1}{2}$ by $\frac{1}{2}$.

$$\frac{1}{2} \cdot 4\frac{1}{2} = \frac{1}{2} \cdot \frac{9}{2} = \frac{9}{4} = 2\frac{1}{4}$$

Answer: $2\frac{1}{4}$ hours

ALTERNATIVE SOLUTION

There is another way to solve this problem. Suppose that the chicken weighs only 4 pounds. Then $\frac{1}{2} \cdot 4 = 2$ (half of 4 is 2). So to solve the real problem, use the same operation: multiply by $\frac{1}{2}$. The calculations are the same as in the first solution.

Sometimes unnecessary information is given. In this case the temperature, 375°, is not needed to solve the problem.

 Practice

Answer each question by solving a simpler problem. Explain your reasoning.

1 Find the sum of this series:

22, 24 , . . . , 68, 70

2 A "going out of business" sale offers folding chairs for $120 a dozen. Mrs. Smith needs to buy chairs for her catering service. How much will she pay for 18 chairs?

3 There are 1,500 students in the senior class at DeWitt Clinton High School. For a class trip, 25 percent of them will go to Washington, D.C. by bus. How many buses will be needed to get them there if each bus holds 47 passengers?

4 A 6-pack of 10-ounce bottles of spring water costs $4.80. A $1\frac{1}{2}$-gallon bottle of the same brand costs $5.75. Which is the better buy? (1 gallon = 128 ounces)

5 A recipe calls for $\frac{3}{4}$ cup of bread stuffing for each pound of turkey. If $9\frac{3}{8}$ cups of stuffing are needed, how much does the turkey weigh?

PROBLEM-SOLVING REVIEW

Solve each problem. Explain the strategy you use.

1 A number is a two-digit multiple of 11. The product of the digits is a perfect square and a perfect cube. Find the number.

2 Ten chairs are placed along the four walls of a room so that each of the walls has the same number of chairs. Show how this can be done.

3 How many numbers are in this set?

2, 5, 8, 11, 14, . . . , 683

4 a Find the missing sums.

1	1
1 + 3	4
1 + 3 + 5	?
1 + 3 + 5 + 7	?
1 + 3 + 5 + 7 + 9	?

b Predict how many odd whole numbers you would have to add to obtain a sum of 169. Explain your reasoning.

5 A dog weighs 12 pounds plus half its weight. How much does the dog weigh?

6 Find the ones digit in 9^{28}.

7 In a dart game, three darts are thrown. All the darts hit the target. What scores are possible?

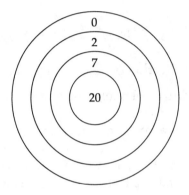

8 A recycling committee was paid $0.14 per pound for $87\frac{1}{2}$ pounds of newspaper and $0.32 per pound for $43\frac{3}{4}$ pounds of aluminum cans. How much money did the committee receive?

9 Maya wants to pin four photographs onto her office bulletin board. What is the least number of pins she will need if a pin goes through every corner of every photograph?

10 The area of a rectangular garden is 40.5 square feet. The garden is twice as long as it is wide. What are the dimensions of the garden?

Chapter 1:

Numeration

Sets of Real Numbers

A **set** is a collection of distinct elements that can be listed within braces { }.

- A set with no elements is known as the **empty set** or the **null set** and can be written as ∅ or { }.
- A set with a specific number of elements such as {1, 2, 3, 4, 5} is a **finite set**. If the set is very large, three dots ... can be used to show "and so on in the same pattern." So {1, 2, 3, ..., 98, 99, 100} is the set of counting numbers from 1 to 100.
- A set that has no last element and continues indefinitely, like {1, 2, 3, 4, ...}, is an **infinite set**.

Counting numbers, or **natural numbers**, are {1, 2, 3, ...} and can be represented on a number line with an arrow to show that the set continues without end.

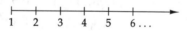

Whole numbers are the counting numbers and 0, {0, 1, 2, 3, ...}. The whole numbers can also be shown on a number line.

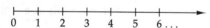

Integers consist of the whole numbers and their negatives, or opposites. {..., −3, −2, −1, 0, 1, 2, 3, ...}. Zero is its own opposite.

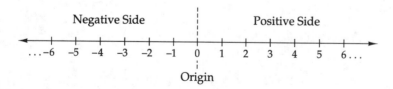

The letters *a* and *b* here are variables. A variable is a letter or symbol that takes the place of a specific unknown quantity.

Rational numbers consist of all numbers that can be expressed as quotients of two integers, $\frac{a}{b}$, where *b*, the denominator, does not equal zero. Every rational number can be represented as a terminating decimal or a repeating decimal.

Terminating decimals, such as 0.25 or −1.00, are rational numbers that have a finite number of nonzero digits to the right of the decimal point. The number 0.25 can also be written 0.25000000000..., but the extra zeros do not tell us anything more about the number. All integers are terminating decimals. For example, −1.00 equals −1.

Repeating decimals, such as −0.333... or 5.7321321321..., are rational numbers. These decimals have an infinite number of nonzero digits to the right of the decimal point in a repeating pattern. The number −0.333... has an infinite number of 3s, and in the number 5.7321321321... the pattern 321 repeats forever. Repeating decimals can be written with a horizontal bar over the digits that define the pattern. The two numbers above can be written as $-0.\overline{3}$ and $5.7\overline{321}$. Note that in the latter number, the bar does not extend over the 7, because the 7 does not repeat.

Irrational numbers are decimal numbers that neither repeat nor terminate. They cannot be expressed as fractions with integers for the numerator and denominator. For example, π ($\pi = 3.1415926...$) and some radicals (such as $\sqrt{2} = 1.41421...$) are irrational numbers.

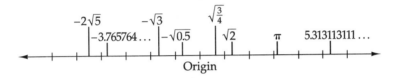

The set of **real numbers** consists of the elements in the set of rational numbers combined with the elements in the set of irrational numbers. The number line is a graphic representation of the set of real numbers. Every real number corresponds to a point on this number line.

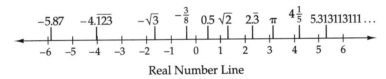
Real Number Line

The relationships between the various sets of numbers are shown in the diagram below. The set of natural numbers falls inside the set of whole numbers, the set of whole numbers is inside the set of integers, and so on. The shaded area has no numbers in it, because every real number is either rational or irrational.

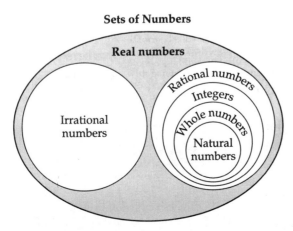

Sets of Real Numbers   15

MODEL PROBLEM

List the numbers in the set {89, 3, 458, 0.89, 0, −3} that are natural numbers.

SOLUTION

The natural numbers are {1, 2, 3,...}. This set would include 89, 3, and 458 from the original set. Zero and −3 are integers, not natural numbers, and 0.89 is a rational number.

Answer: {89, 3, 458}

Practice

1. The set of two-digit natural numbers is:
   (1) finite
   (2) infinite
   (3) empty
   (4) unclear

2. Name the number assigned to point G in this line.

   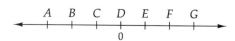

   (1) −3
   (2) 0
   (3) 2
   (4) 3

3. What is the best sequence of names to identify the elements of this set?

   $\{-8, 7, 0.202202220..., 0.\overline{4}\}$

   (1) integer, natural, real, irrational
   (2) integer, whole, irrational, real
   (3) integer, irrational, integer, rational
   (4) integer, real, whole, rational

4. What is the best sequence of names to identify this set of numbers?

   $\left\{1\frac{3}{5}, -10, 0.5\overline{1}, \pi, 17\right\}$

   (1) rational, integer, irrational, irrational, natural
   (2) rational, irrational, rational, irrational, whole
   (3) rational, real, rational, real, natural
   (4) real, integer, irrational, irrational, natural

Exercises 5–14: Identify each number as either a rational or irrational number.

5. 0.17

6. 0.17171717...

7. 0.172222...

8. 0.17117111711117...

9. $\sqrt{52}$

10. π

11. 0

12. $\frac{21}{3}$

13. −5.35

14. $7.\overline{17}$

15. List the elements in the set of the nonnegative integers.

16. Is 0 an irrational number? Explain your answer.

17. How can every whole number be written in the form $\frac{a}{b}$, $b \neq 0$?

18. True or false? Any fraction made up of two natural numbers is a rational number. Explain your answer.

16 Chapter 1: Numeration

Factors and Multiples

The **factors** of a positive number are all the natural numbers that can be divided evenly into that number. For example, the factors of 12 are 1, 2, 3, 4, 6, and 12. Each of these factors divides 12 with no remainder. Since 12 is divisible by 4, 12 is a **multiple** of 4.

A **prime number** is a number greater than 1 that has exactly two different natural numbers as factors: the number 1 and itself. For example, 13 is a prime number. The factors of 13 are 1 and 13, since only $1 \times 13 = 13$. The set of prime numbers from 1 to 25 is {2, 3, 5, 7, 11, 13, 17, 19, 23}.

A **composite number** is a natural number greater than 1 that has more than two factors. The number 12 is a composite number. Its set of six factors is {1, 2, 3, 4, 6, 12}.

A composite number written as the product of its prime factors is called the **prime factorization** of the number. For example, the prime factorization of 60 can be found by using a **factor tree**.

The number 2 is the first prime and is the only even prime number.

The number 1 is neither prime nor composite.

```
        60
       /  \
      2    30
          /  \
         3    10
             /  \
            2    5
```

$60 = 2 \times 2 \times 3 \times 5$

The **greatest common factor (GCF)** of two or more numbers is the largest number that is a factor of each of the given numbers. For example, 4 is the greatest common factor of 8 and 12 because 4 is the largest number that will divide evenly into both 8 and 12.

To Find the GCF

- Write the prime factorization of each number.
- Select the common factors, the factors that appear in both prime factorizations. If a factor appears twice in each prime factorization, select it twice.
- Find the product of the common factors.

 MODEL PROBLEM

Find the GCF of 24 and 60.

SOLUTION

Prime factorization of $24 = 4 \bullet 6 = 2 \bullet 2 \bullet 2 \bullet 3$

Prime factorization of $60 = 4 \bullet 15 = 2 \bullet 2 \bullet 3 \bullet 5$

The common factors are 2, 2, and 3, so the GCF is $2 \bullet 2 \bullet 3 = 12$.

The **least common multiple (LCM)** of two or more numbers is the smallest number that is a multiple of each of the given numbers. For example, 24 is the least common multiple of 8 and 12 because $8 \times 3 = 24$ and $12 \times 2 = 24$.

To Find the LCM

- List the multiples of each given number until a common multiple appears.
- Find the lowest multiple that appears in all the lists.

MODEL PROBLEM

Find the LCM of 9, 18, and 24.

SOLUTION

List the multiples of each number:

Multiples of 9: 9, 18, 27, 36, 45, 54, 63, <u>72</u>, 81, 90, 99, ...

Multiples of 18: 18, 36, 54, <u>72</u>, 90, ...

Multiples of 24: 24, 48, <u>72</u>, 96, 120, ...

The LCM of 9, 18, and 24 is 72.

Practice

1. Which of the following is the prime factorization of 144?
   (1) 1×144
   (2) $9 \times 4 \times 2 \times 2$
   (3) $3 \times 4 \times 3 \times 4$
   (4) $2 \times 2 \times 2 \times 2 \times 3 \times 3$

2. Which statement is true for any natural number x?
   (1) If x is divisible by 70, then it is divisible by 12.
   (2) If x is divisible by 70, then it is divisible by 14.
   (3) If x is divisible by 70, then it is divisible by 16.
   (4) If x is divisible by 70, then it is divisible by 18.

3. $3 \times 3 \times 5 \times y$ could be the prime factorization for which number?
   (1) 11
   (2) 30
   (3) 135
   (4) 335

4. If a movie house gives *one* free ticket to every 60th customer and *two* free tickets to every 160th customer, what customer will be the first to get *three* free tickets?
   (1) 220th
   (2) 360th
   (3) 480th
   (4) 960th

5. What is the least common multiple of natural numbers m and n, if m and n are both prime numbers?
   (1) m
   (2) n
   (3) $m + n$
   (4) mn

6. If an integer N is found on the number line between 40 and 100, inclusive, then how many possible integer values of N are divisible by 4?
   (1) 25
   (2) 21
   (3) 16
   (4) 10

Exercises 7–11: Factor if possible. If the number cannot be factored, state that it is prime.

7 2

8 9

9 11

10 60

11 108

12 Find the greatest common factor of 20 and 63.

13 Find the greatest common factor of 60 and 144.

14 Find the least common multiple of 6, 8, and 9.

15 If set R contains only factors of 32 and set M contains only factors of 24, then what is one number that is contained in set R but *not* in set M?

The Order of Real Numbers

Whenever we compare two real numbers x and y, only *one* of the following can apply:

- $x < y$ or "x is less than y" if x is to the left of y on a number line.

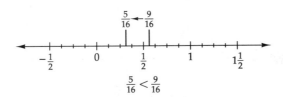

- $x > y$ or "x is greater than y" if x is to the right of y.

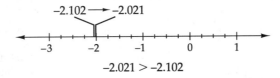

- $x = y$ or "x is equal to y" if x is the same point as y.

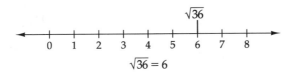

You can remember the meaning of $<$ with this image:

∠ess than

A number sentence with two *greater than* or two *less than* symbols can be used to show where on a number line some quantity may fall. For example, "the square root of 5 is somewhere between 2 and 3" can be written $2 < \sqrt{5} < 3$.

To Compare Decimal Numbers

- Line them up so that the decimal points are in a column.
- Add zeros to the end of the numbers so they are all the same length.
- Compare the numbers as if they were integers, ignoring the decimal points.

MODEL PROBLEM

Put the numbers 1.05, 0.159, 1.5, 0.5 and 0.51 in order from least to greatest.

SOLUTION

Line up the numbers	Make numbers the same length.	Order numbers as if they were integers.
1.05	1.050	0.159
0.159	0.159	0.500
1.5	1.500	0.510
0.5	0.500	1.050
0.51	0.510	1.500

Answer: The numbers in order from least to greatest are 0.159, 0.5, 0.51, 1.05, 1.5

To Compare Fractions

- Find a common multiple of the denominators.
- Build up fractions so they all have the common multiple for a denominator.
- Compare the numerators.

MODEL PROBLEMS

1 Which is greater, $\frac{3}{5}$ or $\frac{4}{7}$?

SOLUTION

$\frac{3}{5} = \frac{3 \times 7}{5 \times 7} = \frac{21}{35}$ $\frac{4}{7} = \frac{4 \times 5}{7 \times 5} = \frac{20}{35}$

Since $21 > 20$, then $\frac{21}{35} > \frac{20}{35}$ and $\frac{3}{5} > \frac{4}{7}$.

You can also compare fractions by cross-multiplying upward.

$7 \times 3 = 21$ $5 \times 4 = 20$

Since $21 > 20$, then $\frac{3}{5} > \frac{4}{7}$.

2 If $\frac{1}{2} < \frac{x}{10} < \frac{4}{5}$, then what is the set of possible integer values for x?

SOLUTION

Express all fractions with the same common denominator. Thus, $\frac{1}{2} = \frac{5}{10}$ and $\frac{4}{5} = \frac{8}{10}$. Now, rewrite the original problem with these equivalent fractions: $\frac{5}{10} < \frac{x}{10} < \frac{8}{10}$.

Since either $\frac{6}{10}$ or $\frac{7}{10}$ could replace $\frac{x}{10}$, the possible integer values for x are 6 and 7.

Answer: {6, 7}

You can also use a calculator to write each fraction as a decimal. Divide the numerator by the denominator, then compare.

Or use the fraction key.

3 Which is smaller, $\frac{7}{8}$ or $\frac{4}{5}$?

SOLUTION

Use a calculator.

$\frac{7}{8} = 7 \div 8 = 0.875$

$\frac{4}{5} = 4 \div 5 = 0.8$

Since $0.8 < 0.875$, then $\frac{4}{5} < \frac{7}{8}$.

1. Which irrational number lies between 0.122 and 0.123?

   (1) 0.1211311141111...
   (2) 0.12133113311133...
   (3) 0.122333...
   (4) 0.123212343212345...

2. Ann agrees to sell her CD player to the person who makes the best offer. Sally offers $550, Mitsi offers $499.98, Doug offers $549.75, and Jean offers $432. To whom should Ann sell the CD player?

   (1) Mitsi
   (2) Sally
   (3) Jean
   (4) Doug

3. The intervals on this number line are equal and the locations of 0 and 1 are shown.

   Point C is:

   (1) $\frac{3}{4}$
   (2) $-\frac{3}{4}$
   (3) $1\frac{1}{4}$
   (4) $-1\frac{1}{4}$

4. Between what two points on this number line is $-1.4141141114...$ located?

   (1) A and B
   (2) B and C
   (3) C and D
   (4) D and E

5. Which statement is *false*?

   (1) $-3 < 9$
   (2) $-4 < -12$
   (3) $-7 < -3 < 5$
   (4) $9 > 0 > -4$

Exercises 6 and 7: Order each set of numbers from least to greatest.

6. $\frac{78}{13}, 5\frac{5}{6}, \frac{1,430}{1,690}, \frac{250}{60}$

7. 0.12, 0.112, 0.2, 0.21, 0.022

8. What is one possible integer value of k for which $\frac{1}{4} < \frac{k}{10} < \frac{1}{3}$ is true?

9. Is there a greatest negative integer? Explain your answer.

10. Is there a natural number that is closest to zero on the number line? Explain your answer.

11. Is there a positive real number that is closest to zero on the number line? Explain your answer.

12. How many natural numbers are in the set of integers between -3 and 4, inclusive?

13. Draw a standard horizontal number line. Show points for 0 and 1 and any other integers that may help you as guides. Graph these three numbers on your line: {1.69, -0.916, 1, $1.\overline{96}$}.

14. Danisha is thinking of a fraction. She challenges you to guess what it is. She tells you that the numerator and the denominator add up to 100 and the fraction is equivalent to $\frac{1}{3}$. What is the fraction?

Converting Between Percents, Fractions, and Decimals

Percent (%) means *out of 100*. Thus, 5% = 5 out of 100 = 5 hundredths = $\frac{5}{100}$ = 0.05.

Although you can use your calculator, memorizing common equivalents will help you to find solutions to problems faster.

Fraction	$\frac{1}{10}$	$\frac{1}{8}$	$\frac{1}{5}$	$\frac{1}{4}$	$\frac{1}{3}$	$\frac{1}{2}$	$\frac{2}{3}$	$\frac{3}{4}$
Decimal	0.1	0.125	0.2	0.25	$0.\overline{3}$	0.5	$0.\overline{6}$	0.75
Percent	10%	$12\frac{1}{2}\%$	20%	25%	$33\frac{1}{3}\%$	50%	$66\frac{2}{3}\%$	75%

Converting To or From a Percent

Operation	Procedure	Examples
Change a percent to a decimal	Divide the number by 100.	$14\% = \frac{14}{100} = 0.14$ $5.6\% = \frac{5.6}{100} = \frac{56}{1,000} = 0.056$ $145\% = \frac{145}{100} = 1.45$
Change a decimal to a percent	Multiply the number by 100.	$0.315 = 0.315 \times 100 = 31.5\%$ $2.35 = 2.35 \times 100 = 235\%$
Change a percent to a fraction or a mixed number	Write the percent as a fraction in hundredths. Reduce the fraction if possible.	$6\% = \frac{6}{100} = \frac{3}{50}$ $375\% = \frac{375}{100} = 3\frac{75}{100} = 3\frac{3}{4}$
Change a fraction or a mixed number to a percent	Use a calculator: First divide the numerator by the denominator. Then change the decimal to a percent.	Change $4\frac{5}{8}$ to a percent. Enter 5 ÷ 8 Display 0.625 So, 0.625 = 62.5% Thus, $4\frac{5}{8} = 4.625 = 462.5\%$

To change a terminating decimal to a fraction, just read the decimal using place-value vocabulary, write it in fraction form, and reduce if necessary. (The denominator of the fraction before reducing always has the same number of zeros as the number of decimal places in the terminating decimal.) For example: 12.44 = 12 and 44 hundredths = $12\frac{44}{100} = 12\frac{11}{25}$.

Changing a Repeating Decimal to a Fraction

Procedure: Steps	Example: 0.066666... (or $0.0\overline{6}$)	Example: 0.73232... (or $0.7\overline{32}$)
1: How many digits repeat?	1 digit (6) repeats	2 digits (32) repeat
2: Write a power of 10 with as many zeros as repeating digits.	1 digit → 1 zero → 10	2 digits → 2 zeros → 100
3: Multiply the decimal by this power of 10.	$0.0\overline{6} \times 10 = 0.\overline{6}$	$0.7\overline{32} \times 100 = 73.2323...$
4: Subtract the original decimal from this product.	0.66666... −0.06666... 0.6	73.23232... − 0.73232... 72.5
5: Subtract 1 from the power of 10 you used in step 2	10 − 1 = 9	100 − 1 = 99
6: Write a fraction. $\frac{\text{difference from step 4}}{\text{difference from step 5}}$	$\frac{0.6}{9}$	$\frac{72.5}{99}$
7: Simplify the fraction.	$\frac{0.6}{9} = \frac{6}{90} = \frac{1}{15}$	$\frac{72.5}{99} = \frac{725}{990} = \frac{145}{198}$

When you work with irrational numbers, especially when you use a calculator, it is important to **estimate** or find a rational approximation for the number. This will help you to judge whether your answer is reasonable.

To round a decimal number, look at the digit to the right of the place you are rounding to. If that number is 5 or greater, add 1 to the last digit you are keeping. If the number is 4 or less, drop off the extra digits without adding.

Examples: 3.47 to the nearest tenth is 3.5.
1.965 to the nearest hundredth is 1.97.
462 to the nearest ten is 460.

 MODEL PROBLEMS

1 Change $9\frac{1}{2}\%$ to a fraction and a decimal.

SOLUTION

Since $9\frac{1}{2}\% = 9.5\%$, then $9.5\% = \frac{9.5}{100} = \frac{95}{1000} = 0.095$

$\frac{95}{1000}$ can be reduced to $\frac{19}{200}$.

Answer: $\frac{19}{200}$ and 0.095

2 Round $\pi = 3.14159265...$ to the nearest hundredth and the nearest thousandth.

SOLUTION

To the right of the hundredths place is a 1, which is less than 4, so drop off the remaining digits. Thus, π to the nearest hundredth is 3.14. To the right of the thousandths place is a 5, which is 5 or greater, so add 1 to the last digit before dropping the remaining digits. Thus, π to the nearest thousandth is 3.142.

Converting Between Percents, Fractions, and Decimals **23**

1 Which fraction represents 0.3%?
  (1) $\frac{3}{10}$ (3) $\frac{3}{1,000}$
  (2) $\frac{3}{100}$ (4) $\frac{1}{300}$

2 0.4% is the same as 1 out of every
  (1) 100 (3) 250
  (2) 125 (4) 1,000

3 $\frac{5}{4}$ is equal to:
  (1) $1\frac{1}{4}\%$
  (2) 25%
  (3) 80%
  (4) 125%

4 Find a fraction that names the same rational number as 0.3232....
  (1) $\frac{32}{100}$ (3) $\frac{10}{33}$
  (2) $\frac{32}{99}$ (4) $\frac{32}{10}$

5 0.005 is equal to:
  (1) $\frac{1}{2}\%$
  (2) 5%
  (3) 50%
  (4) $\frac{1}{200}\%$

6 Write $\frac{5}{12}$ as a repeating decimal.
  (1) 0.4160
  (2) $0.\overline{416}$
  (3) $0.41\overline{6}$
  (4) $0.\overline{0416}$

Exercises 7–9: Write each number in the form $\frac{a}{b}$ where a and b are integers and $b \neq 0$.

7 0.7777...

8 0.31

9 24%

Exercises 10–12: Write each number in decimal form. Round to the nearest thousandth, if necessary.

10 135.9%

11 $\frac{5}{3}$

12 $\frac{1}{5}\%$

Exercises 13–15: Write each number as a percent.

13 $-4\frac{1}{5}$

14 0.8

15 2.3

16 Find a rational approximation for $-\sqrt{21}$ to the nearest hundredth.

17 Find a rational approximation for $\sqrt{5}$ to the nearest thousandth.

18 Write the decimal form for each number: $\frac{4}{9}, \frac{4}{99}, \frac{4}{90}$

19 Put the following numbers in order from least to greatest:
$\frac{5}{6}, \frac{1}{5}, \frac{13}{11}, 1.2, 1.02, \frac{10}{3}$

20 Of 3.14, $\frac{22}{7}$, and $3\frac{14}{99}$, which is the best approximation of π? Explain your reasoning.

24 Chapter 1: Numeration

Exponents and Scientific Notation

An **exponential expression** is a short form for writing a number where a factor appears many times.

$$\text{base} \rightarrow 5^{\overset{\text{exponent}}{\downarrow}3} = 5 \times 5 \times 5 = 125 \leftarrow \text{power}$$

The **exponent** 3 tells how many times the **base** 5 is used as a factor. The exponent raises the base to a **power**.

In the same way, the exponential expression x^3 means $x \cdot x \cdot x$.

Rule	Examples
Any number raised to the first power is that number.	$5^1 = 5$ $30,000^1 = 30,000$
Any number raised to the zero power is always 1, except that 0^0 is undefined.	$5^0 = 1$ $30,000^0 = 1$
For powers of 10, the number of zeros following the 1 is the same as the exponent of 10.	$10,000,000 = 10^7$ $10^4 = 10,000$
For negative powers of integers, the expression becomes a fraction with a numerator of 1 and a denominator of the positive power.	$5^{-3} = \dfrac{1}{5^3} = \dfrac{1}{5 \times 5 \times 5} = \dfrac{1}{125}$

In working with very large or very small numbers, it is often easier to write the numbers in scientific notation. **Scientific notation** is the product of two factors: (a decimal greater than or equal to 1 but less than 10) × (an integer power of 10) or $a \times 10^n$. An example of scientific notation is 8.3×10^4.

You can use a scientific calculator to raise a number to a given power using the y^x key.

$7^3 \rightarrow 7\,\boxed{y^x}\,3 = 343$

Powers of 10

10^4	10,000
10^3	1,000
10^2	100
10^1	10
10^0	1
10^{-1}	0.1
10^{-2}	0.01
10^{-3}	0.001
10^{-4}	0.0001

To Write a Standard Number Greater Than 10 in Scientific Notation

- Move the decimal point to make the number a decimal between 1 and 10.
- Count how many places you moved the decimal point to the left.
- Write the number of places the decimal moved as the exponent.

MODEL PROBLEM

Write 21,400 in scientific notation.

SOLUTION

21,400. =	2.14 ×	10^4
↑	↑	↑
Move the decimal point 4 places to the **left**.	number between 1 and 10	4th power of 10

To Write a Standard Number Between 0 and 1 in Scientific Notation

- Move the decimal point to make the number a decimal between 1 and 10.
- Count how many places you moved the decimal point to the right.
- Write the negative of the number of places the decimal moved as the exponent.

MODEL PROBLEM

Write 0.0000034 in scientific notation.

SOLUTION

0.0000034 =	3.4 ×	10^{-6}
↑	↑	↑
Move the decimal point 6 places to the **right**.	number between 1 and 10	negative 6th power of 10

Note: In scientific notation, if a number is greater than or equal to 10, the exponent of the power of 10 is positive. If a number is greater than or equal to 1 but less than 10, the exponent of the power of 10 is zero. If a number is between 0 and 1, the exponent of the power of 10 is negative.

To change a number in scientific notation to standard form, multiply.

To Multiply by a Power of 10

- To multiply a decimal by a positive power of 10, move the decimal point one place to the right for each power of 10.

    $5.68 \times 10 = 56.8$

    $5.68 \times 10^2 = 568$

    $5.68 \times 10^3 = 5680$ ← add zero as a placeholder

- To multiply a decimal by a negative power of 10, move the decimal point one place to the left for each power of 10.

    $87.25 \times 10^{-1} = 8.725$

    $87.25 \times 10^{-2} = 0.8725$

    $87.25 \times 10^{-3} = 0.08725$ ← add zero as a placeholder

MODEL PROBLEM

Between what two integers does 3.445×10^2 lie?

(1) 3 and 4
(2) 33 and 34
(3) 34 and 35
(4) 344 and 345

SOLUTION

Convert the number from scientific notation into standard form.

$3.445 \times 10^2 = 344.5$

344.5 lies between 344 and 345.

Answer: (4)

Practice

1. Find the expression with the smallest value.
   (1) 5^0
   (2) 1^5
   (3) 0^{10}
   (4) 10^{-1}

2. Express 5.46×10^0 in standard form.
   (1) 546
   (2) 54.6
   (3) 5.46
   (4) 0.546

3. Express 9.19×10^{-5} in standard form.
   (1) 0.0000919
   (2) 0.00000919
   (3) 0.00919
   (4) 919,000

4. In scientific notation 3,450,000 is equal to
   (1) 34.5×10^5
   (2) 345×10^4
   (3) 3.45×10^6
   (4) 3450×10^3

5. Which of the following represents a number between 0.01 and 0.001?
   (1) 3.7×10^0
   (2) 3.7×10^{-1}
   (3) 3.7×10^{-2}
   (4) 3.7×10^{-3}

6. Between what two integers does 4.23×10^{-5} lie?
   (1) 1 and 2
   (2) 0 and 1
   (3) -4 and -5
   (4) -5 and -6

Exercises 7–10: Show how the measurement would be expressed in scientific notation.

7. The distance from the sun to Pluto is about six billion kilometers.

8. The sun is approximately 93 million miles from Earth.

9. The width of a virus is 0.0000001 meter.

10. A diode in a microchip measures 0.00025 centimeter thick.

11. Express 3.76×10^5 in standard form.

12. Express 198,000,000 in scientific notation.

13. Write 30.232×10^6 in scientific notation.

14. Write the fraction $\frac{2}{125}$ in scientific notation.

15. Write 0.0000567×10^{-3} in scientific notation.

16. If $x^6 = x^8$, name all the possible real number values of x that satisfy this equation. Explain your answer.

Exponents and Scientific Notation 27

CHAPTER REVIEW

1. Which of the following numbers is *not* an integer?
   (1) -15
   (2) $\sqrt{64}$
   (3) $\dfrac{-3}{4}$
   (4) 0

2. Which of the following statements is true?
   (1) $-5.4 > -5\dfrac{1}{3}$
   (2) $-5.4 < -5\dfrac{1}{3}$
   (3) $-5.4 > 5\dfrac{1}{3}$
   (4) $5.4 < 5\dfrac{1}{3}$

3. Which of the following sets of numbers is arranged in order, beginning with the smallest?
   (1) $42.4\%, \dfrac{21}{50}, 0.425, \dfrac{266}{625}$
   (2) $\dfrac{266}{625}, 0.425, \dfrac{21}{50}, 42.4\%$
   (3) $\dfrac{21}{50}, \dfrac{266}{625}, 0.425, 42.4\%$
   (4) $\dfrac{21}{50}, 42.4\%, 0.425, \dfrac{266}{625}$

4. What is the correct sequence of words to identify this set of numbers?
   $$\left\{0.8\overline{3}, 10^4, -\dfrac{18}{3}, \sqrt{215}\right\}$$
   (1) natural, real, integer, irrational
   (2) rational, whole, integer, irrational
   (3) irrational, rational, whole, irrational
   (4) irrational, irrational, integer, irrational

5. The set of negative even integers between -3 and 4 has how many elements?

6. Find the least common multiple of 3, 8, 9, and 12.

7. Express $9.2\overline{4}$ as a mixed number.

8. Express the wavelength of a long x-ray, 1.0×10^{-6} cm, in standard notation.

9. In the set {0.1, 0.01, 0.11, 0.011}, which decimal has the greatest value?

Exercises 10–12: Write each number in scientific notation.

10. 0.006

11. 56,200

12. 0.0008722

Exercises 13–15: Write each number as a percent.

13. 0.06

14. $4\dfrac{3}{5}$

15. 10

16. Find the least possible multiple of 4 that is greater than 25

Exercises 17–20: Indicate whether each statement is true or false. If a statement is false, change the symbol of equality or inequality to make it true.

17. $0.12131213121312\ldots > 0.12123123412345\ldots$

18. $-5.335 > -5\dfrac{3}{8}$

19. $0.5 = 0.\overline{5}$

20. $-5.6 > -5.9$

Chapter 2:
Operations With Real Numbers

Fundamental Operations

The four basic **arithmetic operations** are **addition** (+), **subtraction** (−), **multiplication** (×), and **division** (÷). Multiplication can be indicated in several ways:

$a \times b \qquad a \cdot b \qquad (a)(b) \qquad a(b) \qquad ab$

Division can be indicated as:

$a \div b \qquad \dfrac{a}{b} \qquad b\overline{)a} \qquad a/b$

Addition, subtraction, multiplication, and division are called **binary operations** because they are performed on two numbers at a time. (The prefix *bi-* means two.) When we perform a binary operation, we replace any two members of a set—such as the set of real numbers—with exactly one member, or one answer.

Other operations include **powers** and **roots**. Raising a number to a power is a type of multiplication, and extracting a root is related to division.

Properties of Operations

The following table summarizes important properties of arithmetic operations.

Property	Meaning	Examples
Commutative property of addition or multiplication	The order of the numbers does NOT affect the sum or the product.	$5 + 8 = 8 + 5 = 13$ $\dfrac{2}{3} + \dfrac{1}{4} = \dfrac{1}{4} + \dfrac{2}{3} = \dfrac{11}{12}$ $7 \cdot 4 = 4 \cdot 7 = 28$ $3(11) = 11(3) = 33$
Associative property of addition or multiplication	The way the numbers are paired does NOT affect the sum or the product.	$(1 + 2) + 3 = 1 + (2 + 3) = 6$ $\left(\dfrac{1}{2} \times \dfrac{1}{3}\right)\dfrac{1}{5} = \dfrac{1}{2}\left(\dfrac{1}{3} \times \dfrac{1}{5}\right) = \dfrac{1}{30}$ $(1 \times 3)(4) = 1(3 \times 4) = 12$

(continued)

Property	Meaning	Examples
Distributive property	Multiplication can be distributed over addition or subtraction.	$2 \times (3 + 5) = (2 \times 3) + (2 \times 5) = 16$ $7 \times (4 - 1) = (7 \times 4) - (7 \times 1) = 21$
Additive identity	When zero is added to or subtracted from any number, the number remains unchanged.	$18 + 0 = 18$ $18 - 0 = 18$ $0 + 25 = 25$ $25 - 0 = 25$
Additive inverse	The sum of a number and its additive inverse (also called its **opposite**) is zero.	4 and -4 are additive inverses: $4 + (-4) = 0$ $\frac{3}{4}$ and $-\frac{3}{4}$ are additive inverses: $\frac{3}{4} + \left(-\frac{3}{4}\right) = 0$
Multiplicative identity	Any number multiplied by 1 remains unchanged.	$1 \times 15 = 15$ $-7 \times 1 = -7$
Multiplicative inverse or **reciprocal**	The product of any number and its multiplicative inverse (its reciprocal) is 1.	5 and $\frac{1}{5}$ are multiplicative inverses: $5 \times \frac{1}{5} = 1$
Zero product property	The product of zero and any number is zero.	$-8 \times 0 = 0$ $\pi \times 0 = 0$

Note: A number and its reciprocal are either both positive or both negative. For example:

$$6 \times \frac{1}{6} = 1 \qquad \frac{3}{5} \times \frac{5}{3} = 1 \qquad \left(-\frac{1}{6}\right)(-6) = 1 \qquad \left(-\frac{3}{5}\right)\left(-\frac{5}{3}\right) = 1$$

Zero has *no* reciprocal, because $\frac{n}{0}$ is undefined.

MODEL PROBLEMS

For $-\frac{2}{3}$, name the following:

SOLUTIONS

1 Additive inverse

$\frac{2}{3}$, because $-\frac{2}{3} + \frac{2}{3} = 0$

2 Multiplicative identity

1, because 1 is the multiplicative identity for any number

3 Reciprocal

$-\frac{3}{2}$ or $-1\frac{1}{2}$, because $\left(-\frac{2}{3}\right)\left(-\frac{3}{2}\right) = 1$

4 Additive identity

0, because 0 is the additive identity for any number

5 Multiplicative inverse

$-\frac{3}{2}$, because the multiplicative inverse is the reciprocal, found above

 **Practice**

1. The additive inverse of −8 is:
   (1) $-\frac{1}{8}$
   (2) 0
   (3) $\frac{1}{8}$
   (4) 8

2. The reciprocal of $\frac{1}{5}$ is:
   (1) −5
   (2) $-\frac{1}{5}$
   (3) $\frac{1}{25}$
   (4) 5

3. Which statement illustrates the zero product property?
   (1) $\frac{0}{n} \cdot \frac{n}{0} = 1$
   (2) $n^0 = 0$
   (3) $0n = 0$
   (4) $0 - n = -n$

4. Which statement illustrates the distributive property?
   (1) $a(b + c) = ab + ac$
   (2) $a + b + c = c + b + a$
   (3) If $ab = c$, then $a = \frac{c}{b}$.
   (4) $1 \cdot abc = abc$

5. A binary operation is so called because it:
   (1) yields exactly two answers
   (2) is performed on two members of a set
   (3) is performed on members of two different sets
   (4) is a two-step operation

6. *Opposite* means the same as:
   (1) additive inverse
   (2) additive identity
   (3) reciprocal
   (4) zero product

Exercises 7–17: Identify the property illustrated by each statement.

commutative   associative
distributive   additive identity
additive inverse   multiplicative identity
multiplicative inverse

7. $-6(10) = 10(-6)$

8. $1(64) = 64$

9. $6 + 10 = 10 + 6$

10. $6 + 0 = 6$

11. $6 + (10 + 8) = (6 + 10) + 8$

12. $3(4 \times 5) = (3 \times 4) \times 5$

13. $6 \times 8 = 8 \times 6$

14. $6(3 + 4) = (6 \times 3) + (6 \times 4)$

15. $-84 + 84 = 0$

16. $7(31 - 13) = (7 \times 31) - (7 \times 13)$

17. $\frac{p}{q} \cdot \frac{q}{p} = 1$

18. The number 1 is its own multiplicative inverse. Name another number with this property.

19. Which integer has *no* multiplicative inverse?

20. Which number is its own additive inverse?

Fundamental Operations

Closure

If we perform an operation on any number in a set with itself or with any other member of the set and the result is always still in that set, then the set is **closed** under that operation. Here are examples of **closure**:

- For any integers a and b, the sum $a + b$ is an integer.
- For any integers a and b, the difference $a - b$ is an integer.
- For any integers a and b, the product ab is an integer.

Therefore, the integers are closed under addition, subtraction, and multiplication.

However, a set is not closed under an operation if the result is not *always* in the set. For example:

- For all integers a and b, the quotient $a \div b$ is *not* always an integer. For instance, when $a = 5$ and $b = 2$, the quotient $5 \div 2 = 2\frac{1}{2}$, which is *not* an integer.
- Therefore, the integers are *not* closed under division.

Thus we can say that the integers are closed only for the binary operations of addition, subtraction, and multiplication.

> The properties of operations and closure are sometimes called field properties.

MODEL PROBLEM

The set $A = \{-1, 0, 1\}$ is closed under which operation?

(1) addition

(2) subtraction

(3) multiplication

(4) none of the above

SOLUTION

Test each answer option:

(1) Since $1 + 1 = 2$ and 2 is not in set A, this set is NOT closed under addition.

(2) Since $1 - (-1) = 1 + 1 = 2$, the set is NOT closed under subtraction.

(3) Since the product of any number in this set with itself or any other member is only -1, 0, or 1, the set is closed under multiplication.

(4) Since the set is closed under multiplication, this option is obviously false.

> Braces { } indicate a set.

Answer: (3)

Practice

1. The set {0, 1} is closed under:
   (1) multiplication
   (2) division
   (3) addition
   (4) all of the above

2. Which set is closed under division?
   (1) {natural numbers}
   (2) {whole numbers}
   (3) {rational numbers}
   (4) none of the above

3 The set {rational numbers} is *not* closed under:

(1) multiplication
(2) subtraction
(3) extraction of square root
(4) squaring

4. How many of these sets are closed under the given operation?

{0, 2, 4, 6, …} under addition
{0, 1, 2, …} under subtraction
{0, 1, 2} under multiplication
{0, 1} under extraction of square root

(1) exactly one
(2) exactly two
(3) exactly three
(4) all four

Excercises 5–10: Match each set with the operation or operations under which it is closed. An answer may be used more than once or not at all.

Set	Closed Under …
5 {0}	Addition
6 {2, 4, 16, 256, …}	Squaring
7 {… , −3, −2, −1}	Addition, multiplication, and squaring
8 {… , −21, −14, −7, 0, 7, 14, 21, …}	Addition, subtraction, multiplication, and squaring
9 $\left\{\dfrac{1}{3}, \dfrac{1}{9}, \dfrac{1}{27}, \dfrac{1}{81}, \ldots\right\}$	Multiplication, division, and squaring
10 $\left\{\ldots, \dfrac{1}{8}, \dfrac{1}{4}, \dfrac{1}{2}, 1, 2, 4, 8, \ldots\right\}$	Multiplication and squaring

Order of Operations

To **evaluate** or **simplify** a mathematical expression means to carry out the indicated operations as far as possible. In evaluating, it is necessary to follow these rules for the **order of operations**.

- First, simplify any expression within **grouping symbols**: parentheses ( ), brackets [ ], and braces { }. A fraction bar $\dfrac{m}{n}$ also serves as a grouping symbol. When grouping symbols appear within other grouping symbols, work from the inside out.
- Second, evaluate powers (expressions with exponents) and roots (expressions with radicals).
- Third, multiply and divide in order from left to right.
- Last, add and subtract in order from left to right.

To remember order of operations, use the catchword PEMDAS:
<u>P</u>arentheses
<u>E</u>xponents (and roots),
<u>M</u>ultiply and <u>D</u>ivide
<u>A</u>dd and <u>S</u>ubtract

Braces { } are a grouping symbol and also symbolize a set. Notice the context, and don't confuse the two uses!

Fundamental Operations

MODEL PROBLEMS

Simplify each expression.

1 $10 + 8 \div 2$

2 $4^2 + 10 \cdot 3 \div 5 - 10$

3 $5(2^3 + 4) \div \sqrt{36} - 4$

SOLUTIONS

Division comes first: $10 + 8 \div 2 = 10 + 4$
Addition comes next: $10 + 4 = 14$

Answer: 14

Evaluate exponents: $4^2 + 10 \cdot 3 \div 5 - 10 = 16 + 10 \cdot 3 \div 5 - 10$
Multiply and divide in order from left to right:
$16 + 10 \cdot 3 \div 5 - 10 = 16 + 30 \div 5 - 10 = 16 + 6 - 10$
Add and subtract in order from left to right: $16 + 6 - 10 = 22 - 10 = 12$

Answer: 12

$5(2^3 + 4) \div \sqrt{36} - 4$
$= 5(8 + 4) \div 6 - 4$
$= 5(12) \div 6 - 4$
$= 60 \div 6 - 4$
$= 10 - 4 = 6$

Answer: 6

Practice

1 $(21 - 7) \times (15 - 7) - 4 \times 3 =$
  (1) 50
  (2) 75
  (3) 100
  (4) 125

Exercises 2–4: Which choice demonstrates the correct way to evaluate the given statement?

2 $6 + 4 \cdot 8 - 3$
  (1) $(6 + 4) \cdot (8 - 3)$
  (2) $[6 + (4 \cdot 8)] - 3$
  (3) $[(6 + 4) \cdot 8] - 3$
  (4) $6 + [4 \cdot (8 - 3)]$

3 $2^3 + 16 \div 4 + 4$
  (1) $[2^3 + 16] \div (4 + 4)$
  (2) $[(2^3 + 16) \div 4] + 4$
  (3) $2^3 + [16 \div (4 + 4)]$
  (4) $2^3 + [(16 \div 4) + 4]$

4 $6^2 - 2^2 + 5$
  (1) $(6 - 2)^2 + 5$
  (2) $[(6 \cdot 2) - (2 \cdot 2)] + 5$
  (3) $[(6^2) - (2^2)] + 5$
  (4) $6^2 - [(2^2) + 5]$

Exercises 5–20: Evaluate using the order of operations.

5 $16 - 2 \cdot 3$

6 $9 - 3 \cdot 5$

7 $3 + 4 \cdot 2$

8 $7 \cdot 5 - 4 + 3$

9 $-2 \cdot 3^2 + 20$

10 $7 \cdot 5 - 4 \cdot 3 + 2$

11 $20 - 5(5 - 1)$

12 $5 \cdot 10 \div 5 + 2 \cdot 5$

13 $40 - 10^2$

14 $15 - 10 \div 5 \cdot 2 + 4$

15 $-5[7 - 2(-3)]$

16 $18 \div 2 \cdot 3$

17 $3^2(2 - 3) + 5\sqrt{4}$

18 $2 \cdot 3^2 \div 3\sqrt{4}$

19 $(6 - 4)^2 (6 - 4^2)$

20 $\dfrac{7(3 + 1) - 2}{5 \cdot 2 + 2}$

34 Chapter 2: Operations With Real Numbers

Operations With Signed Numbers

Signed numbers (see Chapter 1) are positive (+) and negative (−) numbers. On a horizontal number line, positive numbers are to the right of zero and negative numbers are to the left of zero. On a vertical number line, positive numbers are above zero and negative numbers below zero.

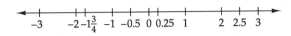

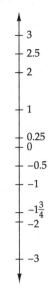

Remember:

- The positive sign + is "understood" and need not be written. Thus *positive three* or +3 can be written simply as 3. When *no* sign is written, a number is always positive.
- The negative sign − must always be written. Thus *negative three* is always written −3.

Absolute Value

In operations with signed numbers, **absolute value** is important. This is the value of a number when we ignore the positive or negative sign. Another way to think of absolute value is that it is a number's distance from zero on the number line, in either direction. The absolute value of any number n is written $|n|$ and is always positive.

Opposites are any two numbers equidistant from zero on the number line in different directions. For any real number n, its opposite is $-n$. An opposite is also called an **additive inverse**.

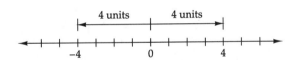

MODEL PROBLEMS

What is the absolute value of each number?   **SOLUTIONS**

1. $|-p|$ if $p > 0$ — p
2. $|35|$ — 35
3. $|xy|$ if $x > 0$ and $y > 0$ — xy
4. $|20 - 25|$ — $|-5| = 5$
5. $|-6|$ — 6

What is the opposite of each number?

6. -6.08 — 6.08
7. $-(p + q)$ — $p + q$
8. $\dfrac{86}{91}$ — $-\dfrac{86}{91}$
9. $3{,}000{,}000$ — $-3{,}000{,}000$

The absolute value of zero is zero:
$$|0| = 0$$

Zero is its own opposite:
$$-0 = 0$$

Practice

1 Which statement is true?
 (1) $7 > |7|$
 (2) $|-5| < 5$
 (3) $|-8| > -2$
 (4) $|-4| < 0$

2 On this number line, A is the opposite of E. What is the value of C?

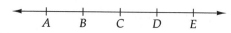

 (1) 0
 (2) 1
 (3) 2
 (4) 3

3 Evaluate: $-|-pq|$ if $p > 0$ and $q < 0$
 (1) pq
 (2) $-pq$
 (3) $p - q$
 (4) 0

4 $|8 - 10| \times |10 - 8| \div |10 - 8| + |8 - 10| =$
 (1) 4
 (2) 3
 (3) -3
 (4) -4

Exercises 5–15: Evaluate:

5 $|-7|$
6 $-|6|$
7 $-|-6|$
8 $2|3|$
9 $2|-3|$
10 $-4|4|$
11 $-4|-4|$
12 $4 + |-2|$
13 $4 - |-2|$
14 $|4 - 7|$
15 $-|6 \cdot 2|$

16 On the number line below, the length of each interval is 2 units and point B is the opposite of point F. What is the value of point G? Explain.

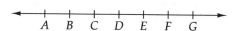

Exercises 17–20: Replace each ? with the symbol that makes the statement true: =, >, or <.

17 $|-6|$? -6
18 $|-8|$? $|6|$
19 $|-6|$? $|6|$
20 $-|-11|$? 11

Adding and Subtracting Signed Numbers

To add and subtract with positive and negative numbers, follow the rules below.

Rules for Addition of Signed Numbers

- To add numbers with *like signs*: Add the absolute values. The sum has the same sign as the given numbers (the addends).

  Examples: $+3 + 2 = +5$ and $-3 + (-2) = -5$.

- To add numbers with *unlike signs*: Subtract the smaller absolute value from the larger absolute value. The sum has the sign of the number with the larger absolute value.

  Example: Add $+3 + (-5)$. First step: $5 - 3 = 2$. Next step: Since -5 has a larger absolute value than $+3$, the sum is negative, -2. Answer: $+3 + (-5) = -2$.

36 Chapter 2: Operations With Real Numbers

MODEL PROBLEMS

1. $7 + (-4) = ?$
2. $-7 + 4 = ?$
3. $-3 + 7 + 5 + 3 + (-8) = ?$

SOLUTIONS

1. $7 + (-4) = 3$
2. $-7 + 4 = -3$
3. An expression such as $-3 + 7 + 5 + 3 + (-8)$ can be computed by grouping the positive and negative integers separately. When opposites appear, they can also be grouped:

   $-3 + 7 + 5 + 3 + (-8) = (7 + 5 + 3) + [-3 + (-8)] = 15 + (-11) = 4$

   $-3 + 7 + 5 + 3 + (-8) = (7 + 5) + (-3 + 3) + (-8) = 12 + 0 + (-8) = 4$

Rules for Subtraction of Signed Numbers

- Subtracting a number is the same as adding its opposite.
  Example: $2 - 5 = 2 + (-5)$.
- Therefore, to subtract signed numbers, change the number being subtracted to its opposite, and then follow the rules for addition: $2 - 5 = 2 + (-5) = -3$.

MODEL PROBLEMS

1. $8 - (-3) = ?$

SOLUTION

$8 - (-3) = 8 + 3 = 11$

Check: $11 + (-3) = 8$

2. $-6 - 5 = ?$

SOLUTION

$-6 - 5 = -6 + (-5) = -11$

Check: $-11 + 5 = -6$

3. A diver is 10 meters below the ocean surface. What is the new depth if the diver descends 2 meters lower?

SOLUTION

Model this situation using a number line:

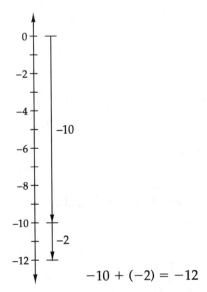

$-10 + (-2) = -12$

Answer: The diver's new depth is 12 meters.

Operations With Signed Numbers

Practice

1 $1{,}364 - 1{,}200 - 64 + 100 =$

  (1) 100
  (2) 150
  (3) 200
  (4) 250

2 If m and p are positive integers, which expression *must* be negative?

  (1) $m - p$
  (2) $p - m$
  (3) $-(m - p)$
  (4) $-(m + p)$

3 If m and p are negative integers, which expression *must* be positive?

  (1) $m - p$
  (2) $p - m$
  (3) $-(m - p)$
  (4) $-(m + p)$

4 $-20 + 14 + (-16) =$

  (1) -40
  (2) -22
  (3) -18
  (4) 18

5 An airplane is flying at an altitude of 6,000 feet. Find its new altitude:

  a after it first ascends 2,500 feet
  b after it then descends 1,500 feet

6 A credit card balance is $452.18. Find the new balance after these things happen: two items costing $26.45 and $15.90 are returned; three items costing $3.17, $16.42, and $57.71 are purchased; and a payment of $100 is made.

7 The temperature is $-6°$F. Find the new temperature:

  a if it falls 12°
  b if it rises 10°

8 Mercury freezes at $-39°$C and boils at 360°C. Find the difference between these two temperatures.

9 Here is the yardage for six plays by a football team: 4 yards, -3 yards, 9 yards, -5 yards, 13 yards, 6 yards. What was the total yardage?

10 Perform the operation $-7 + 4$, modeling it on a number line.

Multiplying and Dividing Signed Numbers

Below are the rules for multiplication and division with positive and negative numbers.

Rules for Multiplication of Signed Numbers

To multiply two signed numbers, multiply the absolute values.
  Then, to give a sign to the product, do the following:

- The product of two *positive* numbers is *positive*. Example: $+2 \cdot (+5) = +10$.
- The product of two *negative* numbers is *positive*. Example: $(-2) \cdot (-5) = +10$.
- The product of one *positive* number and one *negative* number is *negative*.
  Examples: $+2 \cdot (-5) = -10$ and $(-2) \cdot (+5) = -10$.

If two signs are the same, the sign of the product is positive.

If two signs are different, the sign of the product is negative.

MODEL PROBLEMS

SOLUTIONS

1. 13 • (−25) — The two signs are different, so the product is negative: −325

2. −6 • (−4) — The two signs are the same, so the product is positive: 24

3. 183 • (−206) — The two signs are different, so the product is negative: −37,698

4. (−x)(y) — The two signs are different, so the product is negative: −xy

5. The temperature drops 3° each hour for 4 hours. How much does it fall?

   −3 • (+4) = −12

   Answer: The temperature falls 12°.

Rules for Division of Signed Numbers

To divide signed numbers, divide the absolute values.
To give a sign to the quotient, do the following:

- The quotient of two *positive* numbers is *positive*. Example: (+10) ÷ (+2) = +5.
- The quotient of two *negative* numbers is *positive*. Example: (−10) ÷ (−2) = +5.
- The quotient of one *positive* number and one *negative* number is *negative*.

  Examples: +10 ÷ (−2) = −5 and (−10) ÷ (+2) = −5.

If two signs are the same, the quotient is positive.

If two signs are different, the quotient is negative.

MODEL PROBLEMS

Divide:

SOLUTIONS

1. −12 ÷ 2 — The two signs are different, so the quotient is negative: −6
2. −12 ÷ (−2) — The two signs are the same, so the quotient is positive: 6
3. 26 ÷ (−13) — The two signs are different, so the quotient is negative: −2

Practice

1. Which expression does *not* equal −5?
   - (1) −5 + (8 × 2) − (2 × 8)
   - (2) 3 + [2 × (−4)]
   - (3) 3 + [5 × (−2)] + 8
   - (4) [17 × (−2)] + (14 × 2) + 1

2. If a is positive integer and b is a negative integer, which expression *must* be positive?
   - (1) $a - b$
   - (2) $a + b$
   - (3) $a \times b$
   - (4) $a \div b$

Operations With Signed Numbers

3 If *pq* is a negative integer, which expression *must* be positive?
   (1) $-3p^2q^2$
   (2) $-4pq$
   (3) $p + q^2$
   (4) pq^2

4 Given: *mx* is a negative integer. Which *must* also be a negative integer?
   (1) $(-m)(x)$
   (2) $\dfrac{m}{x}$
   (3) $m - x$
   (4) $-m - x$

5 Given: *r* is a positive integer and *s* is a negative integer. Which value is largest?
   (1) $r + s$
   (2) $r - s$
   (3) rs
   (4) $\dfrac{s}{r}$

6 Given: *kx*, *ky*, and *xy* are positive integers. Which statement *must* be true of *k*, *x*, and *y*?
   (1) Exactly two of them are negative.
   (2) They are all positive.
   (3) They are all negative.
   (4) They are either all positive or all negative.

Exercises 7–14: Perform the indicated operations.

7 $2^3 \div (-4) - 2(2 - 5)$

8 $|4(-3)| - |-5|$

9 $3^2 + (-7) - (2^3 - 3)$

10 $12 \div 2 - 4^2 - (-3^2)$

11 $-5 + 3(-6) \div 2 - 1$

12 $8 + \dfrac{12 - 4}{2^2 + 4} - 3$

13 $(-5 \times 2) + [6 \div (-2)] + 5$

14 $(2.5 \times 14) - 30 + [2 \times 5 \div (-2)]$

15 From Monday through Friday, Geri worked for 27 hours earning $7 per hour and for 8 hours earning $9 per hour. How much did she have left if she spent $3 per day for bus fare and $3.50 a day for lunch?

Exercises 16–20: Is each statement true or false? If a statement is false, give a counter example.

16 $|n - m|$ is positive.

17 The difference of two positive numbers is always a positive number.

18 The sum of two negative numbers is never a positive number.

19 The product of a positive number and a negative number is sometimes a positive number.

20 The quotient of two negative numbers is always a positive number.

Operations With Fractions

Adding and Subtracting Fractions

Fractions can be added and subtracted if the denominators are the same (**like**). When the denominators are *not* the same (**unlike**), the two fractions are first written with a common denominator and then added or subtracted. Follow these steps:

- Add or subtract the numerators. The common denominator remains the same.
- Simplify the result if possible.

MODEL PROBLEMS

Find each sum:

1 $\dfrac{1}{8} + \dfrac{3}{8}$

SOLUTION

$\dfrac{1}{8} + \dfrac{3}{8} = \dfrac{1+3}{8} = \dfrac{4}{8} = \dfrac{1}{2}$

2 $\dfrac{8}{15} + \dfrac{7}{9}$

SOLUTION

Step 1. Write the two fractions with 45 as their common denominator (since 45 is the LCM of 15 and 9):

$\dfrac{8}{15} \cdot \dfrac{3}{3} = \dfrac{24}{45}$ and $\dfrac{7}{9} \cdot \dfrac{5}{5} = \dfrac{35}{45}$

Step 2. Add the new numerators. The common denominator stays the same:

$\dfrac{24}{45} + \dfrac{35}{45} = \dfrac{24 + 35}{45} = \dfrac{59}{45}$

Step 3. Simplify the answer:

$\dfrac{59}{45} = 1\dfrac{14}{45}$

To add or subtract positive and negative fractions, follow the rules for addition and subtraction of signed numbers.

 MODEL PROBLEMS

SOLUTIONS

1 $-3\dfrac{4}{5} + \left(-2\dfrac{3}{5}\right)$

$-3\dfrac{4}{5} + \left(-2\dfrac{3}{5}\right) = -\dfrac{19}{5} + \left(-\dfrac{13}{5}\right) = -\dfrac{32}{5} = -6\dfrac{2}{5}$

Or: $-3\dfrac{4}{5} + \left(-2\dfrac{3}{5}\right) = -5\dfrac{7}{5} = -6\dfrac{2}{5}$

2 $\dfrac{5}{8} + \left(\dfrac{-3}{8}\right)$

$\dfrac{5}{8} + \left(\dfrac{-3}{8}\right) = \dfrac{-2}{8} = -\dfrac{1}{4}$

3 $\dfrac{3}{10} - \dfrac{7}{15}$

$\dfrac{3}{10} - \dfrac{7}{15} = \dfrac{9}{30} - \dfrac{14}{30} = \dfrac{-5}{30} = \dfrac{-1}{6}$

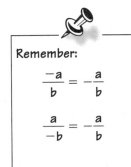

Remember:

$\dfrac{-a}{b} = -\dfrac{a}{b}$

$\dfrac{a}{-b} = -\dfrac{a}{b}$

Operations With Fractions

SOLUTIONS

4 $6\frac{3}{7} - \left(-4\frac{2}{3}\right)$

Add the opposite of $-4\frac{2}{3}$:

$$6\frac{3}{7} - \left(-4\frac{2}{3}\right) = 6\frac{3}{7} + 4\frac{2}{3}$$

Now we use a common denominator, 21, to get:

$$6\frac{9}{21} + 4\frac{14}{21} = 10\frac{23}{21} = 11\frac{2}{21}$$

5 $3\frac{7}{8} + \left(-5\frac{5}{6}\right)$

Use a common denominator, 24, to get:

$$3\frac{7}{8} + \left(-5\frac{5}{6}\right) = 3\frac{21}{24} + \left(-5\frac{20}{24}\right)$$

Next, subtract the smaller absolute value from the larger absolute value:

$$5\frac{20}{24} = 4\frac{44}{24} \text{ so } 5\frac{20}{24} - 3\frac{21}{24} = 4\frac{44}{24} - 3\frac{21}{24} = 1\frac{23}{24}$$

The negative number has the larger absolute value, so give the sum a negative sign:

$$-1\frac{23}{24}$$

Answer: $3\frac{7}{8} + \left(-5\frac{5}{6}\right) = -1\frac{23}{24}$

 Practice

---

1 Acme Shipping charges $6 a pound for packages. You have 3 packages to send, weighing $\frac{3}{5}$ pound, $1\frac{4}{15}$ pounds, and $\frac{23}{30}$ pound. How much will you pay Acme?

(1) $15
(2) $15.80
(3) $16
(4) $16.20

2 You buy $2\frac{1}{2}$ pounds of apples, $\frac{3}{4}$ pound of bananas, $1\frac{2}{5}$ pounds of peaches, and $5\frac{7}{10}$ pounds of watermelon. What is the total weight, in pounds?

(1) $9\frac{1}{2}$
(2) $10\frac{5}{8}$
(3) $10\frac{7}{20}$
(4) $11\frac{13}{30}$

42 Chapter 2: Operations With Real Numbers

3 You need a rope at least 8 feet long. You have four shorter ropes, of the following lengths:

Rope A = $4\frac{2}{3}$ feet	Rope B = $4\frac{2}{5}$ feet
Rope C = $3\frac{3}{4}$ feet	Rope D = $3\frac{2}{3}$ feet

You can tie two or more of these ropes together, but each time you tie you lose $\frac{1}{4}$ foot. Which combination will *not* give you a piece at least 8 feet long?

(1) A and C
(2) B and C
(3) A and D
(4) A, B, and C

4 Which expression does *not* equal 1?

(1) $\frac{1}{2} + \frac{3}{12} + \frac{9}{36}$

(2) $\frac{2}{3} + \frac{2}{6}$

(3) $\frac{3}{8} + \frac{20}{24} - \frac{1}{6}$

(4) $\frac{13}{7} - \frac{12}{14}$

5 Milly, Tilly, Lilly, and Billy are sharing a pizza that has been cut into 8 equal slices. Milly eats 2 slices, Tilly eats 1, Lilly eats 3. What fraction of the pie is left for Billy?

(1) $\frac{3}{8}$

(2) $\frac{1}{2}$

(3) $\frac{1}{8}$

(4) $\frac{1}{4}$

Exercises 6–8: Perform the indicated operation.

6 $10 + \left(-4\frac{1}{3}\right)$

7 $8\frac{2}{3} - \left(-1\frac{1}{4}\right)$

8 $2\frac{1}{8} - 3\frac{5}{6}$

9 On a test, $\frac{1}{4}$ of a class earned A's and $\frac{1}{3}$ earned B's. What part of the class earned either an A or a B?

10 A share of stock gained $\frac{1}{2}$ point, then lost $1\frac{1}{4}$ points, and then gained $2\frac{1}{8}$ points. Find the overall change in its value.

Multiplying and Dividing Fractions

To multiply fractions:

- Find the product of the numerators and the product of the denominators.
- Then simplify the result if possible.

For example:

$$\frac{1}{3} \cdot \frac{2}{3} = \frac{1 \cdot 2}{3 \cdot 3} = \frac{2}{9}$$

$$\frac{2}{3} \cdot \frac{3}{4} = \frac{2 \cdot 3}{3 \cdot 4} = \frac{6}{12} = \frac{1}{2}$$

$$\frac{1}{2} \cdot \frac{2}{5} = \frac{1 \cdot 2}{2 \cdot 5} = \frac{2}{10} = \frac{1}{5}$$

To perform operations with mixed numbers, we often need to write them as improper fractions.

To divide fractions:

- Multiply by the reciprocal of the divisor.
- Simplify the result if possible.

For example:

$$\frac{1}{4} \div \frac{3}{4} = \frac{1}{4} \cdot \frac{4}{3} = \frac{4}{12} = \frac{1}{3}$$

$$\frac{5}{6} \div \frac{3}{8} = \frac{5}{6} \cdot \frac{8}{3} = \frac{40}{18} = \frac{20}{9} = 2\frac{2}{9}$$

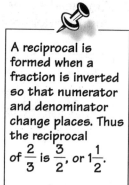

A reciprocal is formed when a fraction is inverted so that numerator and denominator change places. Thus the reciprocal of $\frac{2}{3}$ is $\frac{3}{2}$, or $1\frac{1}{2}$.

To multiply or divide positive and negative fractions, follow the rules for multiplication and division of signed numbers.

MODEL PROBLEMS

Perform the indicated operation.   **SOLUTIONS**

1. $\frac{-2}{15} \cdot \frac{10}{11}$

   $\frac{-2}{15} \cdot \frac{10}{11} = \frac{-20}{165} = \frac{-4}{33}$

2. $-2\frac{4}{7} \cdot \left(-1\frac{5}{16}\right)$

   $-2\frac{4}{7} \cdot \left(-1\frac{5}{16}\right) = -\frac{18}{7} \cdot \left(-\frac{21}{16}\right) = \frac{378}{112} = 3\frac{42}{112} = 3\frac{3}{8}$

3. $\frac{91}{9} \cdot \frac{51}{14}$

   $\frac{91}{9} \cdot \frac{51}{14} = \frac{4,641}{126} = \frac{221}{6} = 36\frac{5}{6}$

4. $\frac{15}{16} \div \frac{3}{20}$

   $\frac{15}{16} \div \frac{3}{20} = \frac{15}{16} \cdot \frac{20}{3} = \frac{300}{48} = 6\frac{12}{48} = 6\frac{1}{4}$

5. $-2\frac{3}{4} \div 5\frac{5}{6}$

   $-2\frac{3}{4} \div 5\frac{5}{6} = \frac{-11}{4} \div \frac{35}{6} = -\frac{11}{4} \cdot \frac{6}{35} = -\frac{66}{140} = -\frac{33}{70}$

6. $-\frac{75}{44} \div -\frac{3}{4}$

   $-\frac{75}{44} \div -\frac{3}{4} = -\frac{75}{44} \cdot -\frac{4}{3} = \frac{300}{132} = \frac{25}{11} = 2\frac{3}{11}$

7. $\frac{9}{20} + \frac{22}{7} \cdot \frac{7}{9}$

   $\frac{9}{20} + \frac{22}{7} \cdot \frac{7}{9} = \frac{9}{20} + \frac{154}{63} = \frac{9}{20} + \frac{22}{9} = \frac{81}{180} + \frac{440}{180} = \frac{521}{180} = 2\frac{161}{180}$

8. $10 \div \frac{1}{90} - \frac{5}{8}$

   $10 \div \frac{1}{90} - \frac{5}{8} = \frac{10}{1} \cdot \frac{90}{1} - \frac{5}{8} = \frac{900}{1} - \frac{5}{8} = 899\frac{3}{8}$

Verbal problems often involve operations with fractions. The next model problems are examples.

44 Chapter 2: Operations With Real Numbers

MODEL PROBLEMS

1. Leo buys $\frac{3}{4}$ pound of ham at $5 a pound, $3\frac{1}{2}$ pounds of cheese at $3 a pound, and $\frac{1}{2}$ pound of salami at $6 a pound. How much change does he get from a $20 bill?

SOLUTION FOR PROBLEM 1

To find the cost of each item, multiply:

$$\frac{3}{4} \times \$5 = \frac{\$15}{4} = \$3.75$$

$$3\frac{1}{2} \times \$3 = \frac{7}{2} \times \$3 = \frac{\$21}{2} = \$10.50$$

$$\frac{1}{2} \times \$6 = \$3$$

To find total cost, add: $3.75 + $10.50 + $3 = $17.25

To find the change, subtract: $20 − $17.25 = $2.75

Answer: He will get $2.75 change.

2. You have a piece of wood $5\frac{1}{4}$ feet long. If you cut it into 7 equal pieces, what will be the length of each piece?

SOLUTION FOR PROBLEM 2

Divide $5\frac{1}{4}$ by 7:

$$5\frac{1}{4} \div 7 = \frac{21}{4} \div 7 = \frac{21}{4} \times \frac{1}{7} = \frac{21}{28} = \frac{3}{4}$$

Answer: Each piece will be $\frac{3}{4}$ foot long. (Or: 9 inches long.)

3. Julio finds $\frac{3}{4}$ of an apple pie on the kitchen table and eats $\frac{1}{3}$ of it. How much of the pie is left?

SOLUTION FOR PROBLEM 3

Step 1. To calculate how much of the whole pie he ate, multiply:

$$\frac{1}{3} \times \frac{3}{4} = \frac{3}{12} = \frac{1}{4}$$

Step 2. To calculate how much pie is left, subtract:

$$\frac{3}{4} - \frac{1}{4} = \frac{2}{4} = \frac{1}{2}$$

Answer: Half of the pie is left.

Practice

1. Pat runs $\frac{2}{3}$ mile. Matt runs one-third as far. What distance does Matt run, in miles?
   (1) $\frac{1}{9}$
   (2) $\frac{2}{9}$
   (3) $\frac{1}{3}$
   (4) 2

2. Bernie's cousin weighs four-fifths as much as Bernie. If Bernie's weight is 135 pounds, what is his cousin's weight?
   (1) 168.75 pounds
   (2) 162 pounds
   (3) 108 pounds
   (4) 101.25 pounds

Operations With Fractions

3 $\dfrac{2}{5} + \left(\dfrac{25}{9} \div \dfrac{1}{3}\right) =$

   (1) $\dfrac{3}{10}$

   (2) $\dfrac{13}{25}$

   (3) $3\dfrac{1}{3}$

   (4) $8\dfrac{11}{15}$

4 A college math professor has to grade 160 exams. He gives $\dfrac{1}{2}$ of the exams to his assistant to grade. Then the assistant gives $\dfrac{1}{4}$ of her exams to a graduate student to grade. How many exams does the graduate student get?

   (1) 20
   (2) 40
   (3) 60
   (4) 70

5 $n \cdot \dfrac{1}{3} \div \dfrac{3}{4} =$

   (1) $\dfrac{n}{4}$

   (2) $\dfrac{n}{3}$

   (3) $\dfrac{3n}{4}$

   (4) $\dfrac{4n}{9}$

6 Which procedure gives the same result as multiplying by $\dfrac{3}{4}$ and then dividing by $\dfrac{3}{8}$?

   (1) dividing by 2
   (2) multiplying by 2
   (3) dividing by 32
   (4) multiplying by $\dfrac{9}{32}$

7 If $\dfrac{3}{4}$ pound of pot roast is recommended per serving, how much should you buy to serve 6 people?

8 To multiply $\dfrac{1}{2} \cdot \dfrac{2}{3}$, Ricky did this:

   $\dfrac{1}{2} \cdot \dfrac{2}{3} = \dfrac{3}{6} \cdot \dfrac{4}{6} = \dfrac{12}{6} = 2$

   Is Ricky's work correct or incorrect? Explain your answer, referring specifically to the rules for operations with fractions. If you think he was *in*correct, redo the problem correctly.

Exercises 9–20: Perform the indicated operations and give each answer in simplest form:

9   $\dfrac{3}{8} \times \dfrac{4}{5}$

10 $3\dfrac{1}{2} \times 1\dfrac{3}{4}$

11 $4 \times 2\dfrac{2}{3}$

12 $\dfrac{3}{4} \div \dfrac{5}{8}$

13 $\dfrac{1}{3} + \dfrac{2}{5} \cdot 10$

14 $\left(\dfrac{2}{3}\right)^2$

15 $\left(\dfrac{1}{2}\right)^3$

16 $\left(\dfrac{1}{3}\right)\left(\dfrac{-4}{5}\right)\left(\dfrac{3}{8}\right)$

17 $\left(-2\dfrac{1}{2}\right)\left(\dfrac{-1}{3}\right)\left(-2\dfrac{1}{4}\right)$

18 $\left(\dfrac{3}{4}\right)^2 - \left(\dfrac{1}{2}\right)^3 + \dfrac{3}{4}$

19 $\dfrac{3}{8} \div \left(\dfrac{5}{6} + \dfrac{2}{3}\right)$

20 $\dfrac{5}{7} \div \dfrac{15}{3} \times \dfrac{11}{13}$

Operations With Decimals

In operations with decimals, we must pay attention to the effect on the decimal point. Where does the decimal point appear in the result? In other words, how many decimal places are there in the result? A calculator can take care of this automatically, but certain rules must be followed when we work with pencil and paper.

Rule for Adding and Subtracting Decimals

- Line up the decimal points in the numbers being added or subtracted and in the sum or difference.

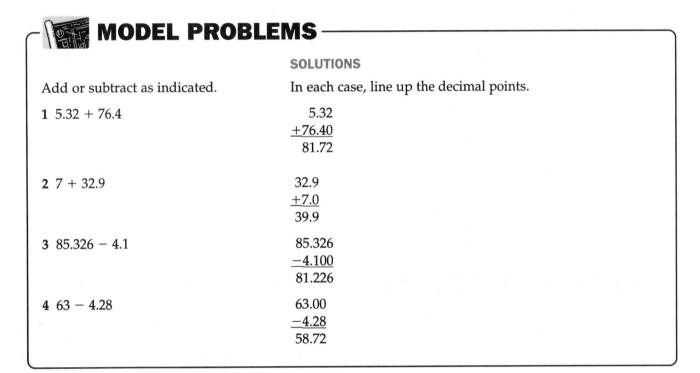

Rule for Multiplying Decimals

- Multiply as with whole numbers. To get the number of decimal places in the product, find the sum of the number of decimal places in the multiplier and the multiplicand.

Operations With Decimals **47**

Rule for Dividing Decimals

- Get rid of the decimal point in the divisor by moving it to the right as many places as necessary to make the divisor a whole number.
- Then move the decimal point in the dividend the same number of places to the right.
- Then divide.

 MODEL PROBLEM

Divide: 13.44 ÷ 0.12

SOLUTION

See the calculation at the right. In the long-division format, the decimal point in the quotient is directly above the *new* position of the decimal point in the dividend.

$$0.12\overline{)13.44.}$$
move 2 places   move 2 places

```
       112.
   12)1344.
       12
       ──
       14
       12
       ──
        24
        24
        ──
         0
```

Operations with decimals are often used in solving verbal problems. The following model problems are examples.

 MODEL PROBLEMS

1. Janet spent $30.38 for fabric that cost $2.48 per yard. How many yards did she buy?

   SOLUTION

   Divide: $30.38 ÷ $2.48 = 12.25

   Answer: She bought 12.25 or $12\frac{1}{4}$ yards.

2. If steak costs $4.95 a pound, and Mrs. Waters buys a 3-pound steak, how much change will she get from a $50 bill?

   SOLUTION

   Step 1. To find total cost, multiply:
   $4.95 × 3 = $14.85

   Step 2. To find her change, subtract:
   $50 − $14.85 = $35.15

   Answer: She will get $35.15 in change.

 Practice

1. Yesterday the high temperature was 92.6°F and the low was 77.8°F. What was the difference between the high and the low?

   (1) 16.6°   (3) 14.4°
   (2) 14.8°   (4) 13.8°

2. A carpenter has fifty 1.5-inch nails and fifty 0.75-inch nails. If all these nails were lined up end to end, what would the length be?

   (1) 112.5 inches   (3) 117.5 inches
   (2) 115 inches   (4) 119 inches

3 Garrett weighed 170 pounds. By dieting and exercising, he lost 0.8 pound a week for 5 weeks. How much did he weigh after the 5 weeks?

   (1) 164 pounds
   (2) 165 pounds
   (3) 166 pounds
   (4) 167 pounds

4 An **odometer** registers miles driven. The odometer on Alyssa's car now reads 2,004.7. What will it read after she drives the following distances?

Day	Miles
Monday	5.6
Tuesday	10.4
Wednesday	7.8
Thursday	11.2
Friday	22.7

   (1) 2,006.2
   (2) 2,045.6
   (3) 2,062.4
   (4) 2,074.7

5 On a 3-day trip, the Lemoine family drove 125.5 miles the first day, 80.4 miles the second day, and 95.7 miles the third day. What was the total distance?

   (1) 3,016 miles
   (2) 301.6 miles
   (3) 300.16 miles
   (4) 3.016 miles

Exercises 6–12: Perform the indicated operations with pencil and paper. Show your work.

6 $9.2 - 2.6$

7 2.8×5.9

8 2.4×0.08

9 0.7×0.6

10 $31.76 \div 4$

11 $0.24 \overline{) 3.948}$

12 $3.573 + 0.00065 + 240 + 6.2$

13 You read 0.15 of a book on Monday and 0.6 of it on Tuesday. How much of the book do you have left to read?

14 A laundry charges $9.50 for the first 12 shirts and $0.55 for each additional shirt. How much will it cost to launder 31 shirts?

15 A library is constructing some new shelves: 4 shelves each 2.75 meters long, and 3 shelves each 3.25 meters long. Will 20 meters of board be just enough, more than enough, or not enough to make all the shelves? If 20 meters is *not* the exact amount required, how much more or less will be needed?

Operations With Exponents

A **power** is the value indicated by a **base** with an **exponent**:

$$\text{Base}^{\text{Exponent}}$$

We say that the exponent *raises the base to a power*. For example:

$$10^2 \quad 4^3 \quad m^n$$

In 10^2, the base is 10, the exponent is 2, and the exponent raises the base to the second power. The expression is read *ten to the second power* or *ten squared*. In 4^3, the base is 4, the exponent is 3, and the exponent raises the base to the third power. This expression is read *four to the third power* or *four cubed*. In m^n the base is m and the exponent is n, and the expression is read *m to the nth power*.

An exponent may be positive, negative, or zero:

- A *positive* whole-number exponent tells us how many times the base is used as a factor. In 10^2, for instance, 10 is used as a factor 2 times. In 4^3, 4 is used as a factor 3 times. In m^n, m is used as a factor n times:

  $10^2 = 10 \cdot 10 = 100 \qquad 4^3 = 4 \cdot 4 \cdot 4 = 64 \qquad m^n = m \cdot m \cdot m \cdot m \ldots$

- A *negative* exponent can be written as a unit fraction with a positive exponent in the denominator:

  $10^{-2} = \dfrac{1}{10^2} = \dfrac{1}{100} \qquad 4^{-3} = \dfrac{1}{4^3} = \dfrac{1}{64} \qquad m^{-n} = \dfrac{1}{m^n}$

- The exponent *zero* has a special meaning. Any number to the zero power equals 1:

  $10^0 = 1 \qquad 4^0 = 1 \qquad 1{,}000{,}000{,}000{,}000^0 = 1 \qquad m^0 = 1$

- The exponent 1 has a special meaning. Any number to the first power equals itself:

  $10^1 = 10 \qquad 4^1 = 4 \qquad 1{,}000{,}000{,}000{,}000^1 = 1{,}000{,}000{,}000{,}000 \qquad m^1 = m$

- The number 1 to any power equals 1:

  $1^0 = 1 \qquad 1^1 = 1 \qquad 1^{153} = 1 \qquad 1^n = 1 \qquad 1^{-n} = \dfrac{1}{1^n} = \dfrac{1}{1} = 1$

Remember:
$x^1 = x$
$x^0 = 1, x \neq 0$
$x^{-n} = \dfrac{1}{x^n}$
$1^n = 1$

Rules for Operations on Terms with Exponents

Operation	Rule	Examples
Addition and subtraction $x^n + x^m = x^n + x^m$ $x^n - x^m = x^n - x^m$	Like bases with unlike exponents cannot be added or subtracted unless they can be evaluated first.	$2^2 + 2^3 = 4 + 8 = 12$ $3^3 - 3^2 = 27 - 9 = 18$ $a^2 + a^3 = a^2 + a^3$ $a^3 - a^2 = a^3 - a^2$
Multiplication $x^n \cdot x^m = x^{n+m}$	To multiply powers of like bases, add the exponents.	$3^4 \times 3^5 = 3^9$ $a^2 \cdot a^3 = a^{2+3} = a^5$ $a^{-4} \cdot a^5 = a^{-4+5} = a^1 = a$
Division $\dfrac{x^n}{x^m} = x^{n-m},\ x \neq 0$	To divide powers of like bases, subtract the exponent of the divisor from the exponent of the dividend.	$\dfrac{4^7}{4^5} = 4^{7-5} = 4^2$ $\dfrac{a^5}{a^2} = a^{5-2} = a^3$ $\dfrac{a^3}{a^8} = a^{3-8} = a^{-5} = \dfrac{1}{a^5}$
Raising a power to a power $(x^m)^n = x^{mn}$	To raise a term with an exponent to some power, multiply the exponents.	$(5^2)^3 = 5^{2 \times 3} = 5^6$ $(a^4)^3 = a^{4 \cdot 3} = a^{12}$
Raising a fraction to a power $\left(\dfrac{x}{y}\right)^n = \dfrac{x^n}{y^n}$	To raise a fraction to a power, raise the numerator and the denominator to that power.	$\left(\dfrac{3}{5}\right)^2 = \dfrac{3^2}{5^2} = \dfrac{9}{25}$ $\left(\dfrac{a}{b}\right)^7 = \dfrac{a^7}{b^7}$
Raising a product to a power $(xy)^n = x^n y^n$	To raise a product to a power, raise each factor to that power.	$(5 \cdot 2)^3 = 5^3 \cdot 2^3 = 1{,}000$ $(4a)^2 = 4^2 \cdot a^2 = 16a^2$ $(ab)^5 = a^5 b^5$

Chapter 2: Operations With Real Numbers

Here are some useful points to remember about working with negative bases:

- $-x^n$ means $-(x^n)$. For example: $-3^2 = -(3^2) = -9$ and $-3x^2 = -3(x^2)$
- $(-x)^n$ means $(-x)(-x)(-x)\ldots$ For example: $(-3)^2 = (-3)(-3) = 9$
- When a negative base is raised to an *even* power, the result becomes positive:
  $(-2)^4 = (-2)(-2)(-2)(-2) = +16$
- When a negative base is raised to an *odd* power, the result remains negative:
  $(-2)^3 = (-2)(-2)(-2) = -8$

Calculator hint: Be sure to use parentheses when you raise a negative base to a power:
$(-8)^2 = 64$
but
$-8^2 = -64$

MODEL PROBLEMS

Perform each operation.

SOLUTIONS

1. $p^0 + p^1 + p^2 + p^5 + 1^p =$ $\quad p + p^2 + p^5 + 2$

2. $5^6 \cdot 5^{-3} =$ $\quad 5^{6+(-3)} = 5^{6-3} = 5^3 = 125$

3. $4^3 \div 4^5 =$ $\quad 4^{3-5} = 4^{-2} = \dfrac{1}{4^2} = \dfrac{1}{16}$

4. $(10^4)^{-1} =$ $\quad 10^{(4)(-1)} = 10^{-4} = \dfrac{1}{10^4} = \dfrac{1}{10,000}$

5. $\left(\dfrac{3p}{2q}\right)^2 =$ $\quad \dfrac{3p \cdot 3p}{2q \cdot 2q} = \dfrac{9p^2}{4q^2}$ Or: $\dfrac{3^2 p^2}{2^2 q^2} = \dfrac{9p^2}{4q^2}$

6. $(4p)^3 =$ $\quad 4^3 \cdot p^3 = 64p^3$

 Practice

1. What is the product of 75^3 and 75^7?
   (1) 75^{10}
   (2) 75^{21}
   (3) 150^{10}
   (4) 150^{21}

2. Which expression is equal to 10,000?
   (1) 100^3
   (2) $(5^4)(2^4)$
   (3) $(10^2)(50^2)$
   (4) $(2)(250^2)$

3. Which expression is *not* equal to the other three?
   (1) $x^2 \cdot x^4$
   (2) $(x^2)^4$
   (3) $x^{-2} \cdot x^8$
   (4) $\dfrac{x^{10}}{x^4}$

4. Which expression is *not* equal to the other three?
   (1) 2^6
   (2) $32(2)^0$
   (3) $2^5 + 2^5$
   (4) $(-2)^6$

5. Which statement is *false*?
   (1) $4^0 = 6^0$
   (2) $3^4 = 9^2$
   (3) $3 \cdot 2^0 = 1$
   (4) $(4^2)^3 = (4^3)^2$

6. If $5^k \times 5^3 = 5^6$, what is the value of k?
   (1) 2
   (2) 3
   (3) 9
   (4) 18

Operations With Exponents **51**

Exercises 7–13: Find each value.

7. $4^3 - 4^2$
8. $4^{-3} + 4^{-2}$
9. $4^3 \cdot 4^2$
10. $4^3 \div 4^2$
11. $3^0 + 6^0$
12. $2^{-1} - 3^{-1}$
13. $(-6)^2$

Exercises 14–18: Simplify.

14. $x^5 \cdot x^{-3}$
15. $(x^3 y^3)^3$
16. $(4x^3)(3x)^2$
17. $\dfrac{x^2}{x^6}$
18. $\left(\dfrac{x^2}{y}\right)^2$
19. Explain why:
    a. $x^2 \cdot x^3 \neq (x^2)^3$
    b. $-4^2 \neq (-4)^2$
20. Find the value of x that makes each statement true:
    a. $(3^3)^{5x} = 3^{30}$
    b. $5^{28} = 5^{3x} \cdot 5^7$
    c. $(6^4 \cdot 6^x) = 6^{24}$

Radical Notation and Operations

Here are some key definitions.

- **Root**: One of two or more equal factors of a number.
- **Square root**: One of two equal factors of a number.
- **Radical sign** $\sqrt{\ }$: Symbol meaning *the square root of*. For example:
  If $\sqrt{a} = x$, then $x \cdot x = a$ or $x^2 = a$.
- **Principal square root**: Every positive number has two square roots. The principal square root is positive. The other root is negative. For example: $3 \cdot 3 = 9$ and $-3 \cdot (-3) = 9$, so $\sqrt{9} = 3$ (principal square root) and -3.

  > A negative number does not have a real-number square root.

- **Radicand**: The number under the radical sign. In $\sqrt{10}$, for example, 10 is the radicand.
- **Coefficient**: A number by which a square root is multiplied. In $3\sqrt{10}$, for example, 3 is the coefficient. The meaning is *three times the square root of ten*.
- **Perfect square**: A number that has a rational square root. For example:

  $\sqrt{9} = 3 \qquad \sqrt{25} = 5 \qquad \sqrt{\dfrac{9}{25}} = \dfrac{3}{5}$

Simplifying Square Roots

A radical is **simplified** when the radicand is the smallest possible positive integer. To simplify a square root:

> Remember:
> $\sqrt{n} \cdot \sqrt{n} = n$
> $(\sqrt{n})^2 = n$
> $\sqrt{n^2} = n$

- Factor the radicand so that one factor is a perfect square. For example:
  $\sqrt{27} = \sqrt{9 \cdot 3}$ (9 is a perfect square)
- Simplify this factor and write its square root as a coefficient:
  $\sqrt{27} = \sqrt{9 \cdot 3} = 3\sqrt{3}$ (the square root of 9 is 3, which becomes the coefficient)

A fractional radicand is considered simplified, or **rationalized**, when there is no radical in the denominator. To rationalize, multiply the numerator and the denominator by the radical you want to eliminate. For example:

$$\frac{3}{\sqrt{7}} = \frac{3}{\sqrt{7}} \cdot \frac{\sqrt{7}}{\sqrt{7}} = \frac{3\sqrt{7}}{7}$$

 MODEL PROBLEMS

SOLUTIONS

1 Simplify: $\sqrt{40}$ $\sqrt{40} = \sqrt{4 \cdot 10} = 2\sqrt{10}$

2 Rationalize: $\dfrac{7}{3\sqrt{2}}$ $\dfrac{7}{3\sqrt{2}} = \dfrac{7}{3\sqrt{2}} \cdot \dfrac{\sqrt{2}}{\sqrt{2}} = \dfrac{7\sqrt{2}}{3 \cdot 2} = \dfrac{7\sqrt{2}}{6}$

 Practice

1 The principal square root of 64 is:
  (1) -8
  (2) 8
  (3) 8 and -8
  (4) not a real number

2 Simplify: $\sqrt{1{,}176}$
  (1) 42
  (2) $7\sqrt{24}$
  (3) $14\sqrt{3}$
  (4) $14\sqrt{6}$

3 Simplify: $\dfrac{3^2}{\sqrt{80}}$
  (1) $\dfrac{\sqrt{80}}{3^{-2}}$
  (2) $\dfrac{4}{9}\sqrt{5}$
  (3) $\dfrac{5}{4}\sqrt{5}$
  (4) $\dfrac{9}{20}\sqrt{5}$

4 $\dfrac{7}{\sqrt{45}} =$
  (1) $\dfrac{7}{9}$
  (2) $\dfrac{3}{7}$
  (3) $\dfrac{7\sqrt{3}}{9}$
  (4) $\dfrac{7\sqrt{5}}{15}$

Exercises 5–10: Simplify.

5 $\sqrt{25}$

6 $\sqrt{864}$

7 $\sqrt{162}$

8 $\dfrac{7}{\sqrt{11}}$

9 $\dfrac{6}{\sqrt{18}}$

10 $\dfrac{19}{\sqrt{19}}$

Radical Notation and Operations

Operations with Square Roots

Rules for Adding and Subtracting Square Roots

- Square roots with like radicands can be added or subtracted.
- If necessary, simplify by factoring so that the radicands are the same.
- Then add or subtract the coefficients.

$x\sqrt{n} + y\sqrt{n}$
$= (x + y)\sqrt{n}$

MODEL PROBLEMS

1 Add: $5\sqrt{7} + 2\sqrt{112}$

SOLUTION

Simplify $2\sqrt{112}$: $\quad 2\sqrt{112} = 2\sqrt{16 \times 7} = 2 \times 4\sqrt{7} = 8\sqrt{7}$

Add the coefficients: $\quad 5\sqrt{7} + 8\sqrt{7} = 13\sqrt{7}$

2 Subtract: $3\sqrt{48} - \sqrt{27}$

SOLUTION

Simplify $3\sqrt{48}$: $\quad 3\sqrt{48} = 3\sqrt{16 \times 3} = 12\sqrt{3}$

Simplify $\sqrt{27}$: $\quad \sqrt{27} = \sqrt{9 \times 3} = 3\sqrt{3}$

Subtract the coefficients: $\quad 12\sqrt{3} - 3\sqrt{3} = 9\sqrt{3}$

Rules for Multiplying Square Roots

- The radicands do *not* need to be the same.
- Write the factors under one radical sign.
- Multiply.
- If possible, simplify the product.

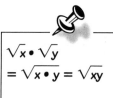

$\sqrt{x} \cdot \sqrt{y}$
$= \sqrt{x \cdot y} = \sqrt{xy}$

MODEL PROBLEMS

Multiply:

SOLUTIONS

1 $\sqrt{3} \cdot \sqrt{6} = \quad \sqrt{3 \cdot 6} = \sqrt{18} = \sqrt{9} \cdot \sqrt{2} = 3\sqrt{2}$

2 $\sqrt{10} \cdot \sqrt{15} = \quad \sqrt{10 \cdot 15} = \sqrt{150} = \sqrt{25 \cdot 6} = 5\sqrt{6}$

Rules for Dividing Square Roots

- The radicands do *not* need to be the same.
- Write the divisor and the dividend as a fraction under one radical sign.
- Simplify the denominator and numerator by factoring.
- If the denominator of the answer contains a radical, rationalize.

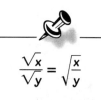

$\dfrac{\sqrt{x}}{\sqrt{y}} = \sqrt{\dfrac{x}{y}}$

MODEL PROBLEM

Divide: $\dfrac{\sqrt{10}}{\sqrt{15}}$

SOLUTION

Rewrite under one radical: $\dfrac{\sqrt{10}}{\sqrt{15}} = \sqrt{\dfrac{10}{15}}$

Simplify: $\sqrt{\dfrac{10}{15}} = \sqrt{\dfrac{2 \times 5}{3 \times 5}} = \sqrt{\dfrac{2}{3}}$

Rationalize: $\sqrt{\dfrac{2}{3}} = \dfrac{\sqrt{2}}{\sqrt{3}} \cdot \dfrac{\sqrt{3}}{\sqrt{3}} = \sqrt{\dfrac{6}{9}} = \dfrac{\sqrt{6}}{3}$

Practice

1 $3\sqrt{48} + 11\sqrt{75} =$
   (1) $33\sqrt{48}$
   (2) $67\sqrt{3}$
   (3) $72\sqrt{5}$
   (4) $81\sqrt{2}$

2 $\sqrt{14} \div \sqrt{6} =$
   (1) $\dfrac{2}{3}$
   (2) $\dfrac{21}{\sqrt{3}}$
   (3) $\dfrac{3}{\sqrt{21}}$
   (4) $\dfrac{\sqrt{21}}{3}$

3 $\sqrt{54} - \sqrt{96} =$
   (1) $-\sqrt{6}$
   (2) $2\sqrt{3}$
   (3) $\sqrt{6}$
   (4) $7\sqrt{6}$

4 $3\sqrt{125} + 6\sqrt{80} =$
   (1) $-29\sqrt{5}$
   (2) $-9\sqrt{5}$
   (3) $39\sqrt{5}$
   (4) 145

5 $\sqrt{8}(\sqrt{13} - \sqrt{117}) =$
   (1) $-4\sqrt{26}$
   (2) $4\sqrt{2}$
   (3) $2\sqrt{13}$
   (4) $2\sqrt{26}$

6 In simplest form, $5\sqrt{12} + 7\sqrt{108} =$
   (1) $12\sqrt{120}$
   (2) $52\sqrt{6}$
   (3) $52\sqrt{3}$
   (4) $20\sqrt{3}$

Exercises 7–20: Perform the indicated operations. Give your answer in simplified radical form.

7 $8\sqrt{12x} + \sqrt{27x}$

8 $4\sqrt{3} - 3\sqrt{3} + 3\sqrt{2}$

9. $3\sqrt{8} - 4\sqrt{3} + \sqrt{18}$

10. $\sqrt{4} + \sqrt{50} - \sqrt{32}$

11. $3\sqrt{6x^2} - 4\sqrt{24x^2} + 2\sqrt{x^2}$

12. $\sqrt{0.5} \cdot \sqrt{32}$

13. $\sqrt{m^2n} \cdot \sqrt{mn^2}$

14. $2\sqrt{5x} \cdot 4\sqrt{3x}$

15. $3\sqrt{7} \cdot 7\sqrt{6}$

16. $\sqrt{2x^3} \cdot 4\sqrt{10x}$

17. $\sqrt{2} \cdot \sqrt{0.02x}$

18. $\dfrac{\sqrt{x^7}}{\sqrt{x^3}}$

19. $\sqrt{\dfrac{x}{4}}$

20. $\dfrac{\sqrt{8x^3}}{\sqrt{2x}}$

Ratios, Rates, and Proportions

Important applications of operations include ratios, rates, and proportions (reviewed in this section) and percentages (reviewed next).

Ratios and Rates

A **ratio** is a comparison of two numbers—called its **terms**—by division. Ratios can be written in several ways. For example, if your school has 4 freshmen to every 3 sophomores, the ratio can be written as:

$$4 \text{ to } 3 \qquad 4:3 \qquad \dfrac{4}{3}$$

The order of the terms is important. When a fraction is used, the numerator is the term before *to* or the colon. The denominator is the term after *to* or the colon.

For calculations involving ratios, the fraction form is most useful. Ratios in fraction form are usually simplified, or "reduced," to lowest terms. Equivalent fractions are equivalent ratios (see "Using Proportions," below).

A **continued ratio** has more than two terms, in a definite order. For example, if the ratio of a to b to c is $2:3:5$, then $\dfrac{a}{b} = \dfrac{2}{3}, \dfrac{b}{c} = \dfrac{3}{5}$, and $\dfrac{a}{c} = \dfrac{2}{5}$.

In general, we use the word *ratio* for a comparison of *like* terms, such as freshmen and sophomores. If the terms are measurements, the unit of measure must be the same. For example, in a ratio of time h spent doing homework to time t spent watching television, h and t should both be in minutes or both be in hours.

A ratio that compares *unlike* terms is called a **rate**. Familiar examples of rates are 55 miles per hour (speed, which compares distance and time) and $10 per hour (pay, which compares money and time). A **unit rate** is a rate written with a denominator of 1. When a given rate has some other denominator, you can find the unit rate by simplifying. For instance:

$$\dfrac{125 \text{ miles}}{2 \text{ hours}} = \dfrac{125}{2} = 62.1 = \dfrac{62.5}{1} \text{ miles per hour}$$

Many real-life problems involve unit prices. A **unit price** is a rate: the rate of cost, or price, per 1 unit of measure, such as $1 per pound. To find the unit price, divide the total cost by the number of units.

Unless a problem states otherwise, "rate" generally means unit rate.

MODEL PROBLEMS

1 Carol's favorite radio station plays 6 new hit songs and 4 oldies every hour. Write a comparison using three ratio notations.

SOLUTION

This problem does not specify order of terms, so we can write the ratio of hits to oldies or oldies to hits.

Words	6 hits to 4 oldies	4 oldies to 6 hits
Colon	6 hits : 4 oldies	4 oldies : 6 hits
Fraction	$\dfrac{6 \text{ hits}}{4 \text{ oldies}} = \dfrac{3 \text{ hits}}{2 \text{ oldies}}$	$\dfrac{4 \text{ oldies}}{6 \text{ hits}} = \dfrac{2 \text{ oldies}}{3 \text{ hits}}$

2 Write a ratio comparing 8 hours to 3 days.

SOLUTION

To make the unit of measure the same, we can convert days to hours or hours to days. (Note that this problem specifies order of terms.)

Converting 3 Days to Hours	Converting 8 Hours to Days
Step 1. There are 24 hours in a day, so multiply: $3 \times 24 \text{ hours} = 72 \text{ hours}$	Step 1. In this case we divide by 24: $8 \text{ hours} \div 24 = \dfrac{8}{24} \text{ day} = \dfrac{1}{3} \text{ day}$
Step 2. Set up the ratio: $\dfrac{8 \text{ hours}}{3 \text{ days}} = \dfrac{8 \text{ hours}}{72 \text{ hours}} = \dfrac{1}{9}$	Step 2. Write the ratio and simplify. $\dfrac{1}{3} \text{ day to 3 days} = \dfrac{1}{3} \div 3 = \dfrac{1}{3} \cdot \dfrac{1}{3} = \dfrac{1}{9}$

3 Your car can go 100 miles on 3 gallons of gasoline. What is the unit rate?

SOLUTION

$$\frac{100 \text{ miles}}{3 \text{ gallons}} = \frac{100}{3} = 33.\overline{3} = \frac{33.\overline{3}}{1}$$

Answer: The rate is $33.\overline{3}$ (or $33\frac{1}{3}$) miles per gallon.

Money is often rounded to the nearest cent.

4 You earn $54 for 4 hours of work. What is the unit rate?

SOLUTION

$$\frac{\$54}{4 \text{ hours}} = \frac{54}{4} = 13.5 = \frac{13.5}{1}$$

Answer: The rate is $13.50 per hour.

5 Which costs less per ounce: 12 ounces of Oatsies for $2.79 or 17.2 ounces of Kornsies for $3.33?

SOLUTION

Find the unit price—the rate per 1 ounce—of each cereal and see which is less:

unit price of Oatsies = $\dfrac{\text{total cost}}{\text{number of units}} = \dfrac{\$2.79}{12} = \$0.2325$, rounded to 23 cents

unit price of Kornsies = $\dfrac{\text{total cost}}{\text{number of units}} = \dfrac{\$3.33}{17.2} = \$0.1936$, rounded to 19 cents

Answer: Kornsies costs less.

Practice

1. John F. Kennedy High School is collecting waste metal and paper for recycling. Five classes bring in the following amounts. Which two classes brought in the same ratio of bags of metal to bags of paper?

Class	Number of bags	
	Metal	Paper
Ms. Ginty's	4	6
Ms. Greenberg's	7	3
Mr. Rondone's	2	3
Mr. Scott's	3	2
Mr. Stanley's	4	3

   (1) Mr. Scott's and Mr. Rondone's
   (2) Ms. Ginty's and Mr. Stanley's
   (3) Mr. Scott's and Mr. Stanley's
   (4) Ms. Ginty's and Mr. Rondone's

2. A candy dish contains p peppermints, s spearmints, and b butterscotch candies. Write an expression for the ratio of spearmints to total number of candies.

   (1) $s : (p + s + b)$
   (2) $s : (p + b)$
   (3) $(p + s + b) : 3$
   (4) $s : b$

3. On each floor of a building, the ratio of offices to windows to doors is $25 : 55 : 33$. What is the ratio of doors to windows?

   (1) 11 to 5
   (2) 11 to 10
   (3) 3 to 5
   (4) 5 to 3

4. Write a ratio comparing 4 feet to 2 yards.

   (1) $4 : 2$
   (2) $4 : 6$
   (3) $12 : 2$
   (4) $12 : 6$

5. Every working day, Angela commutes 35 minutes each way. If she works 7 hours, what is the ratio of her total travel time to her time at work?

   (1) $\dfrac{5}{1}$
   (2) $\dfrac{1}{6}$
   (3) $\dfrac{10}{1}$
   (4) $\dfrac{1}{12}$

6. A school has 500 books for every 25 students. Express this as a unit rate.

   (1) 20 books per student
   (2) 20 students per book
   (3) 500 books per student
   (4) 1 student per 25 books

7. Write two other notations for "7 to 10."

8. A soccer team played 32 games and won 20. Find the ratio and simplify:
   a games won to games lost
   b games won to total number of games

9. A baseball team played 5 games and scored a different number of runs in each game: 3, 4, 1, 0, 5. Express this as a rate.

10. A secretary keyboards a 500-word document in 12 minutes. How many words per minute can he type?

11. Arriving in Freedonia, you cash a $100 traveler's check and get 912.35 freedons (the local currency). What is the exchange rate for:
    a dollars to freedons
    b freedons to dollars

12. Cindy watched a music channel for an hour and a half and counted 5 commercial breaks. She then watched a movie on network TV for 3 hours and counted 13 commercial breaks. Which channel had a higher rate of commercial breaks per hour, and what was its rate?

13. Write as a ratio in simplest form: 750 milliliters to 2 liters.

14. Which costs *more* per ounce: 13.5 ounces of Wheetsies for $3.09 or 20 ounces for $3.99?

15. Which costs *less* per cup: 3 cups of yogurt for $2 or 5 cups for $3?

58 Chapter 2: Operations With Real Numbers

Using Proportions

A **proportion** is a statement that two ratios are equal, or **proportional**. Proportions may be written in several ways. For example:

$$\frac{4}{5} = \frac{8}{10}$$

$$4 : 5 = 8 : 10$$

Four is to five as eight is to ten.
Four-fifths is proportional to eight-tenths.

There is a simple test to determine whether or not two ratios are proportional:

- Two ratios written with colons are equal if and only if the product of the **means** equals the product of the **extremes**:

```
    ┌─Extremes─┐
    ↓          ↓
    4 : 5  =  8 : 10
        ↑  ↑
        └Means┘
```

$5 \cdot 8 = 40$ and $4 \cdot 10 = 40$

- Two ratios in fraction form are equal if and only if their **cross products** are equal:

$$\frac{4}{5} \times \frac{8}{10}$$

$4 \cdot 10 = 40$ and $5 \cdot 8 = 40$

As with ratios, it is often most useful to express proportions as fractions when calculations are necessary. A proportion in fraction form has four terms: a numerator and a denominator for each of two fractions. Proportion problems often involve finding one missing term, or value, when the other three terms are known.

Since cross products are equal, we can use **cross multiplication** to find the missing value in a proportion:

- First, cross multiply the diagonal that does *not* have the variable.
- Then divide by the given number in the other diagonal.

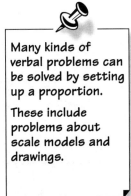

Many kinds of verbal problems can be solved by setting up a proportion.

These include problems about scale models and drawings.

For algebraic solutions to proportion problems, see Chapter 3.

MODEL PROBLEMS

Problems 1–3: In each proportion, find the value of the missing term.

1 $\dfrac{2}{5} = \dfrac{4}{x}$

SOLUTION

Cross multiply: 5 • 4 = 20

Divide: 20 ÷ 2 = 10

Answer: $x = 10$

2 $4 : 5 = s : 15$

SOLUTION

Rewrite in fraction form and cross multiply.

$\dfrac{4}{5} = \dfrac{s}{15}$

Cross multiply: 15 • 4 = 60

Divide: 60 ÷ 5 = 12

Answer: $s = 12$

3 r is to 6 as 3 is to 2

SOLUTION

Rewrite in fractional form and cross multiply.

$\dfrac{r}{6} = \dfrac{3}{2}$

Cross multiply: 6 • 3 = 18

Divide: 18 ÷ 2 = 9

Answer: $r = 9$

4 A map has a scale of 1 inch = 60 miles. If two cities are 93 miles apart, what is the distance between them on the map to the nearest tenth of an inch?

SOLUTION

Set up a proportion: $\dfrac{1 \text{ inch}}{60 \text{ miles}} = \dfrac{x \text{ inches}}{93 \text{ miles}}$

Cross multiply: 93 • 1 = 93

Divide: 93 ÷ 60 ≈ 1.6

Answer: The distance on the map is 1.6 inches.

5 A builder uses 20 bricks to cover 3 square feet of wall. How many square feet can be covered with 2,090 bricks?

SOLUTION

Set up a proportion, being sure to write each ratio in the same way. Here, we use the rate of bricks to square feet:

$\dfrac{20 \text{ bricks}}{3 \text{ square feet}} = \dfrac{2{,}090 \text{ bricks}}{x \text{ square feet}}$

Or:

$\dfrac{20}{3} = \dfrac{2{,}090}{x}$ where x = square feet to be covered with 2,090 bricks

Cross multiply: 3 • 2,090 = 6,270

Divide: 6,270 ÷ 20 = 313.5

Answer: 2,090 bricks will cover 313.5 square feet.

ALTERNATIVE SOLUTIONS

There is more than one way to set up a proportion. Here, our proportion could also be:

$\dfrac{20}{2{,}090} = \dfrac{3}{x}$ or $\dfrac{3}{20} = \dfrac{x}{2{,}090}$ or $\dfrac{2{,}090}{20} = \dfrac{x}{3}$

6 A child's fire truck has two gear belts, one in front and one in back. When the front belt turns 5 times, the rear belt turns 8 times. If the front belt turns 800 times, how many times does the rear belt turn?

SOLUTION

Set up a proportion:

$\dfrac{5 \text{ front turns}}{8 \text{ rear turns}} = \dfrac{800 \text{ front turns}}{x \text{ rear turns}}$ or $\dfrac{5}{8} = \dfrac{800}{x}$

Cross multiply: 8 • 800 = 6,400

Divide: 6,400 ÷ 5 = 1,280

Answer: The rear belt turns 1,280 times.

Practice

1 $\frac{6}{p} = \frac{36}{24}$. Solve for p.
  (1) $p = 1\frac{1}{2}$
  (2) $p = 4$
  (3) $p = 9$
  (4) $p = 144$

2 $3 : x = 2 : 3$. Solve for x.
  (1) $\frac{2}{3}$
  (2) 2
  (3) 3
  (4) $\frac{9}{2}$

3 In Jefferson County, the ratio of Eagle Scouts to Boy Scouts is 1 : 4. If the number of Boy Scouts is 2,400, what is the number of Eagle Scouts?
  (1) 38,400
  (2) 9,600
  (3) 2,400
  (4) 600

4 Which proportion is equivalent to $y : 7 = 5 : 3$?
  (1) $y : 3 = 5 : 7$
  (2) $y : 3 = 7 : 5$
  (3) $5 : y = 3 : 7$
  (4) $5 : y = 7 : 3$

5 According to the directions on the box, $1\frac{1}{3}$ cups of pancake mix and 1 cup water will make enough pancakes for 2 people. How many cups of mix are needed to make pancakes for 5 people?
  (1) $1\frac{7}{8}$ (3) $3\frac{1}{3}$
  (2) $2\frac{2}{3}$ (4) $3\frac{1}{2}$

6 At a book sale, 6 books cost $13. At that rate, how many books could you buy for $32.50?
  (1) 15
  (2) 17
  (3) 19
  (4) 23

7 $\frac{100 \text{ fleas}}{16 \text{ dogs}} = \frac{25 \text{ dogs}}{4 \text{ fleas}}$
  a Explain what is wrong with this proportion.
  b Rewrite it correctly.

8 The wingspan of a jet plane is 22.5 meters. If a model of the plane is built to a scale of 1 centimeter = 3 meters, what is the wingspan of the model?

9 The ratio of boys to girls in a school band is 4 : 3. There are 21 girls in the band. How many members does the band have?

10 In 2 weeks, a family drinks 3 gallons of milk. How many gallons will this family drink in 9 weeks?

11 If $\frac{2}{3}$ of a bucket is filled in 1 minute, how much *more* time will it take to fill the whole bucket?

12 You are making a cookie recipe that calls for 2 cups of flour and 4 eggs. You have $6\frac{1}{2}$ cups of flour, and you want to use it all up. How many eggs will you use?

13 A theater has 1,500 seats. One night, 9 out of every 10 seats were filled. How many people attended that night?

14 The tax bill for a house with an assessed value of $75,620 is $847. At the same rate, find the tax (to the nearest dollar) for a house with an assessed value of $110,000.

15 On a bus, 2 adults can ride for the same price as 3 children. If 1 adult ticket costs $48, what is the price of a child's ticket?

Applications of Percent

Percent (%) means *hundredths* or *out of 100* or *per 100*.

Problem situations that involve applying percents include sales tax, tips, commissions, grades, and percent of increase or decrease.

Computing with Percents

Three basic percent problems are:

- Find a given **percent** of a number.
- Find what percent, or **part**, one number is of another number.
- Find what number, or **whole**, a given number is a certain percent of.

To compute with percents, we use the fraction or decimal equivalent (see Chapter 1). When we use fractions, we can solve percent problems by setting up the following proportion and solving for the unknown quantity—percent, part, or whole:

$$\frac{\text{part}}{\text{whole}} = \frac{\text{percent}}{100}$$

The next model problems are typical examples.

 MODEL PROBLEMS

1. About 65 percent of a person's body weight is water. Find the weight of the water in the body of a 150-pound person.

This problem is asking: What is 65 percent of 150?

SOLUTION

Fraction Method	Decimal Method
Express 65 percent as a fraction: $\frac{65}{100}$.	Convert 65 percent to a decimal:
Think: 65 is the same part of 100 as x is of 150.	$65\% = \frac{65}{100} = 0.65$
Then the proportion is solved for percent:	Multiply:
$\frac{\text{part}}{\text{whole}} = \frac{\text{percent}}{100} \qquad \frac{x}{150} = \frac{65}{100}$	$0.65 \times 150 = 97.5$
Cross multiply: $150 \times 65 = 9{,}750$	
Divide: $9{,}750 \div 100 = 97.5$	

Answer: 65 percent of 150 is 97.5. Therefore, the weight of the water is 97.5 pounds.

62 Chapter 2: Operations With Real Numbers

2 On a 20-question mathematics quiz, Cindy answered 18 questions correctly. What percent of the items did she answer correctly?

This problem is asking: 18 is what percent of 20?

SOLUTION

Fraction Method	Decimal Method
Think: x is the same part of 100 (the whole) as 18 is of 20. Then the proportion is solved for part: $\dfrac{\text{part}}{\text{whole}} = \dfrac{\text{percent}}{100} \qquad \dfrac{18}{20} = \dfrac{x}{100}$ Cross multiply: $100 \times 18 = 1{,}800$ Divide: $1{,}800 \div 20 = 90$	We are finding a part, so the answer will be in decimal form, to be converted to percent. Think: $18 = ? \bullet 20$ Divide: $18 \div 20 = 0.9$ Convert the quotient to a percent: $0.9 = \dfrac{9}{10} = \dfrac{90}{100} = 90$ percent

Answer: 18 is 90 percent of 20. Therefore, Cindy answered 90 percent of the questions correctly.

3 Ninety-five percent of a chorus performed in a concert. If 114 members performed, how many are in the chorus?

This problem is asking: 114 is 95 percent of what number?

SOLUTION

Fraction Method	Decimal Method
Think: 95 is the same part of 100 as 114 is of x. Then the proportion is solved for whole: $\dfrac{\text{part}}{\text{whole}} = \dfrac{\text{percent}}{100} \qquad \dfrac{114}{x} = \dfrac{95}{100}$ Cross multiply: $100 \bullet 114 = 11{,}400$ Divide: $11{,}400 \div 95 = 120$	Convert 95 percent to a decimal: $95\% = \dfrac{95}{100} = 0.95$ Think: $0.95 \bullet ? = 114$ Divide: $114 \div 0.95 = 120$

Answer: 114 is 95 percent of 120. Therefore, the chorus has 120 members.

Note: Percent problems can be solved algebraically using an equation of the form $ax = b$. (See Chapter 3.)

 Practice

1 Keith drove 259 miles of a 500-mile trip. His wife drove the rest of the way. What percent of the trip did his wife drive?

   (1) 25.9%
   (2) 48.2%
   (3) 49.8%
   (4) 51.8%

2 During a fund-raising drive, Kathy, Keri, and Kim raised $10, $15, and $25, respectively. What percent of this money did Kathy raise?

   (1) 5%
   (2) 10%
   (3) 15%
   (4) 20%

Applications of Percent

3. At a summer camp, there are 60 girls and 30 boys. Twenty percent of the girls and 40 percent of the boys passed the swimming test. What percent of the campers passed?

   (1) 24%
   (2) $26\frac{2}{3}\%$
   (3) $36\frac{2}{3}\%$
   (4) 54%

4. Stefan stopped for a drink of water when he had completed 60 percent of a jog. At that point, he had jogged 3 miles. What is the total distance he jogged?

   (1) 2 miles
   (2) 3 miles
   (3) 4 miles
   (4) 5 miles

5. Cassandra paid $35 for a shirt that originally cost $50. What percent of the original price did the shirt cost?

   (1) 40%
   (2) 65%
   (3) 70%
   (4) 85%

6. The sticker price of a car is $15,800. If the sales tax is 8 percent, what is the total cost?

   (1) $15,808
   (2) $17,064
   (3) $17,500
   (4) $18,000

7. After dieting, Devon weighed 180 pounds, which was 90 percent of his original weight. What was his original weight?

8. When a tank is 19 percent full, it contains 136.8 gallons. How much does the full tank hold?

9. Chris is 20 years old and Susan is 50 years old.

   a What percent of Susan's age is Chris?
   b What percent of Chris's age is Susan?

10. A play was attended by 1,071 people. This was 85 percent of the capacity of the auditorium. What is the capacity?

11. Jamila made 51 out of 72 free throws. Find the percentage of free throws made, to the nearest tenth of a percent.

12. A 40 percent discount on a sweater resulted in a saving of $14.40. Find the original price.

13. A bill for a meal is $10.38, including tax. Find the total cost if the customer leaves a 15 percent tip.

14. **Commission** is the amount of money earned for selling a product or service. It is usually a percent of total sales. If a sales representative is paid a monthly salary of $800 plus a 12 percent commission, find his or her income for a month when sales totaled $4,235.50.

15. A test has 125 items. Each correct answer earns 1 point, and 65 percent is a passing grade. If you get 42 answers *wrong*, will you pass? Show your work.

Percent of Increase and Decrease

Percent of increase and **percent of decrease** can be expressed as a ratio:

$$\frac{\text{change in value}}{\text{original value}} \cdot 100$$

Therefore:

$$\text{percent of increase} = \frac{\text{new value} - \text{original value}}{\text{original value}} \cdot 100$$

$$\text{percent of decrease} = \frac{\text{original value} - \text{new value}}{\text{original value}} \cdot 100$$

These formulas can be used to find percent of increase or decrease, original value, or new value when the other two terms are known.

MODEL PROBLEMS

1 If the price of an accordion is marked down from $150 to $120, what is the percent of decrease?

SOLUTION

Percent of decrease is:

$$\left(\frac{\text{original value} - \text{new value}}{\text{original value}}\right) \cdot 100 = \left(\frac{150 - 120}{150}\right) \cdot 100 = \frac{30}{150} \cdot 100 = 20\%$$

2 If a stamp increases in value from $120 to $150, what is the percent of increase?

SOLUTION

Percent of increase is:

$$\left(\frac{\text{new value} - \text{original value}}{\text{original value}}\right) \cdot 100 = \left(\frac{150 - 120}{120}\right) \cdot 100 = \left(\frac{30}{120}\right) \cdot 100 = 25\%$$

Many real-world problems involve percent of increase and decrease. For example, a **markup** is a percent-of-increase situation, and a **discount** is a percent-of-decrease situation.

 Practice

1 If the cost of a No. 2 pencil has increased from 8 cents to 18 cents over the last 10 years, what was the percent of increase?

(1) 175%
(2) 125%
(3) 55.5%
(4) 44.4%

2 You buy a $3.75 book for $3.30. What was the percent discount?

(1) 6.4%
(2) 12%
(3) 13.6%
(4) 45%

3 A CD player costs a store $270. If the markup is 23 percent, what is the selling price?

(1) $293
(2) $332.10
(3) $405
(4) $500

4 Shampoo is on sale: "BUY ONE AND GET THE SECOND AT HALF PRICE!" If you buy 2 bottles, what is the total discount?

(1) 25%
(2) 30%
(3) 50%
(4) 75%

5 The original price of a VCR is $600. It goes on sale with a discount of 25 percent. Then, in a closeout sale, it is discounted 10 percent. What is its final selling price?

(1) $135
(2) $210
(3) $390
(4) $405

6 Ali's salary was cut 20 percent. Later, it was raised 20 percent. After the raise, his pay was:

(1) equal to his original salary
(2) less than it would have been if the raise had come before the cut
(3) 96 percent of his original salary
(4) 120 percent of his original salary

7 Describe what happens when a number is:
   a increased by 100 percent
   b decreased by 100 percent

8 Explain whether or not each of the following is possible:
   a percent of increase greater than 100 percent
   b percent of decrease greater than 100 percent

9 A jacket originally priced at $85 goes on sale for $49.99. What is the percent discount to the nearest percent?

10 The price of a train ticket rises from $15 to $18.30. What is the percent of increase to the nearest percent?

11 A school's enrollment decreases from 4,850 to 3,104. What is the percent of decrease?

12 A balloon ascends from 13,500 feet to 19,008 feet. Its altitude has increased by what percent?

13 What is the total cost of a $28.50 garden rake including an 8 percent sales tax?

14 During an autumn sale, an air conditioner that originally sold for $510 is discounted 25 percent.
   a If you waited for this sale, how much did you save?
   b What is the sale price of the air conditioner?

15 Felipe buys a computer for $2,000 and has it shipped to his home. The sales tax is 6.5 percent. The shipping charge is 8 percent of the pretax price.
   a What is the total cost?
   b How much would Felipe have saved if he had taken the computer home himself instead of shipping it?

CHAPTER REVIEW

1 What is the opposite of $|(-10 - 5)(-10 + 5)|$?
   (1) -75
   (2) -50
   (3) 50
   (4) 75

2 The product of two integers, -1 and -8, is 8. If each integer is increased by 1, what is the new product?
   (1) 18
   (2) 14
   (3) 0
   (4) -18

3 The toll on a bridge is $2 for a car and driver and $0.75 for each additional passenger. If the toll for Marigold's car was $4.25, how many people were riding in it?
   (1) 2
   (2) 3
   (3) 4
   (4) 5

4 $2^3 \cdot 4^3 =$
   (1) 2^9
   (2) 2^{18}
   (3) 8^6
   (4) 8^9

5 $\sqrt{8} \times \sqrt{30} =$
   (1) $4\sqrt{15}$
   (2) $15\sqrt{4}$
   (3) $12\sqrt{20}$
   (4) $10\sqrt{24}$

6 A car travels 165 miles on 6 gallons of gas. You need to find how far it can go on 10 gallons. Which proportion can *not* be used to solve this problem?
   (1) $\dfrac{6}{165} = \dfrac{10}{x}$
   (2) $\dfrac{x}{10} = \dfrac{165}{6}$
   (3) $\dfrac{10}{165} = \dfrac{6}{x}$
   (4) $\dfrac{6}{10} = \dfrac{165}{x}$

66 Chapter 2: Operations With Real Numbers

7 Match each equation with the property it illustrates.

Equations	Properties
(1) $3 + (4 + 5) = (3 + 4) + 5$	(a) Commutative property of multiplication
(2) $5 \times 3 = 3 \times 5$	(b) Additive identity
(3) $8 \times 11 + 8 \times 9 = 8 \times 20$	(c) Multiplicative inverse (reciprocal)
(4) $d + 0 = 0 + d = d$	(d) Commutative property of addition
(5) integer + integer = integer	(e) Associative property of addition
(6) $1 \cdot t = t \cdot 1 = t$	(f) Distributive property
(7) $\left(\dfrac{1}{x}\right)\left(\dfrac{x}{1}\right) = 1, x \neq 0$	(g) Closure
(8) $x + 7 = 7 + x$	(h) Multiplicative identity

8 In football, a team needs 10 yards to make a first down. Suppose that in three plays, your team gains 6 yards, then loses 2 yards, then gains 7 yards. What was the total yardage, and did the team make a first down?
  a Calculate the yardage using signed numbers.
  b Show the situation on a number line.

Exercises 9–17: Simplify, using the order of operations.

9 $10 + 15 \div 5$

10 $12 - 6 \div 3$

11 $3 \times 4 + 8 \div 2 - 6 \div 2$

12 $10 - 48 \div 3 \times 8 + 6$

13 $72 \div 6 - 2 \times 3$

14 $10 - 4[3 - 2(15 \div 3)]$

15 $[60 \div (2 \times 3)] \div 10$

16 $\dfrac{(18 \div 3) + 6}{18 \div (6 \div 3)}$

17 $-42.5 - 2(3^2 - 4.7)(-5)$

18 A speed of 50 miles per hour is equal to about 80 kilometers per hour. Find the speed in kilometers per hour that equals:
  a 55 miles per hour
  b 65 miles per hour

19 A sweatshirt that usually costs $39.95 is on sale at a 15 percent discount. The sales tax is 6 percent. Find to the nearest cent:
  a amount of the discount
  b new price
  c total cost including tax

20 A green log is 50 percent water, by weight. As it ages, it dries out, until it is only 25 percent water. What is the percent decrease in water? Justify your answer with an example.

Now, turn to page 406 and take the Cumulative Review test for Chapters 1–2. This will help you monitor your progress and keep all your skills sharp.

Chapter 3
Algebraic Operations and Reasoning

The Language of Algebra

Any situation in which one or more numbers are unknown can be made into an algebra problem. When an algebra problem is presented in words, part of the job of solving the problem is to rewrite the information using algebraic expressions. **Algebraic expressions** contain numbers, variables, and symbols for the operations. **Algebraic equations** are statements containing two algebraic expressions joined with a sign of equality. **Algebraic inequalities** contain two algebraic expressions joined with a sign of inequality.

Use the following chart to help translate key phrases into mathematical symbols.

Key Words	Symbol	English and Algebra Translations
Signs of Operations		
sum, add, increase, more than, plus, increased by, greater, exceeded by, and	+	f increased by 10 $f + 10$ the sum of x and 9 $x + 9$
difference, subtract, take away, minus, decrease, fewer, less, less than, decreased by, diminished by	−	4 less than y $y - 4$ 32 fewer CDs than Abdul's $A - 32$
multiply, of, product, times	×	the product of 4 and t $4 \times t$ or $4t$ One-half of the pumpkin weighs 3 pounds. $\frac{1}{2}p = 3$
divide, quotient, into, for, per, divided by	÷	500 divided into 4 parts $500 \div 4$ Sue cut a 24-inch board into 8 pieces. $24 \div 8$

Key Words	Symbol	English and Algebra Translations
Sign of Equality		
equals, is equivalent to, is the result of, is	=	Twice a number plus 4 is 14. $2n + 4 = 14$
Signs of Inequality		
is greater than, is more, has more	>	x is greater than 5. $x > 5$
is less than, is fewer, has fewer	<	Pauline and Fredrica together have fewer books than Vicki. $P + F < V$
is greater than or equal to, is at least, has at least	≥	The lecture was at least 4 hours long. $l \geq 4$
is less than or equal to, is at most, has at most	≤	The bank is at most 1.5 miles away. $b \leq 1.5$
is not equal to, is not the same as, cannot equal, does not equal	≠	Today is not the third of the month. $d \neq 3$

Points to Remember

- Before you translate a problem situation into an algebraic expression, be sure you understand what the situation means.
- Define each variable you create with an equals sign.
- When there are two or more unknown quantities, you may need to represent only one of them with a letter variable. Try to express the other unknown quantities in terms of that letter if possible.
- Often, translation assumes that you know certain relationships about money, measurement, or time. If a relationship is not clear to you, you should try making a chart or table of the situation before translating it into an algebraic expression.

 MODEL PROBLEMS

1. If x = Luke's age and $x + 6$ = Nancy's age in years, write an equation for each of the following statements:

   a The sum of Luke's and Nancy's ages is less than 30 years.

   SOLUTION

   $x + (x + 6) < 30$

   b Three times Luke's age equals twice Nancy's age.

   SOLUTION

   $3x = 2(x + 6)$

   c The difference between Nancy's age in 10 years and twice Luke's present age is 4 years.

   SOLUTION

   $[(x + 6) + 10] - 2x = 4$

The Language of Algebra   **69**

2 When 8 is subtracted from 3 times a number, the result is 19. Which of the following equations represents this statement?

(1) $8 - 3x = 19$
(2) $3x - 8 = 19$
(3) $3(x - 8) = 19$
(4) $3(8 - x) = 19$

SOLUTION

Let $x =$ the number.

Then "3 times a number" $= 3x$.

"8 is subtracted from 3 times a number" $= 3x - 8$.

"The result is" translates into an equal sign. Thus, $3x - 8 = 19$.

Answer: (2)

3 Jonas has 20 coins, all nickels and dimes, that have a total value of $1.25. If n represents the number of nickels, which algebraic equation represents this situation?

(1) $20n = 125$
(2) $5n = 1.25$
(3) $5n + 10(20 - n) = 125$
(4) $5n + 10n - 20 = 125$

SOLUTION

Let the number of nickels $= n$.

Then the value of the nickels at five cents per nickel $= 5n$.

Since Jonas has 20 coins in all, the number of dimes he has is 20 minus the number of nickels. Thus, the number of dimes $= 20 - n$.

The value of the dimes at 10 cents per dime $= 10(20 - n)$.

The total amount of money is the sum of the values of the nickels and the dimes. Thus, Jonas's total $= 5n + 10(20 - n)$. If the total value is $1.25, then $5n + 10(20 - n) = 125$ cents.

When you are solving coin problems like this, work in cents so that decimal operations are not required.

Answer: (3)

 Practice

1 Eight years ago, Clyde was 7 years old. Which equation is true if C represents Clyde's age now?

(1) $8 + C = 7$
(2) $7 + C = 8$
(3) $8 - 7 = C$
(4) $C - 8 = 7$

2 Which of the following represents the total cost of x shirts bought at a cost of $(x + 5)$ dollars each?

(1) $(x + 5)$ dollars
(2) $x + (x + 5)$ dollars
(3) $x(x + 5)$ dollars
(4) $x^2 + 5$ dollars

3 If a batch of holiday cookies requires 3 cups of flour, how many cups of flour would be used in baking m batches of cookies?

(1) $m + 3$
(2) $m - 3$
(3) $\dfrac{m}{3}$
(4) $3m$

4 If y represents the tens digit and x the units digit of a two-digit number, then the number is represented by

(1) $y + x$
(2) yx
(3) $10x + y$
(4) $10y + x$

5 George is 15 years old. He is one-third as old as his father. Which equation is true if g represents George's father's age?

(1) $3g = 15$
(2) $g + 3 = 15$
(3) $\dfrac{1}{3}g = 15$
(4) $g - 3 = 15$

Exercises 6–10: Match each statement on the left to the correct translation on the right.

6 The square of a number is greater than one-tenth of the number.

7 A number is greater than twice the number decreased by 10.

8 Twice a number is less than the number and 10.

9 Maria is at least 10 years older than Nadine.

10 A number is ten less than a second number

(a) $n > 2n - 10$
(b) $n = m - 10$
(c) $m \geq n + 10$
(d) $n^2 > \dfrac{n}{10}$
(e) $2n < n + 10$
(f) $n^3 > \dfrac{1}{10}n^2$

Exercises 11–19: Represent each situation with an algebraic equation.

11 Four times a number is nine more than the number.

12 Todd's weight is 16 pounds more than three times his son's weight, w.

13 A number squared, increased by 15, is the same as the square of 1 more than the number.

14 Twice Jacqueline's age is the same as her father's age.

15 The quotient of p divided by r, decreased by the product of p and r, is 5.

16 In 20 years, Tracy's age will be 5 years greater than twice her current age.

17 Jack worked 6 hours on Friday and $x + 2$ hours on Saturday at a constant rate of pay. He earned $92.

18 The square root of half a number is equal to $\dfrac{1}{10}$ the number.

19 From the product of $3x - 2$ and $4x + 3$ subtract double the square of $(2x - 5)$ to get 0.

20 Genna has $12.50 in pennies, dimes, and half-dollars in her piggy bank. She has 5 times as many dimes as half-dollars and 5 times as many pennies as dimes.

  a If Genna has h half-dollars, write an expression for the number of dimes and pennies that she has.

  b Using these expressions, write an equation showing that the total value of all Genna's coins is $12.50.

The Language of Algebra

Evaluating Algebraic Expressions

The group of values that can be substituted for a variable in an algebraic expression is known as the **domain** or **replacement set**. For example, let $6x$ represent the number of soda cans in a certain number, x, of six packs. Here, x can be replaced by any counting number, so the replacement set for x is $\{0, 1, 2, 3, ...\}$. To find the number of sodas in three six-packs, we would let $x = 3$ and substitute 3 in the expression, getting 6×3, or 18. Substituting numbers for variables to find a value for an expression is called **evaluating** the expression.

To Evaluate an Algebraic Expression

- Substitute the given values for the letters (or variables).
- Simplify by using the order of operations.

Some expressions with variables located in the denominator of a fraction or in a radicand ($\sqrt{\ }$) have limits on their replacement sets due to the properties of fractions and radicals. Since division by zero is impossible, a fraction is undefined when its denominator equals 0. Thus, the fraction $\frac{x}{x+7}$ is undefined when $x = -7$. In this case the domain of $\frac{x}{x+7}$ cannot include -7.

Since the square root of a negative number is not real, an algebraic expression with a negative radicand cannot be evaluated as a real number. Thus $2\sqrt{7-x}$ is not a real number when $7 - x$ is negative. In this case the domain cannot include any real numbers greater than 7.

The domain is important in functions. See Chapter 5.

MODEL PROBLEMS

1 Evaluate $a - b(a - x)^2$ if $a = -1$, $b = -2$, and $x = 3$.

SOLUTION

$-1 - (-2)[(-1) - 3)]^2$
$-1 + 2[-4]^2$
$-1 + 2[16]$
$-1 + 32 = 31$

2 If d is an odd integer and e is an even integer, which of the following is an odd integer?

(1) $2d + e$ (3) $de + d$
(2) $2d + 2e$ (4) $3e + 3d + 1$

SOLUTION

The best method to solve a problem involving sets of numbers is to substitute an element from each set. Substitute any even number for e, like 2, and any odd number for d, like 3. Then we have

(1) $2d + e = 2 \times 3 + 2 = 8$
(2) $2d + 2e = 2 \times 3 + 2 \times 2 = 10$
(3) $de + d = 3 \times 2 + 3 = 9$
(4) $3e + 3d + 1 = 3 \times 2 + 3 \times 3 + 1 = 16$

Answer: (3)

3 Which of these values will make the algebraic fraction $\frac{2x+3}{2x-3}$ undefined?

(1) 2
(2) 3
(3) $\frac{2}{3}$
(4) $\frac{3}{2}$

SOLUTION

Substitute the values for x in the denominator to see which would make the denominator equal zero.

(1) $2(2) - 3 = 1$
(2) $2(3) - 3 = 3$
(3) $2\left(\frac{2}{3}\right) - 3 = \frac{4}{3} - 3 = \frac{4}{3} - \frac{9}{3} = -\frac{5}{3}$
(4) $2\left(\frac{3}{2}\right) - 3 = 0$

Answer: (4)

Practice

1. Find the value of $a - (b - c)$ when $a = -3$, $b = 4$, and $c = -5$.
   - (1) -12
   - (2) -4
   - (3) -2
   - (4) 6

2. Find the value of $ab + ac$ when $a = -3$, $b = 4$, and $c = -5$.
   - (1) 27
   - (2) 3
   - (3) -3
   - (4) -27

3. If $x + 3$ is an even integer, which of the following is not an even integer?
   - (1) $2x + 2$
   - (2) $2x + 3$
   - (3) $x - 3$
   - (4) $3x - 1$

4. Which of the fractional expressions has a value greater than 1 if K is an integer greater than 1?
   - (1) $\dfrac{K}{1 + K}$
   - (2) $\dfrac{1}{1 - K}$
   - (3) $\dfrac{K + 1}{K}$
   - (4) $\dfrac{K}{1 - K}$

5. If e is an even integer and d is an odd integer, which of the following is an even integer?
   - (1) $2d^2 + 3e$
   - (2) $de + d^2$
   - (3) $3e^2 + d$
   - (4) $d^2 + 3e$

6. If a and b are prime numbers greater than 2, which of the following expressions must be odd?
   - (1) $a + b - 2$
   - (2) $ab - 1$
   - (3) $ab + 1$
   - (4) $ab + 2$

7. If $m = -3$, then what does $-2m^3$ equal?

8. If $a = \dfrac{2}{3}$, and $b = -\dfrac{3}{2}$, then what is the value of $a^3 + b^2$?

9. If $x = 3$ and $y = -3$, find the value of $x^3 + y^3$.

10. If $r = s = -5$, then what is the value of $r(r + s) - s(r - s)$?

11. What value of r will make $2\sqrt{r}$ undefined?

12. For what values of x will $\dfrac{3(x - 3)^3}{3x}$ be undefined?

13. If $a = 1$ and $b = -20$, find the value of $\dfrac{-a + \sqrt{a^2 - 4b}}{4}$.

14. Given that $s = 4$ and $t = -3$, evaluate the expression $(s^3 - t^3 + s)(t^2 + \sqrt{s})$.

15. If $r = 4$, $p = -1$, and $q = 2$, evaluate $\dfrac{(rp)^2 - pq + 3}{(r - p + q)^2}$.

16. What is the value of $3x^2 - 3x - 3$ as x takes each one of the three following values? Show your work clearly.
    - a. -1
    - b. 0
    - c. 1

Exercises 17–20: Let $x = 3$ and $y = 4$. Fill in each blank in the four statements below with "less than," "equal to," or "greater than."

17. The value of $\dfrac{x}{y - 1}$ is _____ the value of $\dfrac{x}{y}$.

18. The value of $\dfrac{x - 2}{y - 2}$ is _____ the value of $\dfrac{x}{y}$.

19. The value of $\dfrac{x^2 - 1}{2y}$ is _____ the value of $\dfrac{x}{y}$.

20. The value of $\dfrac{x^2}{y^2}$ is _____ the value of $\dfrac{x}{y}$.

Evaluating Algebraic Expressions

Adding and Subtracting Algebraic Expressions

A **term** is an algebraic expression written with numbers, variables, or both and using multiplication, division, or both. A term that has no variables is often called a **constant**. Examples: 5, x, cd, $6mx$, $\dfrac{4x^2y}{-3m^3}$ are all terms. Of these, 5 is a constant.

The **coefficient** of a term is the numerical part of the term. If no coefficient is written, then it is understood to be 1. The coefficient of $6mx$ is 6, the coefficient of $\dfrac{4x^2y}{-3m^3}$ is $-\dfrac{4}{3}$, and the coefficient of cd is 1.

An algebraic expression of exactly one term is called a **monomial**. Examples of monomials include 7, a, and $2x^2$.

Like terms are terms that contain the *same variables* with corresponding variables having the *same exponents*. Terms are separated by plus (+) and minus (−) signs. Examples: $7x^3y^4$ and x^3y^4 are like terms; 5 and 100 are like terms. Y and X are not like terms; n and n^3 are not like terms.

Since algebraic expressions themselves represent numbers, they can be added, subtracted, multiplied, and divided. When algebraic expressions are added or subtracted, they can be combined only if they have like terms.

To Add or Subtract Monomials With Like Terms

- Use the distributive property and the rules of signed numbers to add or subtract the coefficients of each term.
- Write this sum with the variable part from the terms.

MODEL PROBLEMS

SOLUTIONS

1. Add $-2x^3$ and $5x^3$. $\quad -2x^3 + 5x^3 = (-2 + 5)x^3 = 3x^3$

2. Subtract $7mn^2$ from $4mn^2$. $\quad 4mn^2 - 7mn^2 = (4 - 7)mn^2 = -3mn^2$

An algebraic expression of one or more unlike terms is a **polynomial**. **Binomials** are polynomials with *two* unlike terms. $7v + 9$ and $3x^2 - 8y$ are both binomials. **Trinomials** are polynomials with *three* unlike terms. $x^2 - 3x - 5$ and $3a^2bx - 5ax - 2ab$ are both trinomials.

A polynomial with one variable is said to be in **standard form** when it has no like terms and is written in order of descending exponents. For example, $4x + 9 - 5x^2 + 3x^3$ in standard form is $3x^3 - 5x^2 + 4x + 9$. When you are asked to **simplify** a polynomial, you should always write it in standard form.

When you are rearranging terms, keep signs with the terms to their right. Here, the − moves with the $5x^2$.

To Add Polynomials

- Use the commutative property to rearrange the terms so like terms are beside each other.
- Combine like terms.

74 Chapter 3: Algebraic Operations and Reasoning

MODEL PROBLEM

Add $-3x^2 + 4y$ and $5x^3 - 6x^2 - 3y$.

SOLUTION

$(-3x^2 + 4y) + (5x^3 - 6x^2 - 3y)$
$= -3x^2 + 4y + 5x^3 - 6x^2 - 3y$
$= 5x^3 - 3x^2 - 6x^2 + 4y - 3y$
$= 5x^3 + (-3 - 6)x^2 + (4 - 3)y$
$= 5x^3 - 9x^2 + 1y$
$= 5x^3 - 9x^2 + y$

To Subtract Polynomials

- Change the sign of every term in the subtracted polynomial and remove parentheses.
- Combine like terms.

Changing signs is like multiplying each term inside the second parentheses by -1.

MODEL PROBLEM

Subtract $9x^2 - 5x$ from $-4x^2 - 8x$.

SOLUTION

First, rewrite the problem to show the subtraction as $(-4x^2 - 8x) - (9x^2 - 5x)$. Change the signs in the subtracted polynomial and remove the parentheses.

$(-4x^2 - 8x) - (9x^2 - 5x)$
$= -4x^2 - 8x - 9x^2 + 5x$
$= -4x^2 - 9x^2 - 8x + 5x$
$= (-4 - 9)x^2 + (-8 + 5)x$
$= -13x^2 - 3x$

Check.

$(-13x^2 - 3x) + (9x^2 - 5x)$
$= (-13x^2 + 9x^2) + (-3x - 5x)$
$= -4x^2 - 8x$

Practice

1 Simplify $(7x + 6x) - 12x$.

   (1) x
   (2) $2x$
   (3) x^2
   (4) 1

2 Simplify $m - [2 - (2 - m)]$.

   (1) -4
   (2) 0
   (3) $2m - 2$
   (4) $2m$

3 Find the sum of $-3x^2 - 4xy + 2y^2$ and $-x^2 + 5xy - 8y^2$.

   (1) $-2x^2 + xy - 6y^2$
   (2) $-4x^2 + xy - 6y^2$
   (3) $-4x^2 - 9xy - 6y^2$
   (4) $-2x^2 - 9xy + 10y^2$

4 What is the result when $-4x + 6$ is subtracted from $8x + 6$?

   (1) $12x + 12$
   (2) $12x$
   (3) $4x + 6$
   (4) $4x$

5 Simplify $5x - 3y - 7x + y$.

6 What is the result when $10x - 7$ is subtracted from $9x - 15$?

7 What is the result when $3 - 2x$ is subtracted from the sum of $x + 3$ and $5 - x$?

8 Add $2x - 3x^2 - 7$, $3 - 5x - 5x^2$, and $2x^2 + 12 + x - x^2$.

9 Find the sum of $4x^2 - 6x - 3$ and $3x^2 - 5x + 7$.

For Exercises 10–15, remove parentheses and find each sum or difference.

10 $(x^2 + 3x + 1) + (2x^2 - x - 2)$
11 $(-3x^2 + 4x - 8) + (4x^2 + 5x - 11)$
12 $(4x^2 - 4) + (x^2 - x + 4)$
13 $(-3x^2 + 6x + 1) - (4x^2 + 7x - 3)$
14 $(x^2 - x - 9) - (-2x^2 + x + 4)$
15 $(x^2 + 2x - 3) - (x^2 - 4)$

16 What polynomial will produce the sum $6x^2 - x + 2$ when added to $4x^2 - 6x + 3$?

Exercises 17–20: If a taxi ride costs d dollars a mile for the first 5 miles and $d + n$ dollars for each additional mile, write an expression that describes the cost, in dollars, to ride each of the following distances. Combine like terms in your answers.

17 5 miles
18 6 miles
19 7 miles
20 8 miles

Multiplying Algebraic Expressions

Remember:
$x^a \cdot x^b = x^{a+b}$

To Find the Product of Monomials

- Multiply the numerical coefficients using the rule of signs for multiplication.
- Multiply variables of the same base by adding the exponents.
- Multiply these two products together.

 MODEL PROBLEM

Multiply $-4a^2b^3$ and $3a^3b$.

SOLUTION

$(-4a^2b^3)(3a^3b)$
$= (-4 \cdot 3)(a^2 \cdot a^3 \cdot b^3 \cdot b^1)$
$= -12a^5b^4$

To Find the Product of a Polynomial and a Monomial

- First use the distributive property to remove parentheses.
- Then multiply each term of the polynomial by the monomial separately.
- Simplify the product if possible.

Chapter 3: Algebraic Operations and Reasoning

MODEL PROBLEMS

SOLUTIONS

1 $-6(2a^2 - 3a + 1)$ $= -6(2a^2) + (-6)(-3a) + (-6)(1)$
$= -12a^2 + 18a - 6$

2 $3a^2b(a^2 - 2ab - 3b^2)$ $= 3a^2b(a^2) + 3a^2b(-2ab) + 3a^2b(-3b^2)$
$= 3a^4b - 6a^3b^2 - 9a^2b^3$

When we multiply polynomials, each term of one polynomial must multiply each term of the other polynomial. There are three methods for multiplying polynomials: **distributive property**, **columns**, and the **FOIL method**. The FOIL method can be used only to multiply two binomials.

To Use the Distributive Property for Multiplying Polynomials

- Distribute the first polynomial over the terms of the second.
- Solve as before.

 MODEL PROBLEMS

1 Multiply: $(x + 2)(x + 3)$

SOLUTION

Distribute $(x + 2)$ over x and 3.
Distribute x over the first $(x + 2)$ and 3 over the second.
Combine like terms.

$(x + 2)(x + 3)$
$= (x + 2)x + (x + 2)3$
$= x^2 + 2x + 3x + 6$
$= x^2 + 5x + 6$

2 $(2x - 3)(3x^2 - 5x + 4)$

SOLUTION

$(2x - 3)(3x^2 - 5x + 4)$
$= (2x - 3)3x^2 + (2x - 3)(-5x) + (2x - 3)4$
$= 6x^3 - 9x^2 - 10x^2 + 15x + 8x - 12$
$= 6x^3 - 19x^2 + 23x - 12$

Note: The distributive property can also be used this way: Distribute the first term to each term in the second parentheses and then the second term to each term in the second parentheses.

3 $(x + 4)(2x - 3)$

SOLUTION

$(x + 4)(2x - 3)$
$= x(2x) + x(-3) + 4(2x) + 4(-3)$
$= 2x^2 - 3x + 8x - 12$
$= 2x^2 + 5x - 12$

Multiplying Algebraic Expressions

To Use Columns for Multiplying Polynomials

- Set up the multiplication in columns of like terms.
- Multiply as you would large whole numbers.

MODEL PROBLEMS

1 Multiply $x^2 + 2$ and $x + 3$.

SOLUTION

Set up columns of like terms.
Multiply:

 $\quad$ 3 times $(x^2 + 2)$

 $\quad$ x times $(x^2 + 2)$

Add the products in columns:

$$\begin{array}{r} x^2 + 0x + 2 \\ \times \quad\quad x + 3 \\ \hline 3x^2 + 0x + 6 \\ x^3 + 0x^2 + 2x \quad\quad \\ \hline x^3 + 3x^2 + 2x + 6 \end{array}$$

2 Multiply $3x^2 - 5x + 4$ and $2x - 3$.

SOLUTION

$$\begin{array}{r} 3x^2 - 5x + 4 \\ \times \quad\quad 2x - 3 \\ \hline -9x^2 + 15x - 12 \\ 6x^3 - 10x^2 + 8x \quad\quad \\ \hline 6x^3 - 19x^2 + 23x - 12 \end{array}$$

To Use the FOIL Method for Multiplying Binomials

- <u>F</u>   Multiply the first terms.
- <u>O</u>   Multiply the outer terms.
- <u>I</u>   Multiply the inner terms.
- <u>L</u>   Multiply the last terms.
- Combine any like terms.

FOIL stands for <u>F</u>irst, <u>O</u>uter, <u>I</u>nner, <u>L</u>ast.

MODEL PROBLEMS

1 $(x + 3)(x + 4)$

SOLUTION

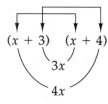

$$\underset{\substack{\text{product of} \\ \text{first terms} \\ x \cdot x}}{x^2} + \underset{\substack{\text{sum of} \\ \text{inner and} \\ \text{outer} \\ 3x + 4x}}{7x} + \underset{\substack{\text{product of} \\ \text{last terms} \\ 3 \cdot 4}}{12}$$

2 $(2a - 3)(5a + 2)$

SOLUTION

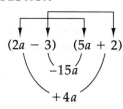

$10a^2 - 11a - 6$

78   Chapter 3: Algebraic Operations and Reasoning

Practice

1. Find the product of $3x^2$ and $2x^3$.
   (1) $5x^5$
   (2) $5x^6$
   (3) $6x^5$
   (4) $6x^6$

2. Find the product: $2x^5y^2 \cdot 5xy^3z^2 \cdot 2yz$
   (1) $9x^6y^6z^3$
   (2) $20x^5y^5z^2$
   (3) $20x^5y^6z^2$
   (4) $20x^6y^6z^3$

3. Find the product: $(2x + 6)(x - 3)$
   (1) $2x^2 + 3x - 18$
   (2) $2x^2 + x - 18$
   (3) $2x^2 - 18$
   (4) $3x + 3$

4. Find the product: $(3a - 2)(5a - 5)$
   (1) $15a^2 - 31a - 10$
   (2) $15a^2 - 25a + 10$
   (3) $15a^2 - 5a + 10$
   (4) $8a^2 - 25a + 7$

5. Find the product of $2x^2y^3$ and $4x^2y^4$.

6. Find the product of $x(-2x)^3$.

7. Find the product of $3x + 5$ and $2x - 7$ and write in standard form.

8. Simplify: $a^3 - 5a^2 - a(3a + 3 - a^3)$.

Exercises 9–17: Simplify the algebraic expressions and write in standard form.

9. $\frac{1}{4}b(12b^3 - 4ab + 16)$

10. $3(4x - 5) - (2x - 10)$

11. $(6x^2 - 2y)(3x^2 - 4y)$

12. $(x + 1)(x^2 + 2x + 1)$

13. $(a - 1)^2 - (a + 1)^2$

14. $0.6 - 0.4(0.3a - 0.01)$

15. $\left(2x + \frac{1}{2}\right)\left(4x + \frac{1}{2}\right)$

16. $\left(\frac{1}{4}x - 9\right)\left(\frac{2}{3}x + 4\right)$

17. $(x^2 + x)(x + 1)$

18. Find the product of $(a - 8)(3a^2 + 2a - 7)$, simplify, and write the answer in standard form. Show your work.

Special Products of Binomials

Certain binomial products appear so often that you should know them on sight. They are useful to have memorized when you are factoring quadratic equations.

Product of a Sum and a Difference: $(x + y)(x - y) = x^2 - y^2$

Distributive Method	Column Method	FOIL Method
Distribute the first term, x, to each term in the second parentheses. Then distribute the second term, y, to each term in the second parentheses. $(x + y)(x - y)$ $= x(x - y) + y(x - y)$ $= x^2 - xy + yx - y^2$ $= x^2 - y^2$	Solution to $(x + y)(x - y)$: $\quad\quad x + y$ $\times \quad\quad x - y$ $\overline{\quad\quad -xy - y^2}$ $\underline{+ x^2 + xy \quad\quad}$ $\quad\quad x^2 \quad\quad -y^2$	**F** Multiply the first terms: $(x)(x) = x^2$ **O** Multiply the outer terms: $(x)(-y) = -xy$ **I** Multiply the inner terms: $(y)(x) = yx$ **L** Multiply the last terms: $(y)(-y) = -y^2$ Product: $x^2 - xy + yx - y^2 = x^2 - y^2$

Square of a Binomial: $(x+y)^2 = x^2 + 2xy + y^2$ and $(x-y)^2 = x^2 - 2xy + y^2$

Distributive Method	Column Method	FOIL Method
To find the first term, square the first term of the binomial: $$x^2$$ To find the middle term, double the product of the two terms of the binomial: $$2xy \text{ or } -2xy$$ To find the third term, square the second term of the binomial: $$(y)(y) = y^2 \text{ or } (-y)(-y) = y^2$$ Products: $$(x+y)^2 = (x+y)(x+y)$$ $$= x^2 + 2xy + y^2$$ $$(x-y)^2 = (x-y)(x-y)$$ $$= x^2 - 2xy + y^2$$	Solution to $(x+y)^2$: $$\begin{array}{r} x+y \\ \times \quad x+y \\ \hline xy + y^2 \\ +\ x^2 + xy \\ \hline x^2 + 2xy + y^2 \end{array}$$ Solution to $(x-y)^2$: $$\begin{array}{r} x-y \\ \times \quad x-y \\ \hline -xy + y^2 \\ +\ x^2 - xy \\ \hline x^2 - 2xy + y^2 \end{array}$$	$(x+y)^2$ or $(x-y)^2$ **F** Multiply the first terms: $$(x)(x) = x^2$$ **O** Multiply the outer terms: $$(x)(y) = xy \text{ or } (x)(-y) = -xy$$ **I** Multiply the inner terms: $$(y)(x) = yx \text{ or } (-y)(x) = -yx$$ **L** Multiply the last terms: $$(y)(y) = y^2 \text{ or } (-y)(-y) = y^2$$ Products: $$(x+y)^2 = x^2 + xy + yx + y^2$$ $$= x^2 + 2xy + y^2$$ $$(x-y)^2 = x^2 - xy - yx + y^2$$ $$= x^2 - 2xy + y^2$$

 MODEL PROBLEMS

SOLUTIONS

1 $(2m - 1)(2m + 1)$ Substitute $2m$ for x and 1 for y in the expression $x^2 - y^2$.
$$(2m)^2 - (1)^2 = 4m^2 - 1$$

2 $(3y + 2)(3y + 2)$ Substitute $3y$ for x and 2 for y in the expression $x^2 + 2xy + y^2$.
$$(3y)^2 + 2(3y)(2) + (2)^2 = 9y^2 + 12y + 4$$

3 $(5x - 6)(5x - 6)$ Substitute $5x$ for x and 6 for y in the expression $x^2 - 2xy + y^2$.
$$(5x)^2 - 2(5x)(6) + (6)^2 = 25x^2 - 60x + 36$$

 When you square a binomial, the first and third terms are always positive. The sign of the middle term is the same as the sign between the terms of the given binomial.

 Practice

1 Which is the simplified form of $(x + 10)(x - 10)$?
  (1) $x^2 + 20x + 100$
  (2) $x^2 + 10x + 100$
  (3) $x^2 - 100$
  (4) $x^2 - 10x + 100$

2 Simplify $(2m + 3)^2$.
  (1) $2m^2 + 6m + 9$
  (2) $2m^2 + 12m + 9$
  (3) $4m^2 + 9$
  (4) $4m^2 + 12m + 9$

3 Simplify: $\left(\dfrac{2}{3}v - 2\right)\left(\dfrac{2}{3}v - 2\right)$

(1) $\dfrac{2}{3}v^2 - 4v + 4$

(2) $\dfrac{4}{9}v^2 - \dfrac{8}{3}v + 4$

(3) $\dfrac{4}{9}v^2 - \dfrac{4}{3}v + 4$

(4) $\dfrac{4}{9}v^2 + \dfrac{4}{3}v - 4$

4 Find the product of $(1.2a + 0.3b)$ and $(1.2a - 0.3b)$.

(1) $0.0144a^2 - 7.2ab + 0.09b^2$
(2) $1.2a^2 + 0.3b^2$
(3) $1.44a^2 - 0.09b^2$
(4) $1.44a^2 + 7.2ab - 0.9b^2$

Exercises 5–12: Multiply the expressions and simplify the result.

5 $(mx + 2)(mx + 2)$

6 $(3x - 3)(3x - 3)$

7 $(12c + 4)(12c - 4)$

8 $(13x + 30)(13x - 30)$

9 $(0.4a - 0.02)^2$

10 $\left(\dfrac{3}{4} + \dfrac{1}{3}x\right)\left(\dfrac{3}{4} - \dfrac{1}{3}x\right)$

11 $(2.1g - 3)(2.1g + 3)$

12 $2x(3x + 1)^2$

Dividing Algebraic Expressions by a Monomial

To Divide a Monomial by Another Monomial

- Divide the numerical coefficients, using the law of signs for division.
- Find the quotient of the variables, using the laws of exponents.
- Multiply the two quotients
- Check the answer by multiplying.

Remember:

$$\dfrac{x^a}{x^b} = x^a \div x^b = x^{a-b}$$

 MODEL PROBLEMS

SOLUTIONS

1 $\dfrac{24x^5}{-3x^2}$

$= \dfrac{24}{-3} \cdot \dfrac{x^5}{x^2} = \dfrac{8}{-1} \cdot x^{5-2} = -8x^3$

Check: $(-3x^2) \times (-8x^3) = 24x^5$

2 $\dfrac{-8a^3b^2}{8a^3b}$

$= \dfrac{-8}{8} \cdot \dfrac{a^3}{a^3} \cdot \dfrac{b^2}{b} = (-1)(1)b = -b$

Check: $(8a^3b)(-b) = -8a^3b^2$

Note: Don't forget! Any expression (except zero) divided by itself is 1.

To Divide a Polynomial by a Monomial

- Divide each term of the polynomial by the monomial, using the distributive property.
- Combine the quotients with correct signs.
- Check by multiplying.

MODEL PROBLEMS

SOLUTIONS

1 $\dfrac{6m^2 - m}{-m}$

$= \dfrac{6m^2}{-m} + \left(\dfrac{-m}{-m}\right) = -6m^{(2-1)} + 1 = -6m + 1$

Check: $-m(-6m + 1) = 6m^2 - m$

2 $\dfrac{9x^5 - 6x^3}{3x^2}$

$= \dfrac{9x^5}{3x^2} + \left(\dfrac{-6x^3}{3x^2}\right) = 3x^3 - 2x$

Check: $3x^2(3x^3 - 2x) = 9x^5 - 6x^3$

3 $\dfrac{4a^2x - 8ax + 12ax^2}{4ax}$

$= \dfrac{4a^2x}{4ax} - \dfrac{8ax}{4ax} + \dfrac{12ax^2}{4ax} = a - 2 + 3x$

Check: $4ax(a - 2 + 3x) = 4a^2x - 8ax + 12ax^2$

Practice

1 Divide: $\dfrac{x^4 y^5 z^2}{x^2 y^3 z}$

(1) $x^2 y^2 z$
(2) $x^6 y^8 z^3$
(3) $x^2 y z^2$
(4) $x^2 y^2 z^2$

2 Divide: $\dfrac{-0.08 a^3 x^2 y^4}{-0.2 a x y^2}$

(1) $0.4 a^2 x y^2$
(2) $0.4 a^3 x^2 y^2$
(3) $-0.016 a^2 x y^2$
(4) $-4 a^3 x^2 y^2$

3 Simplify: $\dfrac{6x^2 + 12x^3}{-6x^2}$

(1) $1 - 2x$
(2) $-1 - 2x$
(3) $-x + 2x^2$
(4) $2x - 1$

4 Simplify: $\dfrac{a^5 x - 2a^4 x^2 + a^3 x^3}{a^2 x}$

(1) $a^5 x - 2a^2 x + a^3 x^3$
(2) $a^3 x - 2a^2 x^2 + ax^3$
(3) $a^3 - 2a^2 x + ax^2$
(4) $a^2 - 2a^2 x + ax$

82   Chapter 3: Algebraic Operations and Reasoning

Exercises 5–20: Simplify the following expressions by dividing by the monomial or constant.

5. $\dfrac{a^4 b^3 c^7}{a^3 c^4}$

6. $\dfrac{21 a^3 b^2}{-7}$

7. $\dfrac{8 x^3 y^2}{-1}$

8. $\dfrac{-18 r^4 c^2}{-3 r c}$

9. $\dfrac{-48 a^4 b^3 c}{-8 a^2 c}$

10. $\dfrac{-44 a^5 x^4 y}{11 a^2 x}$

11. $\dfrac{-100 a^4 d^2 t}{0.25 a d^2 t}$

12. $\dfrac{0.6 a^7 x^2}{0.2 a^2 x}$

13. $\dfrac{p + prt}{p}$

14. $\dfrac{5x^4 + 3x^2}{x^2}$

15. $\dfrac{4a^2 - a}{-a}$

16. $\dfrac{8x^3 - 6xy}{2x}$

17. $\dfrac{12x^2 - 18xy + 24y^2}{6}$

18. $\dfrac{4a^2 + 3b^2 - c^2}{-1}$

19. $\dfrac{9a^5 b^3 - 27 a^2 b^2 + 6 a^3 b^5}{-3ab^2}$

20. $\dfrac{3.4 a^8 b^9 c^{10} - 5.1 a^6 b^2 c^8}{-1.7 a^6 b^2 c^8}$

Operations With Rational Expressions

A **rational expression** is a fraction involving algebraic expressions with a polynomial in the numerator or the denominator or both. For example, $\dfrac{x}{x+7}$, $\dfrac{a+b}{a^2}$, $\dfrac{x^2 + 4xy - y}{x + 5}$, and $\dfrac{x+2}{x^2 + x}$ are all rational expressions. Since rational expressions are simply fractions with variables in them, they behave like fractions under basic operations.

The reciprocal is the fraction turned upside down.

To **multiply rational expressions**, multiply numerators and denominators separately. For example, $\dfrac{3}{x} \cdot \dfrac{2}{5x} = \dfrac{6}{5x^2}$

To **divide two rational expressions**, multiply the dividend by the reciprocal of the divisor. For example, $\dfrac{3}{x} \div \dfrac{2}{5x} = \dfrac{3}{x} \cdot \dfrac{5x}{2} = \dfrac{15x}{2x} = \dfrac{15}{2} = 7\dfrac{1}{2}$

To **add or subtract rational expressions**, change each expression to an equivalent one with a common denominator. When the denominators are the same, the numerators can be added or subtracted. For example, $\dfrac{3}{x} + \dfrac{2}{5x} = \dfrac{15}{5x} + \dfrac{2}{5x} = \dfrac{17}{5x}$

Note: Whenever you work with rational expressions, the limits to the domain of the given equation will still apply to the simplified equation: $\dfrac{x+7}{x+7}$ reduces to 1 only when x is not -7. If x equals -7, $\dfrac{x+7}{x+7}$ is undefined, and therefore will not equal 1.

MODEL PROBLEMS

1 Express $\dfrac{2}{x+1} - \dfrac{1}{x-1}$ as a single fraction. Assume that x does not equal 1 or -1.

SOLUTION

Write the problem.
$$\dfrac{2}{x+1} - \dfrac{1}{x-1}$$

Build up each fraction by multiplying.
$$= \dfrac{2}{x+1} \cdot \dfrac{x-1}{x-1} - \dfrac{1}{x-1} \cdot \dfrac{x+1}{x+1}$$

Distribute numerators.
$$= \dfrac{2x-2}{(x+1)(x-1)} - \dfrac{x+1}{(x+1)(x-1)}$$

Add numerators and combine like terms.
$$= \dfrac{2x-2-(x+1)}{(x+1)(x-1)} = \dfrac{x-3}{(x+1)(x-1)}$$

Multiply denominators and combine like terms.
$$= \dfrac{x-3}{x^2+x-x-1} = \dfrac{x-3}{x^2-1}$$

2 Divide $\dfrac{2}{x+1}$ by $\dfrac{1}{x-1}$. Assume that x does not equal 1 or -1.

SOLUTION

Write the problem.
$$\dfrac{2}{x+1} \div \dfrac{1}{x-1}$$

Multiply the dividend by the reciprocal of the divisor.
$$\dfrac{2}{x+1} \cdot \dfrac{x-1}{1}$$

Multiply numerators and denominators.
$$\dfrac{2(x-1)}{(x+1)1}$$

Distribute.
$$\dfrac{2x-2}{x+1}$$

Practice

1 Multiply $\dfrac{b^3c^7}{a^3d^4} \cdot \dfrac{a^2d^3}{abc^4}$. Assume that none of the variables equal zero.

(1) $\dfrac{b^3c^3}{ad}$

(2) $\dfrac{b^2c^3}{a^2d}$

(3) $\dfrac{b^2c^3}{ad}$

(4) $\dfrac{b^3c^3}{a^4d}$

2 Divide $\dfrac{4x^3y^2}{-9r^4c^2} \div \dfrac{x^2}{-3rc}$. Assume that r and c do not equal zero.

(1) $\dfrac{4x^5y^2}{27r^5c^3}$

(2) $\dfrac{4xy^2}{-3r^4c^2}$

(3) $\dfrac{4xy^2}{3r^3c}$

(4) $\dfrac{4x^5y^2}{-27r^3c}$

84 Chapter 3: Algebraic Operations and Reasoning

3 Simplify: $\dfrac{x}{4+x} - \left(\dfrac{-4}{4+x}\right)$

(1) 1, assuming $x \neq 4$

(2) 1, assuming $x \neq -4$

(3) $\dfrac{x-4}{4+x}$, assuming $x \neq 4$

(4) $\dfrac{x-4}{4+x}$, assuming $x \neq -4$

4 Express $\dfrac{2p}{np^2} + \dfrac{3n}{n^2p}$ as a single fraction. Assume that n and p do not equal zero.

(1) $\dfrac{5}{n^2p^2}$

(2) $\dfrac{5}{np}$

(3) $\dfrac{2n+3p}{n^2p^2}$

(4) $\dfrac{2p+3n}{n^2p^2}$

5 Which expression is equivalent to $\dfrac{a}{b} - \dfrac{1}{a}$? Assume that a and b do not equal zero.

(1) $\dfrac{a-b}{ab}$

(2) $\dfrac{a^2-b}{ab}$

(3) $\dfrac{a-1}{ba}$

(4) $\dfrac{b-a^2}{ba}$

6 Express $\dfrac{x}{x-2} - \dfrac{2}{x+1}$ as a single fraction. Assume that x does not equal 2 or -1.

(1) $\dfrac{x^2+3x-4}{x^2-x-2}$

(2) $\dfrac{x^2+3x-4}{x^2+3x-2}$

(3) $\dfrac{x^2-x+4}{x^2-x-2}$

(4) $\dfrac{x^2-x+4}{x^2-3x-2}$

Exercises 7–12: Assume that no denominators equal zero.

7 Expressed as a single fraction, what is $\dfrac{2}{x-1} + \dfrac{2}{x+1}$?

8 Subtract $\dfrac{7}{x-8}$ from $\dfrac{x}{x+7}$. Show your work.

9 Multiply. $\dfrac{ab}{x} \times \dfrac{x+x^2}{ab}$

10 Divide. $\dfrac{ab}{x} \div \dfrac{x+x^2}{ab}$

11 Add. $\dfrac{ab}{x} + \dfrac{x+x^2}{ab}$

12 Subtract. $\dfrac{ab}{x} - \dfrac{x+x^2}{ab}$

13 Simplify $\dfrac{2x+5}{4} \cdot \dfrac{x}{x+8}$ and state the restrictions on the domain.

14 Simplify $\dfrac{(2x+3)(x-5)}{x(x-5)(x+1)} \cdot \dfrac{x(2x-3)}{(x-1)}$. (Hint: First cancel factors repeated in the numerator and denominator.) What four values for x would make the expression undefined?

Literal Equations and Formulas

Algebraic expressions have no specific value until values are substituted for each variable. Algebraic equations made of algebraic expressions cannot be called true or false until values are chosen from the domain for each variable.

The set of numbers from the domain that makes the equation true is called the **solution set** of the equation.

- If an equation is true for only some values of the variable or variables, it is called a **conditional equation**. Thus, $8x = 16$ is a conditional equation because it is true only when $x = 2$.
- If an equation is true for all possible values of the variable or variables, it is called an **identity**. $4x + 2x = 6x$ is an identity, since any number replacing x will make the equation true.
- If no possible values of the variable or variables make the equation true, it is called a **contradiction**. $1 + x = x$ is a contradiction, since no value of x will make it true.

If no domain is given, it is assumed to be the set of all real numbers.

An equation with more than one variable is called a **literal equation**. For example, $x + y = r$ and $3m = 2n + 1$ are literal equations. A **formula** is a literal equation that expresses a rule about a real-world relationship. For instance, $d = rt$ is a formula comparing distance to rate and time.

For some problems, you may be given a formula. For others, you may be asked to remember a formula or develop one from the problem situation. You may then be given information about all but one of the variables. In all three types of problems, you must then solve for the unknown variable's value.

To Find the Value of a Variable From a Literal Equation or Formula

- Write the equation.
- Substitute the known values of the variables.
- Simplify and solve for the value of the remaining variable.
- State the answer with the correct label.

 MODEL PROBLEMS

1 The formula used to convert from degrees Celsius to degrees Fahrenheit is $F = 1.8C + 32$. The temperature in Montreal is reported at 24°C. Convert that temperature to degrees Fahrenheit.

SOLUTION

Write the formula.	$F = 1.8C + 32$
Substitute known values.	$F = 1.8(24) + 32$
Simplify and solve for F.	$F = 75.2$
Write the answer with the correct label.	$F = 75.2°F$

2 Simple interest is calculated using the formula $I = prt$, where $I =$ simple interest, $p =$ principal, $r =$ annual rate of interest expressed as a decimal or fraction, $t =$ time in years. Compute the simple interest if the initial principal is \$1,000 invested for 3 years at 6% interest.

SOLUTION

Write the formula.	$I = prt$
Substitute known values.	$I = (1,000)(0.06)(3)$
Solve for the variable I.	$I = 60(3) = 180$
Write the answer with the correct label.	$I = \$180$

86 Chapter 3: Algebraic Operations and Reasoning

3 If a slime mold that has been traveling 1 inch per hour has covered a distance of 24 inches, how long has it been moving?

SOLUTION

This relationship is described by the formula $d = rt$. $d = rt$

Substitute known values.   24 in. = (1 in./h)t

Solve for the variable t. $t = \dfrac{24 \text{ in.}}{1 \text{ in./h}}$

Answer with the correct label. t = 24 hours or 1 day

 Practice

1 The centripetal force, F, of a rotating object equals the mass, m, multiplied by the square of its velocity, v, divided by the radius, r, of its path. This formula can be written

(1) $F = m + \left(\dfrac{v}{r}\right)^2$

(2) $F = \dfrac{(mv)^2}{r}$

(3) $F = \dfrac{m + v^2}{r}$

(4) $F = \dfrac{mv^2}{r}$

2 To find the cost of two quarts of milk and three small packages of cream cheese if each quart of milk costs $1.50 and each package of cream cheese costs $0.89, which formula would you use?

(1) $C = m + c$, where $m = 1.5$ and $c = 0.89$
(2) $C = 3m + 2c$, where $m = 1.5$ and $c = 0.89$
(3) $C = 2m + 3c$, where $m = 1.5$ and $c = 0.89$
(4) $C = 2m \bullet 3c$, where $m = 1.5$ and $c = 0.89$

3 Find the number of miles traveled by Mrs. Smith if she drove 40 miles per hour for $\dfrac{3}{4}$ hour and 65 miles per hour for $3\dfrac{1}{2}$ hours.

(1) 485 miles
(2) $332\dfrac{1}{2}$ miles
(3) $257\dfrac{1}{2}$ miles
(4) $109\dfrac{1}{4}$ miles

4 If $S = \dfrac{rm - a}{r - 1}$, what is the value of S when $r = 3$, $m = 15$, and $a = 5$?

(1) 10
(2) 15
(3) 20
(4) 25

5 Find the value of A if $A = \pi dh + \dfrac{1}{2}\pi d^2$ and $\pi = 3.14$, $h = 7$ centimeters, and $d = 10$ centimeters.

(1) $A = 476.99 \text{ cm}^2$
(2) $A = 376.8 \text{ cm}^2$
(3) $A = 230.5 \text{ cm}^2$
(4) $A = 219.8 \text{ cm}^2$

Literal Equations and Formulas

6. The formula used to convert from degrees Fahrenheit to degrees Celcius is $C = \frac{5}{9}(F - 32)$. Find the Celsius temperature when the Fahrenheit temperature is 86 degrees.

   (1) 16°C
   (2) 30°C
   (3) 48°C
   (4) 51°C

7. Write a formula for the cost C in dollars for renting a movie if the cost is $4 for the first 2 days and $1.50 for each additional day.

8. Sales tax is 8%. If the price paid for an item is P, write a formula that can be used to find N, the price of the item before tax.

9. $C = 7N$ is a formula that can be used to find the cost (C) of any number of articles (N) that sell for $7 each.

   a. How will the cost of 9 articles compare to the cost of 3 articles?
   b. If N is doubled, what happens to C?

10. The formula for the volume of a sphere is $V = \frac{4}{3}\pi r^3$. Find the value of V in terms of π when $r = 3$ centimeters.

11. If $S = \frac{a}{1 + r}$, find the value of S when $a = 12$ and $r = 7$.

12. A car rides at a constant rate of 60 miles per hour. How long will it take to travel 105 miles?

Exercises 13–15: Find the value of d in the formula $d = \sqrt{b^2 - 4ac}$, if possible.

13. $a = 3$, $b = 2$, and $c = -\frac{1}{2}$

14. $a = 3$, $b = 4$, and $c = -2$

15. $a = 4$, $b = 3$, and $c = -2$

Exercises 16–19: Find the value of an investment (A) if the value is given by the formula $A = p + prt$.

16. $p = \$600$, $r = 3\%$, and $t = 2$ years

17. $p = \$4,000$, $r = 5\%$, and $t = 8$ years

18. $p = \$1,250$, $r = 4\%$, and $t = 2\frac{1}{4}$ years

19. $p = \$3,500$, $r = 2\%$, and $t = 4.5$ years

20. The thickness of each piece of a certain kind of paper is 0.01 cm. The paper is used in the printing of a book. Each outside cover of the book is 0.2 cm. and the book has 350 pages.

    a. Write a formula for the thickness of the book in meters.
    b. Use your formula to express the width of shelf space in meters needed to store 50 books standing upright.

Solving Linear Equations

The **degree** of a polynomial in one variable is the greatest exponent of that variable. Thus, $5m^4 + 3m - 8$ has degree 4; $x^2 + 1$ has degree 2; and $x + 3$ has degree 1 (because $x = x^1$). An equation or inequality is called **linear** if the expressions on either side of the equals sign have at most a degree of 1.

An equation is like a balanced scale. To keep the equation in balance, *any change made on one side must also be made on the other side*. To solve an equation, you must manipulate the left and right expressions of the equation so that the variable is left alone on one side. This is done by performing opposite or inverse operations to "unwrap" the variable.

A number without a variable, such as 7, has a degree of zero, because $7x^0 = 7 \cdot 1 = 7$.

Once we have a solution, we must test that solution in the equation. This is called checking. **Checking** involves two steps:

- First, the solution must be substituted for the variable *in the original equation*.
- Second, each side of the equation must be simplified. If the two sides of the equation are the same, then the solution was correct.

Solving Equations of the Form $x + a = b$

To undo the addition of a number, add its opposite to both sides of the equation. This is the same as subtracting the number from both sides.

MODEL PROBLEMS

1 Solve the equation $x - 7 = 2$.

SOLUTION

Write the equation.

Add the opposite of -7 to both sides. Arrange in columns and add.

$$\begin{array}{rcr} x - 7 &=& 2 \\ +7 &=& +7 \\ \hline x &=& 9 \end{array}$$

Check. $x - 7 = 2 \rightarrow 9 - 7 = 2$ ✓

Answer: $x = 9$

2 Solve the equation $5 + x = 20$.

SOLUTION

Write the equation.

Add the opposite of 5 to both sides.

$$\begin{array}{rcr} 5 + x &=& 20 \\ -5 &=& -5 \\ \hline x &=& 15 \end{array}$$

Check. $5 + x = 20 \rightarrow 5 + 15 = 20$ ✓

Answer: $x = 15$

 Practice

---

1 If $x + 4\frac{3}{4} = 7$, then $x =$

(1) $11\frac{3}{4}$

(2) $3\frac{3}{4}$

(3) $2\frac{3}{4}$

(4) $2\frac{1}{4}$

2 If $9 = x + 2$, then $x =$

(1) -7
(2) 7
(3) 11
(4) 18

3 If $x - 2.5 = -6$, then $x =$

(1) -8.5
(2) -3.5
(3) 3.5
(4) 8.5

Solving Linear Equations

4 If $x - \$1.25 = \$4.90$, then $x =$
   (1) $3.65
   (2) $3.92
   (3) $6.13
   (4) $6.15

5 To solve $1.4 = x + 8.2$, what number should you add to both sides of the equation?

Exercises 6–10: Solve for the variable.

6 $x - 8 = 12$

7 $x + 5\frac{1}{2} = 3$

8 $x - 1.6 = -3.8$

9 $2 = k + 0.4$

10 $x - \frac{1}{8} = 2\frac{3}{8}$

Solving Equations of the Form $ax = b$

To undo multiplication by a number, do the opposite. That is, divide by that number or multiply by its reciprocal. Remember that the process of multiplying by the reciprocal is the same as dividing by the number.

Consider the equation $3x = 21$. Note that on one side of the equation, x is multiplied by 3. The solution is found by dividing each side of the equation by 3. Multiplying by the reciprocal is another method that can be used to find the solution. Both methods are shown below.

Note: Proportion and conversion problems are equations of the form $ax = b$. For example, the proportion $\frac{x \text{ feet}}{30 \text{ yards}} = \frac{3 \text{ feet}}{1 \text{ yard}}$ is similar to $\frac{1}{30}x = 3$. Proportions with x in the denominator can be inverted. Thus $\frac{30 \text{ yards}}{x \text{ feet}} = \frac{1 \text{ yard}}{3 \text{ feet}}$ has the same solution as $\frac{x \text{ feet}}{30 \text{ yards}} = \frac{3 \text{ feet}}{1 \text{ yard}}$.

MODEL PROBLEMS

1 Solve $12x = 36$ for x.

SOLUTION

Write the equation. $12x = 36$

Divide both sides by 12. $\frac{12x}{12} = \frac{36}{12}$

Simplify. $x = 3$

Check. $12x = 36 \rightarrow 12(3) = 36$ ✓

Answer: $x = 60$

2 Solve the equation $\frac{x}{4} = -5$ for x.

SOLUTION

Write the equation. $\qquad\qquad\qquad\qquad\qquad \frac{x}{4} = -5$

The opposite of division by 4 is multiplication by 4. $\qquad (4)\frac{x}{4} = (4)(-5)$

Simplify each side. $\qquad\qquad\qquad\qquad\qquad x = -20$

Check. $\frac{x}{4} = -5 \rightarrow \frac{-20}{4} = -5$ ✓

Answer: $x = -20$

A term containing a fractional coefficient may be written as a fraction.

$\frac{1}{4}x = \frac{x}{4}$

$\frac{3}{4}x = \frac{3x}{4}$

3 Solve $0.01x = 7$ for x.

SOLUTION

Write the equation. $\qquad\qquad\qquad\qquad\qquad 0.01x = 7$

Rewrite the equation. $\qquad\qquad\qquad\qquad\qquad \frac{1}{100}x = 7$

Multiply both sides by 100. $\qquad\qquad\qquad (100)\frac{1}{100}x = 7(100)$

Simplify. $\qquad\qquad\qquad\qquad\qquad\qquad\qquad x = 700$

Check. $0.01x = 7 \rightarrow 0.01(700) = 7$ ✓

Answer: $x = 700$

 Practice

1 A piece of imported cheese costs $4.77. If the piece weighs $\frac{9}{10}$ of a pound, solve the eqution $\$4.77 = \frac{9}{10}x$ to find how much the cheese costs per pound.

(1) $3.87
(2) $4.29
(3) $5.30
(4) $5.67

2 How long does it take for a car traveling at 45 miles per hour to cover 225 miles?

(1) 265 minutes
(2) 5 hours
(3) 3 hours
(4) 2 hours

3 On a map, 1 inch represents 50 miles. If two cities are 275 miles apart, how many inches apart are they on the map?

(1) 13.75 inches
(2) 5.5 inches
(3) 3.25 inches
(4) 2.25 inches

Solving Linear Equations **91**

4 If a bus travels 152 miles in 4 hours, how long will it take the bus to travel 342 miles at the same rate?

(1) 6.25 hours
(2) 9 hours
(3) 13 hours
(4) 47.5 hours

Exercises 5–15: Solve for the variable and check.

5 $\dfrac{a}{5} = -6$

6 $0.7m = -1.4$

7 $\dfrac{3}{4}x = -27$

8 $-12x = -60$

9 $18 = -2x$

10 $0.06m = 12$

11 $6x = 3$

12 $\dfrac{1}{4}x = 3$

13 $3 = \dfrac{1}{9}x$

14 $\dfrac{2}{3}x = 8$

15 $-0.02x = 6.4$

Solving Equations of the Form $ax + b = c$

On many high school examinations, the solution of equations will usually require more than one operation. In general, in order to "unwrap" or solve multiple-step equations, the operations should be done in reverse order: for equations of the form $ax + b = c$, it is easier to undo the addition first and then undo the multiplication.

MODEL PROBLEMS

1 Solve $-6x + 4 = 34$ for x and check.

SOLUTION

Write the equation.	$-6x + 4 = 34$
Add the opposite of 4 to each side.	$\underline{-4 = -4}$
	$-6x = 30$
Divide each side by -6.	$\dfrac{-6x}{-6} = \dfrac{30}{-6}$
Simplify.	$x = -5$
Check.	$-6x + 4 = 34$
	$-6(-5) + 4 = 34$
	$30 + 4 = 34$
	$34 = 34$ ✓

Answer: $x = -5$

2 Solve: $\frac{2}{7}x + \frac{4}{7} = 2$

SOLUTION

To solve problems with fractional coefficients, it is often easier to multiply both sides of the equation by the least common multiple, thereby eliminating all the fractions.

Multiply every term by the common denominator, 7.	$7 \cdot \frac{2}{7}x + 7 \cdot \frac{4}{7} = 7 \cdot 2$
Simplify. Now there are no fractions.	$2x + 4 = 14$
Add -4 to each side.	$\underline{-4 = -4}$
	$2x = 10$
Divide by 2.	$\frac{2x}{2} = \frac{10}{2}$
Simplify.	$x = 5$
Check.	$\frac{2(5)}{7} + \frac{4}{7} = 2$
	$\frac{10}{7} + \frac{4}{7} = 2$
	$\frac{14}{7} = 2 \checkmark$

Answer: $x = 5$

3 Solve the equation $\frac{5x}{6} + \frac{1}{3} = \frac{5}{12}$.

SOLUTION

Eliminate fractions by multiplying each term by 12 (LCD of 6, 3, and 12).	$12 \cdot \frac{5x}{6} + 12 \cdot \frac{1}{3} = 12 \cdot \frac{5}{12}$
Simplify.	$10x + 4 = 5$
Add -4 to each side.	$\underline{-4 = -4}$
Simplify.	$10x = 1$
Divide each side by 10.	$x = \frac{1}{10}$

Check.

$$\frac{5\left(\frac{1}{10}\right)}{6} + \frac{1}{3} = \frac{5}{12}$$

$$\frac{\left(\frac{1}{2}\right)}{6} + \frac{1}{3} = \frac{5}{12}$$

$$\frac{1}{12} + \frac{1}{3} = \frac{5}{12}$$

$$\frac{1}{12} + \frac{4}{12} = \frac{5}{12} \checkmark$$

(With calculator)

$$\frac{5(0.1)}{6} + \frac{1}{3} = \frac{5}{12}$$

$$\frac{0.5}{6} + \frac{1}{3} = \frac{5}{12}$$

$$0.08\overline{3} + 0.\overline{3} = \frac{5}{12}$$

$$0.41\overline{6} = \frac{5}{12} \checkmark$$

Answer: $x = \frac{1}{10}$

Practice

1 Solve: $7x + 5 = 33$
  (1) $x = 4$
  (2) $x = 5\frac{1}{5}$
  (3) $x = 5\frac{3}{7}$
  (4) $x = 8$

2 Which choice is the solution to $8 = 3x - 4$?
  (1) $x = 1\frac{1}{3}$
  (2) $x = 4$
  (3) $x = 12$
  (4) $x = 36$

3 If $v = V + gt$, what is the value of t when $v = 116$, $V = 20$, and $g = 32$?
  (1) 3,732
  (2) 96
  (3) 64
  (4) 3

4 Which choice is *not* a step you would take to solve the equation $4 - \frac{x}{2} = 0$?
  (1) Add -4 to both sides of the equation.
  (2) Divide both sides by -2.
  (3) Multiply each term by -2.
  (4) Subtract 4 from both sides.

5 To solve $10 - 2x = 18$, what number should you add to both sides of the equation for the first step? For the second step, what number should you divide by?

6 Ms. Healy asked her class to solve the equation $10 + \frac{x}{3} = 23$. Maria multiplied both sides of the equation by 3, then subtracted 10 from both sides. Leon divided both sides of the equation by $\frac{1}{3}$, then subtracted 30 from both sides. Which student was correct? What mistake did the other student make?

Exercises 7–11: Solve for the variable and check.

7 $5y - 7 = 15$

8 $4 - 2x = 14$

9 $\frac{1}{4}m + 5 = -3$

10 $\frac{3}{4}c - 8 = 4$

11 $5x + 6 = 81$

12 A business had $1,395 in sales and $261 in expenses. How much did each of the three partners receive if they divided the profits evenly?

Solving Equations With Like Terms

To solve equations with like terms, move all like terms to the same side of the equals sign and combine them. When many terms are involved, the best method is to simplify both sides of the equation before solving.

Note: In the problems on page 95, the step of checking has been left for the student to complete.

MODEL PROBLEMS

1 Solve for x: $7x - 8 = 10 + 4x$.

SOLUTION

Write the equation.	$7x - 8 =$	$10 + 4x$
Add 8 to each side.	$+ 8 =$	$+ 8$
	$7x =$	$18 + 4x$
Add $-4x$ to each side.	$- 4x =$	$- 4x$
	$3x =$	18
Divide both sides by 3.	$x =$	6

Answer: $x = 6$

2 Willis, Bryan, and Franklin together inherited their aunt's horses after the five oldest mares were sold. The ratio of horses they received, in order, is $1:2:5$. How many horses did each man get if there were originally 101 horses?

SOLUTION

Define all unknowns with one variable, the number of horses in one share, or h.	Willis $= h$ horses Bryan $= 2h$ horses Franklin $= 5h$ horses
Write an equation for the original number of horses.	$h + 2h + 5h + 5 = 101$
Combine like terms.	$8h + 5 = 101$
Add -5 to each side.	$- 5 = - 5$
	$8h = 96$
Divide both sides by 8.	$\dfrac{8h}{8} = \dfrac{96}{8}$
	$h = 12$

Answer: Willis received 12 horses, Bryan received 24 horses, and Franklin received 60 horses.

3 Solve $x + 0.04x = 832$ for x.

SOLUTION

Write the equation.	$x + 0.04x = 832$
Combine like terms.	$1.04x = 832$
Divide each side by 1.04	$\dfrac{1.04x}{1.04} = \dfrac{832}{1.04}$
	$x = 800$

Answer: $x = 800$

4 Neil's telephone service costs $25 per month plus $0.15 for each local call. There is an additional charge for long distance calls. Last month, Neil's telephone bill was $31.50. His long distance charge was $2. How many local calls did he make?

SOLUTION

Let $n =$ number of local calls

Write an equation.	$25 + 0.15n + 2 =$	31.50
Combine like terms.	$27 + 0.15n =$	31.50
Add -27 to each side.	$-27 =$	-27
	$0.15n =$	4.5
Divide both sides by 0.15.	$\dfrac{0.15n}{0.15} =$	$\dfrac{4.5}{0.15}$
	$n =$	30

Answer: Neil made 30 local calls.

Practice

1 If $6x - 4 = 20 - 2x$, then $x =$
 (1) 6
 (2) 4
 (3) 3
 (4) 2

2 If $5x - 2 = 28 - x$, then $x =$
 (1) 6
 (2) 5
 (3) 4
 (4) 3

Solving Linear Equations

3 If $8x - 4 + 7 = 6x + x + 9$, then $x =$
   (1) 6
   (2) 4
   (3) 2
   (4) 0

4 At a school fund raising drive, $350 was collected for washing sedans, coupes, and sport-utility vehicles. The cost of a wash for any type of vehicle was $2. Twice as many coupes were washed as sedans. Four times as many sport-utility vehicles were washed as sedans. How many coupes were washed?
   (1) 175
   (2) 100
   (3) 50
   (4) 25

Exercises 5–18: Solve for the variable and check.

5 $\frac{5}{6}x - \frac{1}{6}x = 10$

6 $\frac{7}{8}x - \frac{1}{8}x = 9$

7 $5x = 15 + 2x$

8 $4x + 6 = 6x$

9 $a - 0.04a = 240$

10 $3x + 10 = 8x$

11 $6x + 3x - 4 = 32$

12 $5x + 2 = 8x - 7$

13 $3x + 5x - x = 56$

14 $9x - 12 = 7x - 6$

15 $5a + 3 = 15 + 2a$

16 $3a - a = a + 8$

17 $7x - 8 - 2x = 64 - 3x$

18 $7x + 6 = 8 - x - 2$

19 Three cousins belong to a school basketball team. The ratio of their field goals in the championship game was 3 to 3 to 4. The total number of field goals made by the cousins was 30. Determine the number made by each cousin. Explain your answer.

20 Three less than a certain number is 12 less than twice the number. Write an equation and solve to find the number.

Solving Equations With Parentheses and Rational Expressions

- To solve equations with parentheses, first use the distributive property to remove the parentheses, then solve as before.
- To solve equations with rational expressions, multiply each term in the equation by the LCD (the least common denominator).
- If the equation is set up like a proportion, then cross multiply, set the two products as equal, and solve.

Note: The step of checking has been left for the student to complete in model problems 4 and 5 on pages 98 and 99.

MODEL PROBLEMS

1 The train stations in New York and Boston are 220 miles apart. A train left New York for Boston at 8:00 A.M. traveling at 65 mph. An hour later, an express train left Boston for New York traveling at 90 mph. At what time did they pass each other?

SOLUTION

When the trains meet, the total distance traveled is the same as the distance between cities, 220 miles. Let t represent the time the express train was traveling and $t + 1$ represent the travel time of the other train.

Use $d = rt$.	$90t + 65(t + 1) = 220$
Remove parentheses.	$90t + 65t + 65 = 220$
Combine like terms.	$155t + 65 = 220$
Add -65 to both sides.	$\underline{\ -65\ \ -65}$
	$155t\ \ \ \ = 155$
Solve.	$t\ \ \ \ = 1$

Check. $90t + 65(t + 1) = 210 \rightarrow 90 + 65(2) = 90 + 130 = 220$ ✓

Answer: The trains passed after the express train had been traveling for one hour, which was at 10:00 A.M.

2 Solve and check: $\dfrac{4a + 3}{3} - \dfrac{7a - 1}{4} = 5$

SOLUTION

Write the equation.	$\dfrac{4a + 3}{3} - \dfrac{7a - 1}{4} = 5$
Multiply each term by the LCD, 12.	$\cancel{12}\left(\dfrac{4a + 3}{\cancel{3}}\right) - \cancel{12}\left(\dfrac{7a - 1}{\cancel{4}}\right) = 12 \cdot 5$
Multiply.	$16a + 12 - 21a + 3 = 60$
Simplify.	$-5a + 15 = 60$
Subtract 15 from both sides.	$\underline{\ -15 = -15}$
	$-5a\ \ \ = 45$
Divide by -5.	$a\ \ \ = -9$

Check. $\dfrac{4a + 3}{3} - \dfrac{7a - 1}{4} = 5 \rightarrow \dfrac{4(-9) + 3}{3} - \dfrac{7(-9) - 1}{4} = \dfrac{-33}{3} - \dfrac{-64}{4} = -11 + 16 = 5$ ✓

Answer: $a = -9$

Solving Linear Equations **97**

3 Solve $\dfrac{2}{3x} + \dfrac{1}{4} = \dfrac{11}{6x} - \dfrac{1}{3}$ for x and check.

SOLUTION

Write the equation.

$$\dfrac{2}{3x} + \dfrac{1}{4} = \dfrac{11}{6x} - \dfrac{1}{3}$$

Multiply every term by the LCD, $12x$.

$$\overset{4}{\cancel{12x}} \cdot \dfrac{2}{\cancel{3x}} + \overset{3}{\cancel{12x}} \cdot \dfrac{1}{\cancel{4}} = \overset{2}{\cancel{12x}} \cdot \dfrac{11}{\cancel{6x}} - \overset{4}{\cancel{12x}} \cdot \dfrac{1}{\cancel{3}}$$

Simplify.

$$4 \cdot 2 + 3x \cdot 1 = 2 \cdot 11 - 4x \cdot 1$$

$$8 + 3x = 22 - 4x$$

Add -8 to each side.

$$\underline{-8 = -8}$$

$$3x = 14 - 4x$$

Add $4x$ to each side.

$$\underline{+4x = + 4x}$$

$$7x = 14$$

Divide each side by 7.

$$\dfrac{7x}{7} = \dfrac{14}{7}$$

Simplify.

$$x = 2$$

Check.

$$\dfrac{2}{3(2)} + \dfrac{1}{4} = \dfrac{11}{6(2)} - \dfrac{1}{3}$$

$$\dfrac{2}{6} + \dfrac{1}{4} = \dfrac{11}{12} - \dfrac{1}{3}$$

$$\dfrac{4}{12} + \dfrac{3}{12} = \dfrac{11}{12} - \dfrac{4}{12}$$

$$\dfrac{7}{12} = \dfrac{7}{12} \checkmark$$

Answer: $x = 2$

4 Solve $\dfrac{4x}{9} = \dfrac{2(x+4)}{3}$.

SOLUTION

Write the equation.

$$\dfrac{4x}{9} = \dfrac{2(x+4)}{3}$$

Use the distributive property.

$$\dfrac{4x}{9} = \dfrac{2x+8}{3}$$

Cross multiply.

$$12x = 9(2x + 8)$$

Use the distributive property.

$$12x = 18x + 72$$

Subtract $18x$ from both sides.

$$\underline{-18x = -18x}$$

$$-6x = 72$$

Divide both sides by -6.

$$\dfrac{-6x}{-6} = \dfrac{72}{-6}$$

Simplify.

$$x = -12$$

Answer: $x = -12$

5 In a group of four consecutive, positive, even integers, the product of the first and the last integers equals the square of the second. Find the integers.

SOLUTION

Consecutive integers can be represented algebraically with one variable. The set $\{x, x + 1, x + 2\}$ is a set of consecutive integers. To represent a set of four consecutive even integers, use $\{x, x + 2, x + 4, x + 6\}$ where x can be any even integer. If x is odd, this same set would represent consecutive odd integers.

Define variables.	first number $= x$ second number $= x + 2$ third number $= x + 4$ last number $= x + 6$
Write the equation.	$x(x + 6) = (x + 2)(x + 2)$
Use FOIL.	$x(x + 6) = x^2 + 4x + 4$
Distribute.	$x^2 + 6x = x^2 + 4x + 4$
Add $-x^2$ to both sides.	$\underline{-x^2 \quad\quad = -x^2}$ $6x = 4x + 4$
Add $-4x$ to both sides.	$\underline{-4x \quad = -4x}$ $2x = 4$
Divide both sides by 2.	$x = 2$

Answer: The numbers are 2, 4, 6, and 8.

 ## Practice

1 Solve $5x - (x + 3) = 7 + 2(x + 2)$ for x.

(1) 3
(2) 4
(3) 6
(4) 7

2 Solve $3x + 2(50 - x) = 110$ for x.

(1) 5
(2) 10
(3) 15
(4) 105

3 $\dfrac{x - 5}{10} = \dfrac{x + 4}{4}$

(1) -40
(2) -10
(3) 10
(4) 40

4 $\dfrac{5}{10x} + \dfrac{1}{3} = \dfrac{3}{5x} + \dfrac{2}{5}$

(1) $-\dfrac{3}{2}$
(2) $-\dfrac{2}{3}$
(3) $\dfrac{2}{3}$
(4) $\dfrac{3}{2}$

5 A cash register has 5 times as many quarters as nickels, two fewer dimes than nickels, and 30 pennies. All together, the cash register contains $8.50 in change. How many nickels are in the cash register?

  (1) 4 nickels
  (2) 5 nickels
  (3) 6 nickels
  (4) 30 nickels

Exercises 6–18: Solve for the variable. Assume that no denominator is equal to zero.

6 $18 - (a - 4) = 9 + (a + 3)$

7 $2x(3x + 1) = 3x(2x + 1) - 2$

8 $(a - 1) - (a + 2) - (a - 3) = a$

9 $\dfrac{x}{5} = \dfrac{x - 3}{2}$

10 $\dfrac{x + 2}{4} = \dfrac{x}{2}$

11 $\dfrac{4x + 5}{6} = \dfrac{7}{2}$

12 $\dfrac{x - 3}{2} = \dfrac{x + 4}{6}$

13 $\dfrac{8}{3x} = \dfrac{4}{4x - 5}$

14 $\dfrac{2x - 3}{4} - \dfrac{x - 2}{3} = 2$

15 $\dfrac{x}{3} - \dfrac{3x}{4} = 5 - \dfrac{5x}{6}$

16 $\dfrac{9}{x} + \dfrac{3}{x} = 4$

17 $\dfrac{17}{2x} - \dfrac{5}{2x} = 12$

18 $1 + \dfrac{1}{2n} + \dfrac{2}{3n} = \dfrac{13}{6n}$

19 Of four consecutive even numbers, 4 times the smallest minus twice the largest is 4. Find the smallest number.

20 The sum of five consecutive integers equals 5.5 times the middle integer. Find the integers.

Solving Literal Equations

Simple literal equations can be solved using the same methods as illustrated in all the preceding examples. When an equation has more than one variable, we can solve for any variable using these methods, treating the other variables like constants.

 MODEL PROBLEMS

1 Solve for R in $I = \dfrac{E}{R}$.

SOLUTION

Write the equation. $I = \dfrac{E}{R}$

Multiply both sides by R to eliminate the fraction. $R \cdot I = \dfrac{E}{R} \cdot R$

$RI = E$

Divide both sides by I. $\dfrac{RI}{I} = \dfrac{E}{I}$

$R = \dfrac{E}{I}$

2 Solve the equation $a + c = b - a$ for a.

SOLUTION

Write the equation. $a + c = b - a$

Combine the a terms.
$\quad +a = +a$
$\quad 2a + c = b$

Isolate the a.
$\quad -c = -c$
$\quad 2a = b - c$

$\dfrac{2a}{2} = \dfrac{b - c}{2}$

$a = \dfrac{b - c}{2}$

100 Chapter 3: Algebraic Operations and Reasoning

3 Solve the equation $m(n + r) = s$ for n.

SOLUTION

Write the equation.	$m(n + r) = s$
Use the distributive property.	$mn + mr = s$
Add $-mr$ to both sides.	$\underline{ -mr = -mr}$
	$mn = s - mr$
Divide both sides by m.	$\dfrac{mn}{m} = \dfrac{s - mr}{m}$
Simplify.	$n = \dfrac{s - mr}{m}$

Answer: $n = \dfrac{s - mr}{m}$ or $\dfrac{s}{m} - r$

4 Solve for p in $A = p + prt$

SOLUTION

Write the equation.	$A = p + prt$
Use the distributive property.	$A = p(1 + rt)$
Divide both sides by $(1 + rt)$.	$\dfrac{A}{(1 + rt)} = \dfrac{p(1 + rt)}{(1 + rt)}$
Simplify.	$\dfrac{A}{(1 + rt)} = p$

Answer: $p = \dfrac{A}{1 + rt}$

Practice

1 Solve the volume formula, $V = lwh$, for h.
  (1) $h = Vlw$
  (2) $h = V - lw$
  (3) $h = V + lw$
  (4) $h = \dfrac{V}{lw}$

2 Solve $c = ax + b$ for x.
  (1) $x = \dfrac{c - b}{a}$
  (2) $x = ca - b$
  (3) $x = \dfrac{cb}{a}$
  (4) $x = a(c - b)$

3 Solve for x in the equation $\dfrac{ax}{b} = \dfrac{bx}{a}$. Assume that a and b are not equal and do not equal zero.
  (1) $x = \dfrac{a^2}{b^2}$
  (2) $x = 0$
  (3) x can equal any real number as the equation is an identity.
  (4) x has no answer as the equation is a contradiction.

4 Solve the equation $p(q + r) = 1$ for q.
  (1) $q = 1 - p - r$
  (2) $q = \dfrac{1}{p + r}$
  (3) $q = \dfrac{1 - pr}{p}$
  (4) $q = \dfrac{1 - p}{r}$

Exercises 5–13: Solve for x.

5 $ax = 7$

6 $mx = g$

7 $2a + x = 6$

8 $ax + n = m$

9 $nx = 4j - 5x$

10 $R = \dfrac{a(x + b)}{3}$

11 $5x - 8a = 3x + 7a$

12 $\dfrac{1}{x} - \dfrac{1}{a} = \dfrac{1}{b}$

13 $\dfrac{ax}{c} - b = \dfrac{c}{b}$

Exercises 14–19: Solve for the indicated variable.

14 $B = \dfrac{P}{R}$ for R

15 $S = \dfrac{\pi r^2 A}{90}$ for A

16 $S = c + g$ for g

17 $R = \dfrac{gs}{g+s}$ for g

18 $fx - gx = h$ for x

19 $2x = ax + 7$ for x

20 The formula for the total amount of money owed on a long-term loan is $A = p + prt$.

  a Solve for t (time in years)
  b Find t when p (principal) = $7,400, A (total amount) is $9,176, and r (rate) is 8%.

Linear Inequalities in One Variable

An **inequality** consists of two or more expressions joined by a sign of inequality. The **signs of inequality** are < (less than), > (greater than), ≤ (less than or equal to), ≥ (greater than or equal to), and ≠ (not equal to).

A **linear inequality in one variable** is an inequality of the first degree that contains only one variable. Examples of linear inequalities in one variable are $x - 2 > 2x$ and $5x + 2 \neq 4$.

A **compound inequality** is formed by joining two inequalities with *and* or *or*. Examples of compound inequalities are "x is greater than 3 *or* less than 0" ($x > 3$ or $x < 0$) and "2 is less than y *and* y is less than or equal to 4" ($2 < y$ and $y \leq 4$).

Unless a restriction is stated, the domain of any inequality is the real number line. The solution set is the set of all numbers in the domain that make the inequality true. This set can be represented on a number line.

An open circle marking a value on the number line indicates that the number is *not* included in the solution set.

Compound inequalities joined with *and* can be combined. "$2 < y$ and $y \leq 4$" is the same as $2 < y \leq 4$. There is no shortcut for phrases with *or*.

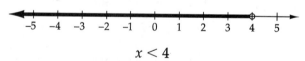

$x < 4$

A closed circle marking a value on the number line indicates that the number is *included* in the solution set.

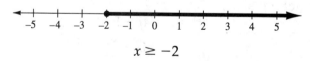

$x \geq -2$

102 Chapter 3: Algebraic Operations and Reasoning

Properties of Inequalities

In each case below, assume that a, b, and c are real numbers.

Property	Description	Examples
Comparison property of numbers	For all numbers a and b, exactly one of the following statements is true: $a < b$, $a = b$, or $a > b$.	For 2 and 3, only $2 < 3$ is true.
Addition property of inequality	If $a > b$, then $a + c > b + c$.	$4 > 1$, so $4 + 2 > 1 + 2$ $6 > 3$
	If $a < b$, then $a + c < b + c$.	$-3 < 1$, so $-3 + 5 < 1 + 5$ $-2 < 6$
Subtraction property of inequality	If $a > b$, then $a - c > b - c$.	$2 > -5$, so $2 - 4 > -5 - 4$ $-2 > -9$
	If $a < b$, then $a - c < b - c$.	$0 < 3$, so $0 - 4 < 3 - 4$ $-4 < -1$
Multiplication property of inequality	If $a > b$ and c is positive ($c > 0$), then $ca > cb$.	$3 > 1$, so $2 \cdot 3 > 2 \cdot 1$ $6 > 2$
	If $a < b$ and c is positive ($c > 0$), then $ca < cb$.	$5 < 6$, so $3 \times 5 < 3 \times 6$ $15 < 18$
	If $a > b$ and c is negative ($c < 0$), then $ca < cb$.	$9 > 4$, so $-1 \times 9 < -1 \times 4$ $-9 < -4$
	If $a < b$ and c is negative ($c < 0$), then $ca > cb$.	$3 < 8$, so $-3 \times 3 > -3 \times 8$ $-9 > -24$
Division property of inequality	If c is positive ($c > 0$) and $a > b$, then $\frac{a}{c} > \frac{b}{c}$.	$12 > 6$, so $\frac{12}{3} > \frac{6}{3}$ $4 > 2$
	If c is positive ($c > 0$) and $a < b$, then $\frac{a}{c} < \frac{b}{c}$.	$4 < 8$, so $\frac{4}{2} < \frac{8}{2}$ $2 < 4$
	If c is negative ($c < 0$) and $a > b$, then $\frac{a}{c} < \frac{b}{c}$.	$3 > 2$, so $\frac{3}{-1} < \frac{2}{-1}$ $-3 < -2$
	If c is negative ($c < 0$) and $a < b$, then $\frac{a}{c} > \frac{b}{c}$.	$3 < 6$, so $\frac{3}{-3} > \frac{6}{-3}$ $-1 > -2$

To solve an inequality, follow the same steps that apply to solving equations, with one very important exception: *When you multiply or divide by a negative number, you must change the direction of the inequality.*

Therefore $-3x > 3$, when divided by -3, becomes $x < -1$.

To check the answer for an inequality, substitute a number from the graph of the solution set.

Linear Inequalities in One Variable

MODEL PROBLEMS

1 Graph the the following: **SOLUTIONS**

  a $x \neq 1$

All real numbers except 1 are included in the solution set.

  b $(x \leq -1)$ or $(x > 4)$

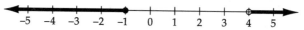

The word *or* means that the solution set is all the values that satisfy at least one of the inequalities.

  c $(x > 2)$ and $(x \leq 5)$

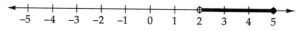

The word *and* means that the solution set is all the values that satisfy both inequalities.

2 Solve and graph $x + 3 \geq 6$.

SOLUTION

$$x + 3 \geq 6$$
$$x + 3 - 3 \geq 6 - 3$$
$$x \geq 3$$

3 Solve and graph $1 - 4x < 13$.

SOLUTION

$$1 - 4x < 13$$
$$1 - 1 - 4x < 13 - 1$$
$$-4x < 12$$

$$\frac{-4x}{-4} > \frac{12}{-4}$$ *Note the change in the inequality!*

$$x > -3$$

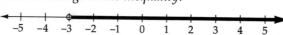

4 Solve the inequality $\frac{2x}{5} - \frac{x}{2} > \frac{9}{10}$, and graph the solution set.

SOLUTION

The LCD of 5, 2, and 10 is 10.

$$10\left(\frac{2x}{5} - \frac{x}{2}\right) > 10\left(\frac{9}{10}\right)$$

$$4x - 5x > 9$$
$$-x > 9$$
$$(-1)(-x) < (-1)9$$ *Note the change in the inequality!*
$$x < -9$$

5 The members of a school booster club are creating buttons for sale at basketball games. The machine to make the buttons costs $45. The material needed to make each button costs 20 cents. The buttons will be sold for $1 each. How many buttons must be sold to make a profit of at least $100?

SOLUTION

Let number of buttons sold = b.

Then the income from sales = $\$1.00(b)$, or simply b dollars.

The expense of making buttons = $0.20b$ dollars.

The total expenses = $45 + 0.2b$ dollars.

Profit is equal to the income from sales minus the expenses. $b - (45 + 0.2b) \geq 100$
Profit is at least $100.

Distribute -1 over the parentheses. $b - 45 - 0.2b \geq 100$

Combine like terms. $0.8b - 45 \geq 100$

Add 45 to both sides. $0.8b \geq 145$

Divide both sides by 0.8. $b \geq \dfrac{145}{0.8}$

$b \geq 181.25$

Replace 181.25 with the nearest greater whole number, $b \geq 182$
since they cannot sell pieces of buttons.

Answer: They must sell at least 182 buttons to make a profit of at least $100.

 Practice

1 Which of the following is a member of the solution set of $-6 < x \leq -1$?

(1) -1
(2) -6
(3) 3
(4) -7

2 Which inequality is represented by this graph?

(1) $4 \leq x < -3$
(2) $-3 \leq x \leq 4$
(3) $-3 < x \leq 4$
(4) $-3 \leq x < 4$

3 Which graph represents the solution set of $2x > 6$?

(1) 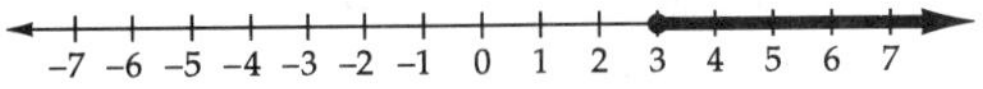
(2) 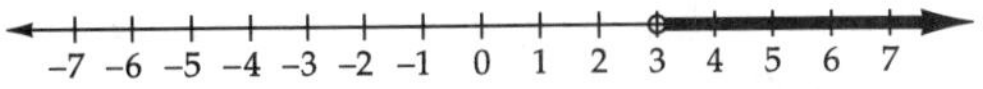
(3) 
(4)

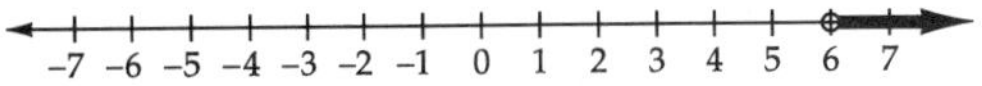

4 Which graph shows the solution to $(x < 3)$ and $(x > -1)$?

(1) 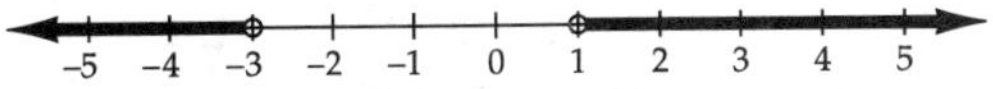
(2) 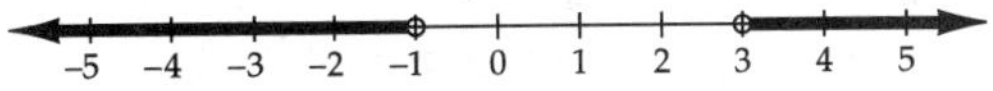
(3) 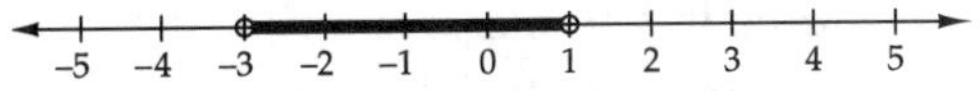
(4)

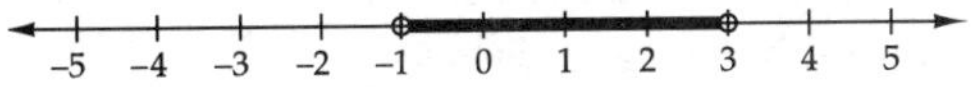

5 Find the solution set for $\frac{1}{2}x + 4 \geq 4$ if the domain or replacement set for x is $\{-2, 0, 2, 3\}$.

(1) $\{-2, 0\}$
(2) $\{0\}$
(3) $\{2, 3\}$
(4) $\{0, 2, 3\}$

6 The solution set of $-5(x - 4) + 3x > -16$ is

(1) $x > 18$
(2) $x > -6$
(3) $x < 6$
(4) $x < 18$

Exercises 7–12: Solve each inequality and graph the solution set.

7 $x - 3 > 1$

8 $4j + 1 \geq 25$

9 $1.5 - 3x \neq -4.5$

10 $(c \geq -1)$ or $(c < -4)$

11 $(2y + 4 > 0)$ and $(y < 1)$

12 $3(x + 2) + 11 > 20$

13 If the replacement set for x is $\{-4, -3, -2, -1, 0, 1, 2, 3, 4\}$, what is the solution set for $-2x > 6$?

14 What are the integer values in the solution set of $3 \leq x < 7$?

15 Find the whole number that makes the following sentence true when substituted for x: $(3x > 6)$ and $(x < 4)$.

16 Solve the following compound inequality and graph the solution set:

$(9 - 4j > 17)$ or $(2j + 1 > 1)$

17 Andrew used vocabulary lists to practice for the SAT. He studied for 5 days, each day increasing the number of new words he knew by 5. After 5 days, he believed that he knew at least 75 new words. What is the smallest number of words he knew on the first day?

18 Joyce must move a shipment of books from the lobby to the fifth floor. The sign in the elevator says "Maximum weight: 1,000 pounds." Each box of books weighs 60 pounds.

a Find the maximum number of boxes the elevator can lift.

b If Joyce, who weighs 115 pounds, must ride up with the boxes, then what is the maximum number of boxes she can put in the elevator?

Solving Systems of Linear Equations

Two or more linear equations with the same variables form a **system of linear equations**, sometimes called **simultaneous equations**. A **solution** to a system is any pair of values (x, y) that makes all equations true. For example, the pair $(2, 4)$ means $x = 2$ and $y = 4$. $(2, 4)$ is a solution to the system of equations $y = 2x$ and $y = x + 2$.

There are three types of linear systems based on the number of solutions.

Inconsistent System	Dependent System	Consistent System
No common solution.	An infinite number of common solutions.	One common solution.
Equations $x + y = 8$ and $x + y = 5$ have no common solution. It is impossible for the sum of the same numbers to equal both 8 and 5.	Equations $x + y = 8$ and $3x + 3y = 24$ form a dependent system. The second equation is formed by multiplying each term of the first equation by 3.	Equations $x + y = 8$ and $x - y = 2$ are neither inconsistent nor dependent. Their solution is found in a model problem on page 107.

Any system of linear equations will fall into one of these three categories. The majority of systems you will be asked to solve will have just one common solution. The complete solution of dependent systems is the same as the solution of either original linear equation. These solutions are discussed in Chapter 5.

There are two methods for solving systems of equations algebraically: the **addition-subtraction method** and the **substitution method**. Remember that the solution to a system of equations is a value for *each variable*, not just one.

To Solve Systems of Equations with the Addition-Subtraction Method

- If necessary, rewrite the equations in the form $ax + by = c$. This is the **standard form** of a linear equation.
- Decide which variable you want to eliminate.
- Multiply one or both equations by constants, if necessary, so that the coefficients of the variable you want to eliminate are opposites.
- Set the equations up in columns. Align like terms.
- Add the columns. This will result in a single equation in one variable.
- Solve the resulting equation.
- Substitute the resulting value into either *original* equation.
- Solve the resulting equation.
- Check by substituting both values into *all original equations*.

Note: If both variables cancel out when the column are added, the system of equations is either inconsistent or dependent, so a single solution cannot be found.

 MODEL PROBLEMS

1 Solve the system of equations $x + y = 8$ and $x - y = 2$.

SOLUTION

The equations are already in standard form, and the y terms are already opposites.

Set up the equations in columns.

$$x + y = 8$$
$$x - y = 2$$

Add the columns.

$$2x = 10$$

Solve for x.

$$\frac{2x}{2} = \frac{10}{2}$$
$$x = 5$$

Substitute 5 for x in the first equation.

$$(5) + y = 8$$

Solve for y.

$$-5 = -5$$
$$y = 3$$

Check. Substitute 5 for x and 3 for y in the original equations.

$$x + y = 8 \qquad\qquad x - y = 2$$
$$(5) + (3) = 8 \qquad (5) - (3) = 2$$
$$8 = 8 \checkmark \qquad\qquad 2 = 2 \checkmark$$

Answer: (5, 3)

2 Find the solution of the system $x - 2y = 1$ and $x + y = 10$.

SOLUTION

The coefficients of x are the same.
Write the first equation. $\quad x - 2y = 1$
Multiply each term by -1. $\quad (-1)x + (-1)(-2y) = (-1)1$
Simplify. $\quad -x + 2y = -1$
Set up the equations in columns and add.

$$-x + 2y = -1$$
$$x + y = 10$$
$$\overline{3y = 9}$$

Solve. $\quad y = 3$
Replace y in either equation.
$$x + y = 10$$
$$x + 3 = 10$$
$$x = 7$$

Check by substituting 3 for y and 7 for x in both original equations.

$\quad x - 2y = 1 \qquad x + y = 10$
$\quad 7 - 2(3) = 1 \qquad 7 + 3 = 10$ ✓
$\quad 7 - 6 = 1$ ✓

Answer: $(x, y) = (7, 3)$

3 Solve the system $5x - 2y = 10$ and $2x + y = 31$.

SOLUTION

More work is involved to make the y coefficients cancel.
Write the second equation. $\quad 2x + y = 31$
Multiply each term by 2. $\quad (2)2x + (2)y = (2)31$
Simplify. $\quad 4x + 2y = 62$
Set up the equations in columns and add.

$$5x - 2y = 10$$
$$4x + 2y = 62$$
$$\overline{9x = 72}$$
$$x = 8$$

Solve.
Replace x in either equation.
$$2x + y = 31$$
$$2(8) + y = 31$$
$$16 + y = 31$$
$$y = 15$$

Check by substituting 15 for y and 8 for x in both original equations.

$\quad 5x - 2y = 10 \qquad 2x + y = 31$
$\quad 5(8) - 2(15) = 10 \qquad 2(8) + (15) = 31$
$\quad 40 - 30 = 10$ ✓ $\qquad 16 + 15 = 31$ ✓

Answer: $(x, y) = (8, 15)$

To Solve Systems of Equations With the Substitution Method

- Solve one of the equations for one of the variables.
- Substitute the resulting algebraic expression in the second equation.
- Solve the second equation for the second variable.
- Substitute the resulting value in either original equation.
- Solve for the first variable.
- Check by substituting both values in the equation not used in step 4.

Chapter 3: Algebraic Operations and Reasoning

MODEL PROBLEMS

1 Solve this system of equations: $3x - 4y = 2$ and $x = 14 - 2y$

SOLUTION

Since the solutions are the same for each equation, $x = 14 - 2y$ is true in both equations.

Write the first equation.	$3x - 4y = 2$
Substitute $14 - 2y$ for x.	$3(14 - 2y) - 4y = 2$
Distribute the 3.	$42 - 6y - 4y = 2$
Combine like terms.	$42 - 10y = 2$
Add $10y$ to both sides.	$42 = 10y + 2$
Subtract 2 from both sides.	$40 = 10y$
Solve.	$4 = y$

Substitute 4 for y in either equation
$x = 14 - 2y$
$x = 14 - 2(4)$
$x = 14 - 8 = 6$

Check by substituting 6 for x and 4 for y in both original equations.

$3x - 4y = 2$ $x = 14 - 2y$
$3(6) - 4(4) = 2$ $6 = 14 - 2(4)$
$18 - 16 = 2$ ✓ $6 = 14 - 8$ ✓

Answer: $(x, y) = (6, 4)$

2 Henrietta has seven bills, all tens and twenties, that total $100 in value. How many of each bill does she have?

SOLUTION

Let x represent the number of tens and y represent the number of twenties.

Henrietta's number of bills can be shown by $x + y = 7$.

The cash value of the bills can be shown by $10x + 20y = 100$.

Now we have the system $x + y = 7$ and $10x + 20y = 100$.

Write the first equation. $x + y = 7$
Solve for x.
$$\begin{array}{r} -y = -y \\ \hline x = 7 - y \end{array}$$

Write the second equation. $10x + 20y = 100$
Substitute $(7 - y)$ for x. $10(7 - y) + 20y = 100$
Solve.
$70 - 10y + 20y = 100$
$70 + 10y = 100$
$$\begin{array}{r} -70 = -70 \\ \hline 10y = 30 \\ y = 3 \end{array}$$

Substitute 3 for y in one of the original equations and solve for x.
$x + y = 7$
$x + (3) = 7$
$$\begin{array}{r} -3 = -3 \\ \hline x = 4 \end{array}$$

To check, substitute 3 for y and 4 for x in each equation.

$10x + 20y = 100$ $x + y = 7$
$10(4) + 20(3) = 100$ $3 + 4 = 7$ ✓
$40 + 60 = 100$ ✓

Check the original problem.

4 bills + 3 bills = 7 bills

$(4 \times 10) + (3 \times 20) = 40 + 60 = 100$

Answer: Henrietta has 4 ten-dollar bills and 3 twenty-dollar bills.

Note: When you are solving word problems, be sure that your answer fits the problem, not just your equations.

1. You want to solve the system below by eliminating y. You multiply the first equation by 7. By what number should you multiply the second equation?

   $2x - 5y = 1$

   $-3x + 7y = -3$

   (1) -7
   (2) -5
   (3) 5
   (4) 7

2. What is the solution to this system of equations?

   $2x + y = 7$

   $3x - y = 3$

   (1) $(4, 1)$
   (2) $(2, -1)$
   (3) $(2, 3)$
   (4) $(3, 2)$

3. Prestige Parking charges $5 for the first hour or any part of that hour and $3 per hour for each additional hour or part of an hour. Paradise Parking charges $8 for the first hour or any part of that hour and $1.50 per hour for each additional hour or part of an hour. For how many hours of parking will the charge be the same in both garages?

   (1) From 1 hour and 1 minute to 2 hours
   (2) From 2 hours and 1 minute to 3 hours
   (3) From 3 hours and 1 minute to 4 hours
   (4) From 4 hours and 1 minute to 5 hours

4. A circus act has 3 times as many elephants as acrobats. Jorge noticed that all together, there were 56 legs in the circus ring. How many elephants were in the show?

   (1) 14 elephants
   (2) 12 elephants
   (3) 9 elephants
   (4) 4 elephants

5. What is the solution to this system of equations?

   $2x + y = 12$

   $x + 2y = 9$

6. In the vote for school president, 476 seniors voted for one or the other of the two candidates. The candidate who won had a majority of 94. How many seniors voted for the winner?

7. The Gateway Arch in St. Louis is 86 feet taller than the Washington Monument. If the height of the Gateway Arch is 369 feet more than one-half the height of the Washington Monument, how tall is each structure?

8. The sum of two numbers is 78. Their difference is 18. Find the numbers.

9. The factory foreman makes $9 more per hour than Janet, the most senior worker. If one dollar is subtracted from the foreman's rate of pay, the resulting amount is $\frac{3}{2}$ what Janet makes. Find the rate of pay for Janet and the foreman.

10. Carlos and Sam together unpacked cartons for 3 hours at a rate of 8 cartons per hour. During that time, Carlos unpacked twice as many cartons as Sam. How many cartons did each of them unpack?

11. Sandralee ordered a frankfurter and soda. Her cost was $3.75. Maria ordered two frankfurters. Her cost was $4.50. What was the cost of the soda?

12. A student was given the following two equations and asked to determine if they were equivalent: $n + \frac{m}{9} = \frac{1}{4}$ and $9n + m - \frac{1}{4} = 2$. Show how they are or are not equivalent. Explain your answer clearly.

13. Temperatures measured in degrees Celsius and degrees Fahrenheit are related by the formula $F = 1.8C + 32$ where F represents degrees Fahrenheit and C represents degrees Celsius. Jacques has a quick approximation: "Double the Celsius temperature and add 30 degrees." For what Celsius and Fahrenheit temperatures is his method exact?

14. Solve the following system of equations.

    $3x + 12 = y$

    $8x - 2y = 40$

CHAPTER REVIEW

1. The distance in feet, S, an object will fall equals the product of 16 and the time, t, squared, in seconds. $S =$
   (1) $16t^2$
   (2) $16 + t^2$
   (3) $\sqrt{16t}$
   (4) $\dfrac{16}{t^2}$

2. Christine will rent a bicycle the next time she goes to the boardwalk. She must leave a $25 deposit. When she returns the bike she will get a refund on her deposit and she will be charged $5 per hour or part of an hour. If she keeps the bicycle for 3 hours and 15 minutes, how much does she spend?
   (1) $15
   (2) $20
   (3) $40
   (4) $45

3. Mr. Schmidtmann took 55 minutes to drive into the city and back. He took 5 minutes less for the return trip than for the drive into the city. How long did his return trip take?
   (1) 20 minutes
   (2) 25 minutes
   (3) 30 minutes
   (4) 35 minutes

4. Express $\dfrac{8a^2}{-3b} + \dfrac{3}{4a}$ as a single fraction. Assume that $a \neq 0$ and $b \neq 0$.
   (1) $\dfrac{9b - 32a^3}{12ab}$
   (2) $\dfrac{8a^2 + 3}{-3b + 4a}$
   (3) $\dfrac{-8a^2 + 3}{12ab}$
   (4) $\dfrac{32a^2 + 9b}{12ab}$

5. What polynomial must be added to $7a^2 - 3a + 8$ to get $3a^2 + 5a - 9$?

6. Simplify: $3(a - b - c) + 2(b - c - a)$

7. Simplify: $\dfrac{15a^3b^2 - 10a^2b^3}{-5a^2b^2}$

Exercises 8–11: Solve for x.

8. $2(4x - 1) = 3(2x + 1)$

9. $5(x + 2) = 6 + 3(2x - 1)$

10. $\dfrac{3x - 5}{2} + \dfrac{5x + 1}{4} = 2x$

11. $\dfrac{x}{2} + \dfrac{x}{3} - \dfrac{x}{4} = 21$

Exercises 12–14: Solve for x and graph the solution set.

12. $6 - 4x > 8 + 5x$

13. $0.5(3 - 8x) < 10(1 - 0.5x)$

14. $\dfrac{x}{2} - \dfrac{x}{6} \leq \dfrac{7}{3}$

15. The relationship between the velocity of an object, the time it is in motion, and the distance it travels is $d = vt$. If an object travels 120 meters in 40 seconds, what is the velocity of the object in meters per second?

16. If the replacement set for x is $\{-3, -2, -1, 0, 1, 2, 3\}$, write the solution set for $-3x < 0$.

17. If $x = 10$ and $y = 3$, then what is the value of $(2x + 7) + y$?

18. Solve for f in the equation $ab(d + f) = f$.

19. In a county math league, 520 students from 100 different teams compete. Each team is required to have at least 5 members. What is the largest possible number of members on any one team?

20. The sophomore class is planning to sell school mascot puppets to help cover the cost of a class trip. A local supplier will charge $76.25 for the design and $2.25 for each puppet ordered. The class plans to sell the puppets for $5. How many puppets must be sold to earn at least $2,000?

Now turn to page 408 and take the Cumulative Review test for Chapters 1–3. This will help you monitor your progress and keep all your skills sharp.

Chapter 4
Factoring and Quadratic Equations

Factoring

In working with algebraic fractions, fractional equations, formulas, and quadratic equations, we often need to find two or more **factors** whose product is a given expression. This process is called **factoring**. Factoring is the reverse of multiplication:

- In multiplication, the factors are given and we must find the product. For example, the product of $4m(r + 5)$ is $4mr + 20m$, and the product of $(x + 3)(x - 2)$ is $x^2 + x - 6$.
- In factoring, the product is given and we must find the factors. For example, factoring $3x + 6$ gives $3(x + 2)$, and factoring $x^2 + x - 6$ gives $(x + 3)(x - 2)$.

In algebra, factoring means breaking a polynomial into two or more parts, its factors. When these factors are multiplied, the product is the original polynomial. Three **types of factoring** in algebra are:

Type 1: Factoring out the greatest common monomial.
Type 2: Factoring a binomial that is the difference of two squares.
Type 3: Factoring trinomials.

We will use two **special products** of polynomials (see Chapter 3) as shortcuts in factoring:

If a polynomial cannot be factored, it is called prime. For example, $a^2 + b^2$ is prime.

Special Product 1	Special Product 2
$(x + y)(x - y) = x^2 - y^2$	$(x + y)(x + y) = x^2 + 2xy + y^2$
	$(x - y)(x - y) = x^2 - 2xy + y^2$

Common Monomial Factors

To factor out the **greatest common monomial factor** (GCF), follow the steps in these model problems.

112

MODEL PROBLEMS

1 Factor: $2x^2 + 2x$

SOLUTION

Step 1. Examine the polynomial to find like factors that appear in each term. Look for the largest number and the variable with the highest exponent that will divide evenly into each term of the polynomial:

$2, x$

Step 2. Expressed as a product, $2x$, this is the greatest common monomial factor (GCF). Write the GCF outside a pair of parentheses:

$2x(\ )$

Step 3. Divide each term by the GCF:

$2x^2 \div 2x = x$
$2x \div 2x = 1$

Step 4. Write the results as a polynomial inside the parentheses:

$2x(x + 1)$

Check: Multiply.

$2x(x + 1)$
$= (2x \cdot x) + (2x \cdot 1)$
$= 2x^2 + 2x$ ✓

Column method:

$\quad\quad x + 1$
$\times \quad\quad 2x$
$\overline{\ \ 2x^2 + 2x\ }$ ✓

2 Factor: $12abc^2 - 24a^2c$

SOLUTION

Step 1. Find like factors: $12, a, c$

Step 2. Set up parentheses: $12ac(\ )$

Step 3. Divide:

$12abc^2 \div 12ac = bc$
$-24a^2c \div 12ac = -2a$

Step 4. Fill in the parentheses: $12ac(bc - 2a)$

Check: Multiply.

$12ac(bc - 2a)$
$= (12ac \cdot bc) + [12ac \cdot (-2a)]$
$= 12abc^2 + (-24a^2c)$
$= 12abc^2 - 24a^2c$ ✓

Column method:

$\quad\quad bc - \ 2a$
$\times \quad\quad 12ac$
$\overline{\ 12abc^2 - 24a^2c\ }$

To check factors, multiply. The product should be the given expression.

Here are some more examples:

Expression	Factors
$x^3y + 4xy^2$	$xy(x^2 + 4y)$
$6a^3 + 12a^2 + 3a$	$3a(2a^2 + 4a + 1)$
$15a^2x - 3x$	$3x(5a^2 - 1)$
$6cy - 6cx^2$	$6c(y - x^2)$
$3x(4x - 1) + 5(4x - 1)$	$(4x - 1)(3x + 5)$

 **Practice**

1. Which expression is prime?
   - (1) $x^2 + 2x$
   - (2) $3x^2 + 9$
   - (3) $x + 1$
   - (4) $2x^2 + 6x - 10$

Exercises 2 and 3: In which case is the greatest common monomial factored out correctly?

2. $6x^3 - 9x^2 - 12x$
   - (1) $3x^2(6x - 9 - 12)$
   - (2) $3x(2x^2 - 3x - 4)$
   - (3) $x(6x^2 - 9x - 12)$
   - (4) $3(2x^3 - 3x - 4x)$

3. $5a^2bc - 5ab^2c - 5abc$
   - (1) $5abc(abc)$
   - (2) $5abc(a - b - 5)$
   - (3) $5abc(a - b - 1)$
   - (4) $5abc(a - b - c)$

4. In which case is the greatest common monomial factored out *incorrectly*?
   - (1) $27x^3 - 6x^2 + 15x = x(27x^2 - 6x + 15)$
   - (2) $x^2(a - 3b) + x^2(a - 3b) = 2x^2(a - 3b)$
   - (3) $21xyz - 35axy = 7xy(3z - 5a)$
   - (4) $2g^2 + 2g = 2g(g + 1)$

Exercises 5–15: Factor each expression.

5. $5a - 5b$
6. $2x + 6y$
7. $ax + ab$
8. $gd - gr$
9. $ar^2 + 5a$
10. $10x^2y + 15xy^2$
11. $a(a - 7) + 9(a - 7)$
12. $x^2(x + 2) + 7(x + 2)$
13. $30x^5y^5 - 20x^3y^9 + 10x^3y^5$
14. $y(y - 1) + 2(y - 1)$
15. $8rta - 6rtb + 10rtc$

Factoring the Difference of Two Squares

In a second type of factoring, we need to factor a binomial of the form $x^2 - y^2$. We use special product 1 (see page 112), the product of binomials that are the sum and the difference of the same terms:

$$(x + y)(x - y) = x^2 - y^2 \quad (x + 6)(x - 6) = x^2 - 36 \quad (4 + x)(4 - x) = 16 - x^2$$

Certain products occur so often that you should recognize them at sight, and this product is one of them. It can be calculated using the distributive property or the FOIL method (see Chapter 3).

Special product 1 gives us an important rule:

- Two binomials of the form $(x + y)(x - y)$ produce the difference of two squares, $(x^2 - y^2)$.
- Therefore, to factor this special product, the difference of two squares, write two binomials that are the sum and difference of the square roots of the two terms.

 FOIL means First, Outer, Inner, Last— the order for multiplying the terms of two binomials.

MODEL PROBLEMS

Factor the following:

SOLUTIONS

1. $a^2 - b^2$ $(\sqrt{a^2} + \sqrt{b^2})(\sqrt{a^2} - \sqrt{b^2}) = (a + b)(a - b)$

2. $16x^2 - 25$ $(\sqrt{16x^2} + \sqrt{25})(\sqrt{16x^2} - \sqrt{25}) = (4x + 5)(4x - 5)$

3. $4y^2 - 25x^2$ $(\sqrt{4y^2} + \sqrt{25x^2})(\sqrt{4y^2} - \sqrt{25x^2}) = (2y + 5x)(2y - 5x)$

4. $\frac{1}{4}a^2b^2 - x^4$ $\left(\sqrt{\frac{1}{4}a^2b^2} + \sqrt{x^4}\right)\left(\sqrt{\frac{1}{4}a^2b^2} - \sqrt{x^4}\right) = \left(\frac{1}{2}ab + x^2\right)\left(\frac{1}{2}ab - x^2\right)$

5. $x^4 - 16$ $(\sqrt{x^4} - \sqrt{16})(\sqrt{x^4} + \sqrt{16}) = (x^2 - 4)(x^2 + 4) = (x - 2)(x + 2)(x^2 + 4)$

Practice

Exercises 1 and 2: Factor the binomial.

1. $x^2y^2 - 100 =$
   - (1) $10xy(-10xy)$
   - (2) $(x + y + 10)(x + y - 10)$
   - (3) $(xy - 10)(xy - 10)$
   - (4) $(xy + 10)(xy - 10)$

2. $\frac{9}{16}y^2 - 36 =$
   - (1) $\left(\frac{9}{4}y + 4\right)\left(\frac{9}{4}y - 4\right)$
   - (2) $\left(6y + \frac{3}{4}\right)\left(6y - \frac{3}{4}\right)$
   - (3) $\left(\frac{3}{4}y + 6\right)\left(\frac{3}{4}y - 6\right)$
   - (4) $\left(\frac{3}{4}y + 9\right)\left(\frac{3}{4}y - 9\right)$

3. Which binomial is factored correctly?
   - (1) $\frac{x^2}{4} - \frac{y^2}{9} = \left(\frac{x}{2} + \frac{y}{3}\right)\left(\frac{y}{3} - \frac{x}{2}\right)$
   - (2) $b^2 - 0.01 = (b - 0.1)(b - 0.1)$
   - (3) $49 - 0.81x^2y^2 = (7 + 0.09xy)(7 - 0.09xy)$
   - (4) $4a^6 - 9b^{10} = (2a^3 + 3b^5)(2a^3 - 3b^5)$

4. Which binomial is factored *incorrectly*?
   - (1) $y^2 - 1 = (y + 1)(y - 1)$
   - (2) $\frac{4}{9}a^2 - \frac{9}{25}y^2 = \left(\frac{2}{3}a + \frac{3}{5}y\right)\left(\frac{2}{3}a - \frac{3}{5}y\right)$
   - (3) $9y^2 - 16 = (3y + 4)(3y - 4)$
   - (4) $x^2y^2 - z^2 = (x^2y + 1)(x^2y - 1)$

Exercises 5–20: Factor each binomial.

5. $a^2 - 49$
6. $m^2 - 64$
7. $49x^2 - y^2$
8. $36 - x^2$
9. $16x^2 - 1$
10. $r^2 - s^2$
11. $x^2 - y^4$
12. $a^2 - 4b^2$
13. $9a^2 - 16y^2$
14. $81x^4 - 16y^8$
15. $49 - 4m^2$
16. $25a^2 - 81b^2$
17. $y^2 - c^2d^2$
18. $s^2t^2 - k^2n^2$
19. $x^4 - 0.0016$
20. $81 - x^4$

Factoring   **115**

Factoring Trinomials

A third type of factoring involves trinomials. Two cases of factoring trinomials are:

- Perfect trinomal squares
- General quadratic trinomials

Perfect Trinomial Squares   In factoring a perfect trinomial square, we use special product 2 (see page 112), the square of a binomial:

$(x + y)^2 = (x + y)(x + y) = x^2 + 2xy + y^2$

$(x - y)^2 = (x - y)(x - y) = x^2 - 2xy + y^2$

Methods of squaring a binomial are reviewed in Chapter 3. One method is FOIL. Special product 2 lets us identify a perfect trinomial square:

- The first and last terms must be perfect squares and must be positive (+).
- The middle term must be twice the product of the square roots of the first and last terms.
- The middle term can be positive (+) or negative (−). The sign of the middle term (+ or −) is the same as the sign between the terms of the binomials.

Special product 2 also gives us a rule for factoring a perfect trinomial square:

- If the middle term is positive, the factors will be of the form $(x + y)(x + y)$.
- If the middle term is negative, the factors will be of the form $(x - y)(x - y)$.
- The x term in the factors is the square root of the first term in the given trinomial.
- The y term in the factors is the square root of the last term in the given trinomial.

MODEL PROBLEMS

Factor these perfect-square trinomials:

1 $x^2 + 6x + 9$

SOLUTION

The middle term is positive, so the factors will have the form $(x + y)(x + y)$.

The x term in the factors is $\sqrt{x^2} = x$.

The y term in the factors is $\sqrt{9} = 3$.

Answer: $(x + 3)(x + 3)$ or $(x + 3)^2$

2 $x^2 - 10x + 25$

SOLUTION

The middle term is negative, so the factors will have the form $(x - y)(x - y)$.

The x term in the factors is $\sqrt{x^2} = x$.

The y term in the factors is $\sqrt{25} = 5$.

Answer: $(x - 5)(x - 5)$ or $(x - 5)^2$

3 $4x^2 + 12x + 9$

SOLUTION

The middle term is positive, so the factors will have the form $(x + y)(x + y)$.

The x term in the factors is $\sqrt{4x^2} = 2x$.

The y term in the factors is $\sqrt{9} = 3$.

Answer: $(2x + 3)(2x + 3)$ or $(2x + 3)^2$

General Quadratic Trinomials A general quadratic trinomial has the form:

$ax^2 + bx + c$

It may be the product of two different binomials or the same binomials. (When it is the product of the same binomials, it is a perfect trinomial square, reviewed above.)

To factor a general quadratic trinomial, we use trial and error, testing all pairs of possible binomial factors until the product is the given trinomial. However, we can shorten this process by applying FOIL, the factoring method, and certain rules:

- If all three terms in the given trinomial are positive (+), both factors have the form $(x + y)$.
- If only the middle term of the trinomial is negative (−), both factors have the form $(x - y)$.
- If the last term is negative (−), one factor has the form $(x + y)$ and the other $(x - y)$. (In this case the middle term can be positive or negative.)
- When the coefficient of the first term is 1, we need to test only the factors of the last term. In this case, the sum of the factors of the last term must equal the coefficient of the middle term.

Another approach to factoring is to use **field properties** (see "Properties of Operations" and "Closure" in Chapter 2). This method is demonstrated in Model Problem 6 below.

> Here we assume that the first term ax^2, is positive. If it is negative, see page 121.

MODEL PROBLEMS

Factor each quadratic trinomial.

1 $x^2 + 7x + 12$

SOLUTION

Since all the signs of the given trinomial are positive, both factors will have the form $(x + y)$.

Here the coefficient of the first term is 1, so we try only factors of 12: 1 • 12, 2 • 6, and 3 • 4:

Try $(x + 1)(x + 12) = x^2 + 13x + 12$ (wrong)

Try $(x + 2)(x + 6) = x^2 + 8x + 12$ (wrong)

Try $(x + 3)(x + 4) = x^2 + 7x + 12$ ✓

Shortcut: The sum of the factors of 12 must be 7. Only one pair meets this condition: 3, 4.

2 $x^2 - 10x + 24$

SOLUTION

Only the middle term is negative, so both factors will have the form $(x - y)$.

Since the coefficient of the first term is 1, test just the factors of 24. In this case, at least one of these factors must be negative, since their sum must be −10. Because their product is +24, *both* must be negative. Thus the possible factors are: −1 • (−24), −2 • (−12), −3 • (−8), and −4 • (−6).

Only one pair (−4 and −6) gives a sum of −10.

Answer: Therefore, the factors must be $(x - 4)(x - 6)$.

3 $m^2 - m - 2$

SOLUTION

Here, the last term is negative, so the factors will have the form $(x + y)(x - y)$.

The coefficient of the first term is 1, so test only the factors of the last term. Since the last term is -2, one factor must be negative and the other positive: $1 \cdot (-2)$ or $-1 \cdot 2$.

The coefficient of the middle term is -1, and only the pair $1, -2$ will add up to -1.

Answer: The factors are $(m + 1)(m - 2)$.

4 $x^2 + 3x - 18$

SOLUTION

The factors will have the form $(x + y)(x - y)$.

We need to test only the factors of the last term. One must be positive and the other negative. The possible factors are $1 \cdot (-18), -1 \cdot 18, 2 \cdot (-9), -2 \cdot 9, 3 \cdot (-6)$, and $-3 \cdot 6$.

Of these, only the pair $-3, 6$ will add up to the middle coefficient, 3.

Answer: The factors are $(x + 6)(x - 3)$.

5 $2x^2 + 9x - 35$

SOLUTION

The factors will have the form $(x + y)(x - y)$.

Here the coefficient of first term is 2. Therefore we must consider the factors of both 2 and -35. Factors of 2: $1 \cdot 2$ and $2 \cdot 1$

Factors of -35: $1 \cdot (-35), -1 \cdot 35, 5 \cdot (-7)$, and $-5 \cdot 7$

We can eliminate $1 \cdot (-35)$ and $-1 \cdot 35$, since these would yield sums that are too great.

Use trial and error with the remaining choices, applying FOIL to check. For example:

Try $(x - 5)(2x + 7) = 2x^2 + 7x - 10x - 35 = 2x^2 - 3x - 35$ (wrong)

Try $(2x - 7)(x + 5) = 2x^2 + 10x - 7x - 35 = 2x^2 - 3x - 35$ (wrong)

Try $(2x - 5)(x + 7) = 2x^2 + 14x - 5x - 35 = 2x^2 + 9x - 35$ ✓

Answer: $(2x - 5)(x + 7)$

With experience, you will get better and better at guessing and testing.

6 Apply field properties to factor $x^2 + 8x + 15$.

SOLUTION

$x^2 + 8x + 15$	Given quadratic trinomial
$x^2 + (5 + 3)x + (5 \cdot 3)$	$5 + 3 = 8$ and $5 \cdot 3 = 15$
$x^2 + 5x + 3x + (5 \cdot 3)$	Distributive property
$(x^2 + 5x) + [3x + (5 \cdot 3)]$	Associative property
$[(x \cdot x) + (x \cdot 5)] + [(3 \cdot x) + (3 \cdot 5)]$	Commutative property
$x(x + 5) + 3(x + 5)$	Distributive property (multiplication over addition)
$(x + 3)(x + 5)$	Distributive property (multiplication over addition)

Answer: $(x + 3)(x + 5)$

 **Practice**

1 Which expression is a perfect trinomial square?
  (1) $x^2 - 5x + 4$
  (2) $x^2 + \frac{1}{2}x + \frac{1}{4}$
  (3) $4x^2 - 4x + 1$
  (4) $x^2 + 7x + \frac{4}{49}$

2 Factor: $7x^2 + 9x - 10$
  (1) $(7x + 10)(x - 1)$
  (2) $(7x - 5)(x + 2)$
  (3) $(7x + 2)(x - 5)$
  (4) $(7x - 1)(x + 10)$

3 Factor: $2x^2 - x - 3$
  (1) $(2x + 1)(x + 3)$
  (2) $(2x + 3)(x - 1)$
  (3) $(2x - 3)(x + 1)$
  (4) $(2x - 3)(x - 1)$

4 Which expression is prime?
  (1) $x^2 + 4x + 3$
  (2) $x^2 + 5x + 6$
  (3) $x^2 + 6x + 8$
  (4) none of the above

Exercises 5–20: Factor each trinomial if possible.

5 $x^2 - 10x + 25$
6 $x^2 - 7x + 12$
7 $x^2 + 2x - 8$
8 $x^2 - 8x - 20$
9 $11x^2 + 12x + 1$
10 $11x^2 - 12x + 1$
11 $x^2 - 5x + \frac{25}{4}$
12 $3y^2 - 4y - 4$
13 $x^2 + 2x + 1$
14 $x^2 + 3x + 2$
15 $4x^2 - 20x + 25$
16 $2x^2 + 5x + 6$
17 $16x^2 + 40x + 25$
18 $a^4 + 2a^2b^2 + b^4$
19 $6x^2 - 11x + 5$
20 $4x^2 - 39x + 81$

Factoring Completely

In general, to factor a polynomial completely:

- Find any common monomial factors.
- If possible, find the binomial factors of the remaining terms, and continue until there are no more possible factors.
- Last, write all the factors as a product.

Flowchart for Factoring Completely

First: Remove common monomial factors.

Second: If the items that remain are

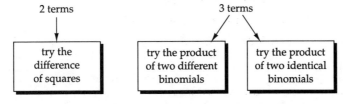

Factoring

MODEL PROBLEMS

Factor completely:

SOLUTIONS

1. $ax^2 - a$

   $a(x^2 - 1)$
   $a(x - 1)(x + 1)$

2. $9a^4 - 36b^4$

   $9(a^4 - 4b^4)$
   $9(a^2 - 2b^2)(a^2 + 2b^2)$

3. $x^4 - 3x^3 - 40x^2$

   $x^2(x^2 - 3x - 40)$
   $x^2(x - 8)(x + 5)$

4. $ax^5 - a^5x^5$

   $ax^5(1 - a^4)$
   $ax^5(1 - a^2)(1 + a^2)$
   $ax^5(1 - a)(1 + a)(1 + a^2)$

Practice

1. In factoring completely, the first step is:
   (1) Write the factors as a product.
   (2) Find common monomial factors.
   (3) Use the FOIL method.
   (4) Find the bionomial factors.

Exercises 2–4: Factor the given expression completely.

2. $2x^2 - 72y^2$
   (1) $2(x + 6y)(x - 6y)$
   (2) $(\sqrt{2}x + 6\sqrt{2}y)(\sqrt{2}x - 6\sqrt{2}y)$
   (3) $2x(x - 6y)^2$
   (4) $2(x - 6y)^2$

3. $2x^2 - 8x - 10$
   (1) $2(x - 5)(x - 1)$
   (2) $2(x - 5)(x + 1)$
   (3) $(2x - 10)(x + 1)$
   (4) $(x - 10)(2x + 1)$

4. $x^3 - 8x^2 + 16x$
   (1) $(x^2 - 8)^2$
   (2) $(x - 8)(x^2 + 2)$
   (3) $x(x - 4)^2$
   (4) $x(x - 8)(x + 2)$

Exercises 5–20: Factor each expression completely.

5. $5x^2 - 20$
6. $ab^4 - ax^4$
7. $3x^2 + 12x + 12$
8. $9x^3 - 9x$
9. $3x^2 - 75$
10. $6a^2 - 6a^4$
11. $x^3 - x^2 - 2x$
12. $a^3 - a$
13. $x - 25x^3$
14. $ax^3 - 36ax$
15. $25x^2 - 100y^2$
16. $2a^4 - 32b^4$
17. $9x^2 + 18xy + 9y^2$
18. $5x^2 - 25x + 30$
19. $a - ar^2$
20. $x^2(x + 2) - 9(x + 2)$

A Special Case: Factoring Out −1

If the first term of a general quadratic trinomial is negative, we can factor the trinomial more easily by first factoring out −1. This method is reviewed here.

When the numerator and denominator of a fraction are the same, we divide and get 1 as the quotient. When they are opposites, we get −1.

$\dfrac{n}{n} = 1$	$\dfrac{-n}{-n} = 1$	$\dfrac{n}{-n} = -1$	$\dfrac{-n}{n} = -1$
$\dfrac{3}{3} = 1$	$\dfrac{-7}{-7} = 1$	$\dfrac{4}{-4} = -1$	$\dfrac{-6}{6} = -1$

The examples above are obvious, but something like $\dfrac{x-7}{7-x} = -1$ is not so obvious. To get a clearer picture, factor out −1 from either the numerator or the denominator so that $x - 7$ is the same as $-1(7 - x)$. Thus:

$$\frac{x-7}{7-x} = \frac{-1(7-x)}{(7-x)} = -1$$

MODEL PROBLEMS

In each problem, factor out −1 and then simplify.

1 Factor −1 out of the denominator: $\dfrac{2x - 5}{5 - 2x}$

SOLUTION

$$\frac{2x-5}{5-2x} = \frac{(2x-5)}{-1(2x-5)} = \frac{1}{-1} = -1$$

2 Factor −1 out of the numerator: $\dfrac{7 - x}{x^2 - 49}$

SOLUTION

$$\frac{7-x}{x^2-49} = \frac{-1(x-7)}{(x+7)(x-7)} = \frac{-1}{x+7}$$

3 Factor −1 out of the denominator: $\dfrac{6xy - 2x^2}{x - 3y}$

SOLUTION

$$\frac{6xy-2x^2}{x-3y} = \frac{2x(3y-x)}{x-3y} = \frac{2x(3y-x)}{-1(3y-x)} = \frac{2x}{-1} = -2x$$

4 Factor −1 out of the numerator: $\dfrac{3 - 6x}{2x^2 + 5x - 3}$

SOLUTION

$$\frac{3-6x}{2x^2+5x-3} = \frac{3(1-2x)}{(2x-1)(x+3)} = \frac{-1(3)(2x-1)}{(2x-1)(x+3)} = \frac{-3}{x+3}$$

5 Use factoring out −1 to factor this trinomial: $-x^2 - x + 6$

SOLUTION

$$-x^2 - x + 6 = -1(x^2 + x - 6) = -1(x + 3)(x - 2)$$

1 Which is equivalent to -1?

(1) $\dfrac{x^2 - 4}{(2 - x)^2}$ (3) $\dfrac{4x - 6y}{6y - 4x}$

(2) $\dfrac{x^2 - xy}{y - x}$ (4) $\dfrac{2x - y}{2y - x}$

Exercises 2–4: For each given expression, factor out -1 and simplify.

2 $\dfrac{x^2 - xy}{y^2 - xy}$

(1) $-\dfrac{x}{y}$

(2) $\dfrac{xy - x^2}{y^2 - xy}$

(3) $-\dfrac{y - x}{x - y}$

(4) $\dfrac{2x}{y}$

3 $\dfrac{8 - 2x}{x^2 - 16}$

(1) $\dfrac{-2}{x + 4}$

(2) $\dfrac{4 - x}{x - 8}$

(3) $\dfrac{2}{4 - x}$

(4) $\dfrac{-1}{x - 2}$

4 $\dfrac{x^2 + x - 56}{64 - x^2}$

(1) $\dfrac{7 - x}{x + 8}$

(2) $-\dfrac{x - 7}{x - 8}$

(3) $\dfrac{7 - x}{8 - x}$

(4) $-\dfrac{x + 7}{x + 8}$

Exercises 5–18: Simplify each fraction.

5 $\dfrac{5 - z}{z - 5}$

6 $\dfrac{a - b}{b - a}$

7 $\dfrac{x - 6y}{6y - x}$

8 $\dfrac{x - y}{2y - 2x}$

9 $\dfrac{x^2 - 9}{3 - x}$

10 $\dfrac{2x - 6}{9 - 3x}$

11 $\dfrac{7 - a}{3a - 21}$

12 $\dfrac{x - 7}{49 - x^2}$

13 $\dfrac{x^2 - 4x}{16 - x^2}$

14 $\dfrac{a^2 + 2a - 15}{3 - a}$

15 $\dfrac{3 - 2x}{2x^2 + 3x - 9}$

16 $\dfrac{4 - x}{x^2 - 8x + 16}$

17 $\dfrac{9 - a^2}{a^2 - 4a + 3}$

18 $\dfrac{8 - x}{x^2 + x - 72}$

Exercises 19 and 20: Factor each trinomial by first factoring out -1.

19 $4x - x^2 - 21$

20 $7x - 2x^2 - 6$

122 Chapter 4: Factoring and Quadratic Equations

Solving Factorable Quadratic Equations

As reviewed above, a **quadratic trinomial** has the form:

$ax^2 + bx + c$

This expression can also be described as a **second-degree polynomial in one variable**: it has one variable x whose highest power is 2.

An equation involving a polynomial of the second degree in one variable is called a **quadratic equation**. This section reviews solutions to factorable quadratic equations.

Standard Form

The **standard form of a quadratic equation** is:

$ax^2 + bx + c = 0$

where a, b, and c are real numbers and $a \neq 0$.

When a quadratic equation is written in standard form, all of the terms on one side are arranged in descending order of exponents, and the other side is zero. For example:

$x^2 + x - 6 = 0 \qquad x^2 - 5 = 0$

MODEL PROBLEMS

Write in standard form:

SOLUTIONS

1. $x^2 + 3x = 4$ → $x^2 + 3x - 4 = 0$
2. $2x + x^2 + 1 = 0$ → $x^2 + 2x + 1 = 0$
3. $10 - 13x + 3x^2 = 0$ → $3x^2 - 13x + 10 = 0$
4. $9 + 6x + x^2 = 0$ → $x^2 + 6x + 9 = 0$
5. $x^2 + x + x + 1 = 0$ → $x^2 + 2x + 1 = 0$
6. $3x + 4 = x^2$ → $-x^2 + 3x + 4 = 0$ (Or: $x^2 - 3x - 4 = 0$)

Practice

1. If $n \neq 0$, which quadratic equation is written in standard form?
   (1) $x + x^2 = n$
   (2) $-x^2 - x - n = 0$
   (3) $x + x^2 + n = 0$
   (4) $x^2 - x = n$

2. If these quadratic equations are written in standard form, in which is it true that $a = 1$, $b = 2$, and $c = 3$?
   (1) $2x + x^2 = 3$
   (2) $3 + 2x = -x^2$
   (3) $2x = 3 + x^2$
   (4) $1 + 2x + 3x^2 = 0$

3 Which sentence represents a quadratic equation that can be expressed in standard form?

(1) A number squared is equal to the sum of zero and the same number squared.
(2) There are three consecutive even integers such that one-third the sum of the greater two is equal to the least of the three.
(3) The square of the difference between a number and one is thirty-six.
(4) One divided by a number is equal to one divided by the sum of twice the number and one.

4 Which of the following can be written as a quadratic equation in standard form?

(1) $x^2 + 3x - 5 = x^2 + x + 7$
(2) $x(x + 3) = 5x(1 - x^2)$
(3) $x^2 - 9 = \dfrac{1}{x}$
(4) $3(x^2 + 1) = (x + 5)(x - 5)$

Exercises 5–9: Write each quadratic equation in standard form.

5 $x(x + 2) = 8$
6 $x^2 = 35 - x - 15$
7 $x(x - 5) = 24$
8 $x^2 - 7x = -2x + 6$
9 $x(2x - 7) = 3x^2 - 8$

10 The following examples can *not* be written as quadratic equations in standard form. For each, explain why not:

a $(x + 1)(x - 1) = x^2$
b $x(x + 1)(x - 1) = 0$
c $x + 1 = 5x - 5$

Quadratic Trinomial Equations

Second-degree polynomial expressions can often be factored to get two **first-degree**, or **linear**, expressions—that is, expressions in which the highest power of x is 1. This gives us a strategy for solving quadratic equations.

Suppose that the product of two expressions is 0. Using the zero product property of multiplication, we know that *at least one* of these expressions equals zero:

If a and b are real numbers, then $ab = 0$ if and only if $a = 0$ or $b = 0$.

Therefore, to solve a quadratic equation, we factor the quadratic trinomial to get two linear binomials. We can do this by reversing the FOIL process and using the rules for factoring (see above and Chapter 3).

- Write the quadratic equation in standard form.
- Factor the quadratic expression.
- Set each linear factor equal to zero (by the zero product property).
- Solve each linear equation to find the solution set, called the **roots**.
- Check each value in the original quadratic equation.

MODEL PROBLEMS

Solve each quadratic equation and check your solution.

1 $x^2 + 5x + 4 = 0$

SOLUTION

Factor the left side of the equation:

$x^2 + 5x + 4 = (x + 1)(x + 4)$

Solve the equation:

$x^2 + 5x + 4 = (x + 1)(x + 4) = 0$

At least one factor must equal zero: $x + 4 = 0$ or $x + 1 = 0$.

Set each factor equal to zero and solve each equation:

$x + 4 = 0 \quad\quad x + 1 = 0$
$\quad x = -4 \quad\quad\quad x = -1$

Check:

$x = -4$	$x = -1$
$x^2 + 5x + 4 = 0$	$x^2 + 5x + 4 = 0$
$(-4)^2 + 5(-4) + 4 = 0$	$(-1)^2 + 5(-1) + 4 = 0$
$16 - 20 + 4 = 0$	$1 - 5 + 4 = 0$
$0 = 0$ ✓	$0 = 0$ ✓

Answer: $x = -4$ or $x = -1$. The solution set is $\{-4, -1\}$.

2 $x^2 - 8x + 16 = 0$

SOLUTION

Factor the left side of the equation:

$x^2 - 8x + 16 = (x - 4)(x - 4) = 0$

Set the factor equal to zero and solve:

$x - 4 = 0$
$\quad x = 4$

Check:

$x^2 - 8x + 16 = 0$
$(4)^2 - 8(4) + 16 = 0$
$16 - 32 + 16 = 0$
$0 = 0$

Answer: $x = 4$. The solution set is $\{4\}$.

Solving Factorable Quadratic Equations

3 $x^2 + 2x = 15$

SOLUTION

Rewrite the equation in standard form:
$$x^2 + 2x - 15 = 0$$

Factor the left side of this equation. (Look for two numbers whose product is -15 and whose sum is 2.)
$$x^2 + 2x - 15 = (x + 5)(x - 3) = 0$$

At least one factor must equal zero, so $x + 5 = 0$ or $x - 3 = 0$. Solve both equations:

$x + 5 = 0 \qquad x - 3 = 0$
$\quad x = -5 \qquad\quad x = 3$

Check:

$x = -5$	$x = 3$
$x^2 + 2x = 15$	$x^2 + 2x = 15$
$(-5)^2 + 2(-5) = 15$	$(3^2) + 2(3) = 15$
$25 - 10 = 15$	$9 + 6 = 15$
$15 = 15$ ✓	$15 = 15$ ✓

Answer: $x = -5$ or $x = 3$. The solution set is $\{-5, 3\}$.

4 $3x^2 + 8x + 4 = 0$

SOLUTION

Factor the left side of the equation:
$$3x^2 + 8x + 4 = (3x + 2)(x + 2) = 0$$

At least one factor must equal zero, so $3x + 2 = 0$ or $x + 2 = 0$. Solve both equations:

$3x + 2 = 0 \qquad x + 2 = 0$
$\quad x = -\dfrac{2}{3} \qquad\quad x = -2$

Check:

$x = -\dfrac{2}{3}$	$x = -2$
$3x^2 + 8x + 4 = 0$	$3x^2 + 8x + 4 = 0$
$3\left(-\dfrac{2}{3}\right)^2 + 8\left(-\dfrac{2}{3}\right) + 4 = 0$	$3(-2)^2 + 8(-2) + 4 = 0$
$3\left(\dfrac{4}{9}\right) - \dfrac{16}{3} + 4 = 0$	$3(4) - 16 + 4 = 0$
$\dfrac{4}{3} - \dfrac{16}{3} + 4 = 0$	$12 - 16 + 4 = 0$
$-\dfrac{12}{3} + 4 = 0$	$-4 + 4 = 0$
$-4 + 4 = 0$	$0 = 0$ ✓
$0 = 0$ ✓	

Answer: $x = -\dfrac{2}{3}$ or $x = -2$. The solution set is $\left\{-\dfrac{2}{3}, -2\right\}$.

5 $4x^2 + 24x + 20 = 0$

SOLUTION

Factor the left side of the equation:

$$4x^2 + 24x + 20 = 4(x + 5)(x + 1)$$

Eliminate the constant factor by dividing both sides of the equation:

$$\frac{4(x + 5)(x + 1)}{4} = \frac{0}{4}$$

$$(x + 5)(x + 1) = 0$$

At least one factor must equal zero, so $x + 5 = 0$ or $x + 1 = 0$. Solve both equations:

$x + 5 = 0 \qquad x + 1 = 0$
$x = -5 \qquad x = -1$

Check:

$x = -5$	$x = -1$
$4x^2 + 24x + 20 = 0$	$4x^2 + 24x + 20 = 0$
$[(4)(-5)^2] + [(24)(-5)] + 20 = 0$	$[(4)(-1)^2] + [(24)(-1)] + 20 = 0$
$(4 \cdot 25) + (-120) + 20 = 0$	$(4 \cdot 1) + (-24) + 20 = 0$
$100 - 120 + 20 = 0$	$4 - 24 + 20 = 0$
$-20 + 20 = 0$	$-20 + 20 = 0$
$0 = 0$ ✓	$0 = 0$ ✓

Answer: $x = -5$ or $x = -1$. The solution set is $\{-5, -1\}$.

Note: Quadratic equations in standard form can also be solved with the **quadratic formula**, substituting the values of a, b, and c:

$$x = \frac{-b \pm \sqrt{b^2 - 4ac}}{2a}$$

See the appendix on how to use a calculator to evaluate this formula.

In some cases, factoring can be used to solve a higher-order equation by expressing it as a factor and a quadratic expression. If there is a common monomial factor that contains a variable, that factor may be zero and give you a root. Do *not* eliminate it by division.

The next model problem is a typical example. It involves a **cubic equation**, an equation of the form:

$$ax^3 + bx^2 + cx + d = 0$$

where a, b, c, and d are real numbers and $a \neq 0$. A factorable cubic equation has three real roots.

MODEL PROBLEM

Solve, and check your solution: $a^3 = 4a^2 + 5a$

SOLUTION

Write the cubic equation in standard form:

$a^3 - 4a^2 - 5a = 0$

Factor the left side to get a monomial and a quadratic expression:

$a(a^2 - 4a - 5) = 0$

Then factor completely:

$a(a - 5)(a + 1) = 0$

Set each factor equal to zero and solve the resulting equations:

$a = 0 \qquad a - 5 = 0 \qquad a + 1 = 0$
$a = 0 \qquad a = 5 \qquad a = -1$

Check:

$a = 0$	$a = 5$	$a = -1$
$a^3 = 4a^2 + 5a$	$a^3 = 4a^2 + 5a$	$a^3 = 4a^2 + 5a$
$0^3 = 4(0^2) + 5(0)$	$5^3 = 4(5^2) + 5(5)$	$(-1)^3 = 4(-1)^2 + 5(-1)$
$0 = 0 + 0$ ✓	$125 = 4(25) + 25$	$-1 = 4(1) + (-5)$
	$125 = 100 + 25 = 125$ ✓	$-1 = 4 - 5 = -1$ ✓

Answer: $a = 0, a = 5,$ or $a = -1$. The solution set is $\{0, 5, -1\}$.

"Solve the equation"—as in the model problems above—is a typical problem situation. But problems involving quadratic equations may take other forms and therefore may call for different strategies. Look at the next model problems.

MODEL PROBLEMS

1. One root of the equation $x^2 + 3x + k = 0$ is -1.
   a. Find the value of k.
   b. Find the second root.

SOLUTION

a. The value $x = -1$ makes the equation true, so substitute -1 for x:

$(-1)^2 + 3(-1) + k = 0$
$1 - 3 + k = 0$
$-2 + k = 0$
$k = 2$

Answer: The value of k is 2.

128 Chapter 4: Factoring and Quadratic Equations

b Substitute 2 for k in the given equation:
$$x^2 + 3x + 2 = 0$$
Factor:
$$(x + 2)(x + 1) = 0$$
Set each factor equal to zero and solve the resulting equations:
$$x + 2 = 0 \quad\quad x + 1 = 0$$
$$x = -2 \quad\quad x = -1$$
Check the second root:
$$x^2 + 3x + 2 = 0$$
$$(-2)^2 + 3(-2) + 2 = 0$$
$$4 + (-6) + 2 = 0$$
$$-2 + 2 = 0$$
$$0 = 0 \checkmark$$

Answer: The second root is -2.

2 Write an equation of the form $ax^2 + bx + c = 0$ such that the roots (the solution set) will be $\{-2, 7\}$.

SOLUTION

Work backward: Let $x = -2$ and $x = 7$

Rewrite the equations: $x + 2 = 0$ and $x - 7 = 0$

Write the expressions on the left as factors, in parentheses, and set the product equal to 0:
$$(x + 2)(x - 7) = 0$$
Multiply, using FOIL: $x^2 - 5x - 14 = 0$

Check:

$x = -2$	$x = 7$
$x^2 - 5x - 14 = 0$	$x^2 - 5x - 14 = 0$
$(-2)^2 - 5(-2) - 14 = 0$	$7^2 - 5(7) - 14 = 0$
$4 + 10 - 14 = 0$	$49 - 35 - 14 = 0$
$0 = 0 \checkmark$	$0 = 0 \checkmark$

ALTERNATIVE SOLUTION

Sum of the roots $= -\dfrac{b}{a}$

Product of the roots $= \dfrac{c}{a}$

You can always assume that $a = 1$. Then:

$\dfrac{-b}{a} = -b$, so $-b = -2 + 7 = 5$, and $b = -5$

$\dfrac{c}{a} = c = -2 \cdot 7 = -14$

If $b = -5$ and $c = -14$, then the equation is $x^2 - 5x - 14 = 0$.

Answer: The equation is $x^2 - 5x - 14 = 0$.

Note: In Model Problem 2, the answer could also be any equivalent equation.

Practice

1. Solve: $x^2 - 13x - 30 = 0$
   (1) $\{-15, -2\}$
   (2) $\{-15, 2\}$
   (3) $\{15, -2\}$
   (4) $\{15, 2\}$

2. Solve: $x^2 + 5x = 6$
   (1) $\{-1, 6\}$
   (2) $\{1, -6\}$
   (3) $\{2, -3\}$
   (4) $\{2, 3\}$

3. For which equation is the solution set $\{3, 4\}$?
   (1) $x^2 + 3x + 4 = 0$
   (2) $x^2 - 7x + 12 = 0$
   (3) $x^2 + 12x + 7 = 0$
   (4) $x^2 - 9x = 16$

4. For which equation is the solution set $\{-8, 2\}$?
   (1) $x^2 + 6x - 16 = 0$
   (2) $x^2 - 6x - 10 = 0$
   (3) $x^2 - 12x - 64 = 0$
   (4) $x^2 - 16x + 6 = 0$

Exercises 5–13: Solve each equation and check your solution.

5. $x^2 - 10x + 25 = 0$
6. $x^2 + 22x + 57 = 0$
7. $x^2 + 14x + 40 = 0$
8. $x^2 - 11x + 18 = 0$
9. $2x^2 + 9x + 7 = 0$
10. $5x^2 - 26x + 5 = 0$
11. $3x^2 + 8x + 5 = 0$
12. $2x^2 + 32x - 114 = 0$
13. $3x^2 + 14x + 8 = 0$

Exercises 14–17: In these problems, solution sets are given. For each solution set, write an equation of the form $ax^2 + bx + c = 0$.

14. $\{4, -2\}$
15. $\{5, 8\}$
16. $\{-3, -3\}$
17. $\{-11, 3\}$

Items 18–20: In each of these problems, an equation and one of its roots are given. Find:

a. the value of k
b. the second root

18. 5 is a root of $x^2 - 7x + k = 0$
19. 5 is a root of $x^2 - 3x + k = 0$
20. 7 is a root of $x^2 - 3x = -k$

Incomplete Quadratic Equations

Recall that a quadratic equation is an equation of the form $ax^2 + bx + c = 0$ where a, b, and c are real numbers and $a \neq 0$.

An **incomplete quadratic equation** is a quadratic equation in which $b = 0$ or $c = 0$ or both $b = 0$ and $c = 0$. In general, the following three forms represent incomplete quadratic equations.

- Constant is missing: $ax^2 + bx = 0$
- Linear term is missing: $ax^2 + c = 0$
- Constant and linear term are missing: $ax^2 = 0$

Note that each type can be written in standard ("complete") form:

Incomplete Form	Standard Form
$ax^2 + bx = 0$	$ax^2 + bx + 0 = 0$
$ax^2 + c = 0$	$ax^2 + 0x + c = 0$
$ax^2 = 0$	$ax^2 + 0x + 0 = 0$

Solving Quadratic Equations When the Constant Is Missing To solve an incomplete quadratic of the form $ax^2 + bx = 0$:

- Rewrite the equation, if necessary, so that all terms with the unknown are on one side of the equals sign, leaving zero on the other side.

  If the equation is fractional, first multiply by the LCD to clear the equation of all fractions, and then rewrite.

  If there are parentheses, remove them, and then rewrite.

- Combine like terms, if any.
- Factor completely.
- Set each factor equal to zero and solve the resulting equations.
- Check both solutions.

MODEL PROBLEMS

1 Solve and check your solution: $x^2 = 6x$

SOLUTION

Step 1. Rewrite: $x^2 - 6x = 0$

We omit step 2, since there are no like terms to combine.

Step 3. Factor: $x(x - 6) = 0$

Step 4. Set each factor equal to zero and solve:

$x = 0 \qquad x - 6 = 0$
$x = 0 \qquad x = 6$

Step 5. Check:

$x = 0$	$x = 6$
$x^2 = 6x$	$x^2 = 6x$
$0^2 = 6(0)$	$6^2 = 6(6)$
$0 = 0$ ✓	$36 = 36$ ✓

Answer: $x = 0$ or $x = 6$. The solution set is $\{0, 6\}$.

2 Solve for x and check: $ax^2 + bx = 0$

SOLUTION

We do not need to rewrite (step 1), and there are no like terms to combine (step 2), so we go on to steps 3, 4, and 5.

Step 3. Factor: $x(ax + b) = 0$

Step 4. Set each factor equal to zero and solve:

$x = 0 \qquad ax + b = 0$
$\qquad\qquad ax = -b$
$\qquad\qquad x = \dfrac{-b}{a} \text{ or } -\dfrac{b}{a}$

Step 5. Check:

$x = 0$	$x = -\dfrac{b}{a}$
$ax^2 + bx = 0$	$ax^2 + bx = 0$
$a(0^2) + b(0) = 0$	$a\left(-\dfrac{b}{a}\right)^2 + b\left(-\dfrac{b}{a}\right) = 0$
$0 + 0 = 0$	$a\left(\dfrac{b^2}{a^2}\right) + b\left(-\dfrac{b}{a}\right) = 0$
$0 = 0$ ✓	$\dfrac{b^2}{a} + \left(-\dfrac{b^2}{a}\right) = 0$
	$0 = 0$ ✓

Answer: $x = 0$ or $x = -\dfrac{b}{a}$.

The solution set is $\left\{0, -\dfrac{b}{a}\right\}$.

Solving Quadratic Equations When the Linear Term Is Missing To solve incomplete quadratics of the form $ax^2 + c = 0$:

- Rewrite, if necessary, so that all terms with the unknown are on one side of the equation and a constant is on the other side.

    If necessary, before rewriting multiply by the LCD to clear the equation of all fractions; also, remove any parentheses.

- Combine like terms, if any.
- Divide both sides of the equation by the coefficient of the x^2 term.
- Extract the square root of both sides of the equation, writing the symbol $\pm$ before the square root of the known quantity. If possible, simplify.
- Check each root.

> Remember: Square roots of equals are equal

MODEL PROBLEMS

1 Solve and check your solution: $4x^2 - 27 = x^2$

SOLUTION

Step 1. Rewrite: $4x^2 - x^2 = 27$

Step 2. Combine like terms: $3x^2 = 27$

Step 3. Divide both sides by 3: $x^2 = 9$

Step 4. Take the square root of both sides and simplify:
$$\sqrt{x^2} = \pm\sqrt{9}$$
$$x = \pm 3$$

Step 5. Check:

$x = 3$	$x = -3$
$4x^2 - 27 = x^2$	$4x^2 - 27 = x^2$
$4(3)^2 - 27 = (3)^2$	$4(-3)^2 - 27 = (-3)^2$
$4(9) - 27 = 9$	$4(9) - 27 = 9$
$36 - 27 = 9$	$36 - 27 = 9$
$9 = 9$ ✓	$9 = 9$ ✓

ALTERNATIVE SOLUTION

Sometimes, as with $4x^2 - 27 = x^2$, an equation can be solved by factoring.

Rewrite so that all terms are on one side:
$4x^2 - x^2 - 27 = 0$

Combine: $3x^2 - 27 = 0$

Divide by 3: $x^2 - 9 = 0$

Factor: $(x + 3)(x - 3) = 0$

Set each factor equal to zero:
$x + 3 = 0$ or $x - 3 = 0$

Solve: $x = -3$ and $x = 3$

Answer: $x = 3$ or $x = -3$.

The solution set is $\{3, -3\}$.

2 Solve and check: $4x^2 = 49$

SOLUTION

Divide both sides by 4: $x^2 = \dfrac{49}{4}$

Take the square root of both sides and simplify:

$$x = \pm\sqrt{\dfrac{49}{4}} = \pm\dfrac{7}{2}$$

Check:

$x = \dfrac{7}{2}$	$x = -\dfrac{7}{2}$
$4x^2 = 49$	$4x^2 = 49$
$4\left(\dfrac{7}{2}\right)^2 = 49$	$4\left(-\dfrac{7}{2}\right)^2 = 49$
$4\left(\dfrac{49}{4}\right) = 49$	$4\left(\dfrac{49}{4}\right) = 49$
$49 = 49$ ✓	$49 = 49$ ✓

Answer: $x = \dfrac{7}{2}$ or $x = -\dfrac{7}{2}$.

The solution set is $\left\{\dfrac{7}{2}, -\dfrac{7}{2}\right\}$.

3 Solve for x and check: $9x^2 - m^2 = 0$

SOLUTION

Rewrite: $9x^2 = m^2$

Divide by 9: $x^2 = \dfrac{m^2}{9}$

Take the square root of both sides and simplify: $x = \pm\sqrt{\dfrac{m^2}{9}} = \pm\dfrac{m}{3}$

Check:

$x = \dfrac{m}{3}$	$x = -\dfrac{m}{3}$
$9x^2 - m^2 = 0$	$9x^2 - m^2 = 0$
$9\left(\dfrac{m}{3}\right)^2 - m^2 = 0$	$9\left(-\dfrac{m}{3}\right)^2 - m^2 = 0$
$9\left(\dfrac{m^2}{9}\right) - m^2 = 0$	$9\left(\dfrac{m^2}{9}\right) - m^2 = 0$
$m^2 - m^2 = 0$	$m^2 - m^2 = 0$
$0 = 0$ ✓	$0 = 0$ ✓

Answer: $x = \dfrac{m}{3}$ or $x = -\dfrac{m}{3}$. The solution set is $\left\{\dfrac{m}{3}, -\dfrac{m}{3}\right\}$.

Solving Quadratic Equations When the Constant and Linear Term Are Missing

To solve a quadratic equation of the form $ax^2 = 0$, follow this simple rule:

If $ax^2 = 0$, then $x = 0$.

MODEL PROBLEM

Solve: $5x^2 = 0$

SOLUTION
Divide both sides by 5: $x^2 = 0$

Solve: $\sqrt{0} = 0$, so x must equal 0.

Answer: $x = 0$

Practice

1 Solve: $x^2 - 4x = 5x$

(1) $\{-5, 0, 1\}$
(2) $\{-5, 1\}$
(3) $\{0, 9\}$
(4) $\{-9, 0, 9\}$

2 Solve: $\dfrac{x}{2} = \dfrac{2}{x}$

(1) $\{2\}$
(2) $\{0, 2\}$
(3) $\{-2, 2\}$
(4) $\{-2, 0, 2\}$

Solving Factorable Quadratic Equations

3 For which equation is the solution set {0, 1}?
  (1) $x^2 - x = 0$
  (2) $x^3 - x = 0$
  (3) $x^2 - 1 = 0$
  (4) $x - 1 = 0$

4 For which equation is the solution set $\left\{-\dfrac{6}{5}, \dfrac{6}{5}\right\}$?
  (1) $\dfrac{12}{5x} = \dfrac{x}{60}$
  (2) $\dfrac{6-x}{4} = \dfrac{5}{6+x}$
  (3) $\dfrac{6}{x} = \dfrac{25x}{6}$
  (4) $5x^2 + 6x = 0$

Exercises 5–19: Solve and check your solution.

5 $x^2 + 2x = 0$
6 $x^2 = 5x$
7 $4x^2 = 28x$
8 $3x^2 = -5x$
9 $8a^2 = 2a$
10 $3x^2 - 4x = 5x$
11 $\dfrac{x^2}{5} = \dfrac{x}{15}$
12 $5x(x-6) - 8x = 2x$
13 $3x^2 + 25 = 25 - 15x$
14 $3x^2 = 6ax$ for x
15 $\dfrac{2x^2}{25} = 18$
16 $16x^2 - 400 = 0$
17 $x^2 - 16 = 48$
18 $(x-6)(x+6) = 28$
19 $a^2x^2 = 49$ for x

20 Solve each formula for the variable indicated:
  a $A = \pi r^2$ for r
  b $S = at^2$ for t
  c $A = 4\pi r^2$ for r
  d $E = mc^2$ for c
  e $K = \dfrac{1}{2}mv^2$ for v
  f $F = \dfrac{mv^2}{r}$ for v
  g $s = \dfrac{1}{2}gt^2$ for t
  h $V = \pi r^2 h$ for r

Fractional Quadratic Equations

Recall that to solve a fractional equation such as $\dfrac{4x}{9} = \dfrac{2(x+4)}{3}$, we cross multiply and solve for x:

$$18(x+4) = 12x$$
$$18x + 72 = 12x$$
$$18x - 12x = -72$$
$$6x = -72$$
$$x = -12$$

134   Chapter 4: Factoring and Quadratic Equations

Sometimes cross multiplying results in a quadratic equation. To solve a fractional quadratic equation:

- Cross multiply only if there is just one fraction on each side of the equals sign.
- For any other fractional equation, clear the fractions by multiplying each term by the lowest common denominator (LCD).
- Rewrite the equation in standard form, with all terms on one side of the equals sign: $ax^2 + bx + c = 0$.
- Factor and solve.
- Check your solution set.

MODEL PROBLEMS

1 Solve and check your solution:

$$\frac{x+5}{3} = \frac{10}{x-8}$$

SOLUTION

Cross multiply: $(x+5)(x-8) = 30$

Use FOIL: $x^2 - 8x + 5x - 40 = 30$

Combine terms and write the equation in standard form: $x^2 - 3x - 70 = 0$

Factor: $(x+7)(x-10) = 0$

Solve:

$x + 7 = 0 \quad x - 10 = 0$
$x = -7 \quad\quad x = 10$

Check: The check must be done with the *original* equation.

$x = -7$	$x = 10$
$\dfrac{x+5}{3} = \dfrac{10}{x-8}$	$\dfrac{x+5}{3} = \dfrac{10}{x-8}$
$\dfrac{-7+5}{3} = \dfrac{10}{-7-8}$	$\dfrac{10+5}{3} = \dfrac{10}{10-8}$
$\dfrac{-2}{3} = \dfrac{10}{-15}$ or $-\dfrac{2}{3} = -\dfrac{10}{15}$	$\dfrac{15}{3} = \dfrac{10}{2}$
$-\dfrac{2}{3} = -\dfrac{2}{3}$ ✓	$5 = 5$ ✓

Answer: $x = -7$ or $x = 10$. The solution set is $\{-7, 10\}$.

2 Solve and check your solution: $2 + \dfrac{5}{n} = \dfrac{12}{n^2}$

SOLUTION

Multiply each term by the LCD, n^2:

$$2(n^2) + \dfrac{5}{n}(n^2) = \dfrac{12}{n^2}(n^2)$$

Simplify: $2n^2 + 5n = 12$

Write in standard form: $2n^2 + 5n - 12 = 0$

Factor: $(2n - 3)(n + 4) = 0$

Solve:

$$2n - 3 = 0 \qquad n + 4 = 0$$
$$n = \dfrac{3}{2} \qquad\quad n = -4$$

Check:

$n = \dfrac{3}{2}$	$n = -4$
$2 + \dfrac{5}{n} = \dfrac{12}{n^2}$	$2 + \dfrac{5}{n} = \dfrac{12}{n^2}$
$2 + \left(5 \div \dfrac{3}{2}\right) = 12 \div \left(\dfrac{3}{2}\right)^2$	$2 + \dfrac{5}{-4} = \dfrac{12}{(-4)^2}$
$2 + \dfrac{10}{3} = \dfrac{48}{9}$	$\dfrac{8}{4} - \dfrac{5}{4} = \dfrac{12}{16}$
$\dfrac{16}{3} = \dfrac{16}{3}$ ✓	$\dfrac{3}{4} = \dfrac{3}{4}$ ✓

Answer: $n = \dfrac{3}{2}$ or $n = -4$. The solution set is $\left\{\dfrac{3}{2}, -4\right\}$.

Note: In checking, decimal equivalents can also be used.

Practice

1 In which case will cross multiplying result in a quadratic equation?

(1) $\dfrac{x}{7} = \dfrac{4}{x - 3}$

(2) $\dfrac{x}{7} = \dfrac{x - 3}{4}$

(3) $\dfrac{x}{x - 3} = \dfrac{4}{7}$

(4) $\dfrac{7}{x} = \dfrac{4}{x - 3}$

2 In which case do we need to use the LCD?

(1) $\dfrac{2x}{3} = \dfrac{6}{x}$

(2) $\dfrac{x}{2} - 2 = \dfrac{6}{x}$

(3) $\dfrac{x}{2} - \dfrac{8}{x} = 0$

(4) each of the above

3 Find the solution set: $\dfrac{2x+3}{6x+1} = \dfrac{1}{2x}$

(1) $\left\{\dfrac{1}{2}, -\dfrac{1}{2}\right\}$

(2) $\{2, -2\}$

(3) $\{1, -1\}$

(4) $\left\{\dfrac{1}{3}, -\dfrac{1}{3}\right\}$

4. Find the solution set: $\dfrac{2}{x} = \dfrac{3}{x^2 - 1}$

(1) $\left\{-\dfrac{1}{2}, 2\right\}$

(2) $\left\{\dfrac{1}{2}, -2\right\}$

(3) $\left\{\dfrac{1}{2}, -\dfrac{1}{2}\right\}$

(4) $\{1, -1\}$

Exercises 5–20: Solve each equation and check your solution.

5 $\dfrac{x}{5} = \dfrac{3}{x+2}$

6 $\dfrac{8}{x} = \dfrac{x+2}{3}$

7 $\dfrac{3x}{4} = \dfrac{x^2}{8}$

8 $\dfrac{7x}{3} = \dfrac{x^2}{6}$

9 $\dfrac{x+2}{2} = \dfrac{1}{x+3}$

10 $\dfrac{x+4}{3} = \dfrac{3}{x+4}$

11 $\dfrac{x-2}{2} = \dfrac{3}{x+3}$

12 $x - \dfrac{16}{x} = 6$

13 $\dfrac{x}{3} + \dfrac{9}{x} = 4$

14 $\dfrac{x}{2} + 1 = \dfrac{12}{x}$

15 $\dfrac{x}{3} - 1 = \dfrac{6}{x}$

16 $x - \dfrac{35}{x} = -2$

17 $x - \dfrac{5}{x} = 4$

18 $\dfrac{7}{x} - 5 = 2x$

19 $\dfrac{6}{x} + x = \dfrac{11}{2}$

20 $\dfrac{2}{x} + 4 = x - \dfrac{3}{x}$

Solving Quadratic-Linear Pairs Algebraically

A **quadratic-linear pair** is a system of one quadratic equation and one linear equation with the same two variables x and y. In such a system the quadratic equation has the form:

$y = ax^2 + bx + c$

where a, b, and c are real numbers and $a \neq 0$.

The solutions to a quadratic-linear system are the ordered pairs of values (x, y) that satisfy both equations. As reviewed in Chapter 3, a system of two linear equations can be solved algebraically by the **substitution method** or the **addition method** (also called the **elimination method**). These methods can be adapted to solve a quadratic-linear pair.

Substitution Method

- Solve the linear equation for one of the variables, x or y. (Solving for either variable will work. Common sense will usually indicate which is better.)
- Substitute this expression for the appropriate variable in the quadratic equation.
- Solve the quadratic equation.
- Substitute the solutions found in step 3 in the linear equation and solve.
- Check by substituting the ordered pairs in both given equations.

For graphic solutions to quadratic-linear pairs, see Chapter 5.

 MODEL PROBLEM

1 Solve by the substitution method:

$y = x^2 - 4x + 4$

$2y = x + 4$

SOLUTION

Since the x term is squared in the quadratic equation, we solve the linear equation for y:

$2y = x + 4$

$y = \frac{1}{2}x + 2$

Substitute this expression for y in the quadratic equation:

$y = x^2 - 4x + 4$

$\frac{1}{2}x + 2 = x^2 - 4x + 4$

To clear the fraction, multiply each term by 2. Then write in standard form.

$x + 4 = 2x^2 - 8x + 8$

$0 = 2x^2 - 9x + 4$

Or:

$2x^2 - 9x + 4 = 0$

Solve by factoring: $2x^2 - 9x + 4 = (2x - 1)(x - 4) = 0$. Then:

$2x - 1 = 0 \quad x - 4 = 0$

$x = \frac{1}{2} \quad\quad x = 4$

Substitute both x values in the linear equation to find the corresponding y values.

$x = \frac{1}{2}$	$x = 4$
$2y = x + 4$	$2y = x + 4$
$2y = \frac{1}{2} + 4 = 4\frac{1}{2} = \frac{9}{2}$	$2y = 4 + 4$
	$2y = 8$
$y = \frac{9}{2} \div 2 = \frac{9}{4}$	$y = 4$

138   Chapter 4: Factoring and Quadratic Equations

Check: Substitute both ordered pairs in both given equations.

$(x, y) = \left(\dfrac{1}{2}, \dfrac{9}{4}\right)$		$(x, y) = (4, 4)$	
$y = x^2 - 4x + 4$	$2y = x + 4$	$y = x^2 - 4x + 4$	$2y = x + 4$
$\dfrac{9}{4} = \left(\dfrac{1}{2}\right)^2 - 4\left(\dfrac{1}{2}\right) + 4$	$2\left(\dfrac{9}{4}\right) = \dfrac{1}{2} + 4$	$4 = 4^2 - (4 \cdot 4) + 4$	$2(4) = 4 + 4$
$\dfrac{9}{4} = \dfrac{1}{4} - 2 + 4$	$\dfrac{18}{4} = 4\dfrac{1}{2}$	$4 = 16 - 16 + 4$	$8 = 8$ ✓
$2\dfrac{1}{4} = 2\dfrac{1}{4}$ ✓	$4\dfrac{1}{2} = 4\dfrac{1}{2}$ ✓	$4 = 4$ ✓	

Answer: $(x, y) = \left(\dfrac{1}{2}, \dfrac{9}{4}\right)$ or $(4, 4)$.

Addition (or Elimination) Method

- Add the two equations to eliminate one variable, y. You may need to add an opposite (subtract). If necessary, first write an equivalent equation for one or both of the original equations.
- The sum will be a quadratic equation in one variable, x. Solve it.
- Substitute the solutions in the original linear equation and solve.
- Check by substituting the ordered pairs in the given equations.

MODEL PROBLEM

Solve by the addition (elimination) method:
$y = x^2 - 4x + 4$
$2y = x + 4$

SOLUTION

To eliminate y, multiply all terms in the quadratic equation by 2. Multiply the linear equation by -1 and add:

$y = x^2 - 4x + 4 \quad \rightarrow \quad 2y = 2x^2 - 8x + 8$
$2y = x + 4 \quad \rightarrow \quad \underline{-2y = - x - 4}$
$ 0 = 2x^2 - 9x + 4$

Now proceed as in the Model Problem on page 138: Solve this quadratic equation, substitute the values of x in the linear equation to solve it, and check the ordered pairs by substitution in the given equations.

Shortcut: When the y term is the same or equivalent in both equations, set the other expressions equal to each other, simplify (write in standard form), and solve.

Solving Quadratic-Linear Pairs Algebraically

1. Solve:

   $y = 3 - x^2$
   $y = -x - 3$

   (1) $(x, y) = (-3, 0)$ or $(2, -5)$
   (2) $(x, y) = (-1, 4)$ or $(6, -9)$
   (3) $(x, y) = (-2, -1)$ or $(3, -6)$
   (4) $(x, y) = (0, -3)$ or $(1, -4)$

2. Solve:

   $y = x^2 - 8x + 7$
   $y = -x - 3$

   (1) $(x, y) = (-5, -2)$ or $(-2, 1)$
   (2) $(x, y) = (-2, 1)$ or $(4, -7)$
   (3) $(x, y) = (1, -4)$ or $(7, 10)$
   (4) $(x, y) = (2, -5)$ or $(5, -8)$

3. If $7x - y = 10$ and $y = x^2 + 3x - 10$, which is a possible value for x?

   (1) $x = \dfrac{10}{7}$
   (2) $x = 2$
   (3) $x = 4$
   (4) $x = 5$

4. If $x^2 - y = 15$ and $x + y = 15$, which is a possible value for x?

   (1) $x = 0$
   (2) $x = 5$
   (3) $x = 10$
   (4) $x = 15$

5. When Alice and Valerie tried to solve the following system of equations, Alice said there were an infinite number of real solutions and Valerie said there were no real solutions. Which student, if either, is correct? Explain.

   $y = 2x^2 - 2x + 3$
   $y = -2x - 3$

Exercises 6–10: Solve each system and check your solution.

6. $y = -\dfrac{1}{2}x^2 + 3x$

   $2y - x = 0$

7. $y = -x^2 + 6x - 1$

   $y = x + 3$

8. $y = \dfrac{1}{2}x^2 + 3x - 2$

   $2y = 8x - 1$

9. $y = -\dfrac{1}{2}x^2 + 3x - \dfrac{1}{2}$

   $y = \dfrac{5}{4}x - 6$

10. $y = \dfrac{1}{2}x^2 - 5x + 2$

    $x - 2y = 24$

Verbal Problems Involving Quadratics

Factoring polynomial equations is a tool for solving word problems. In this section, quadratics are applied in number problems and then in real-world problems.

Number Problems

A common problem situation involves finding the value of a described number or numbers. The next model problems, solved alegbraically, are typical.

Remember:
$n, n + 1, n + 2, \ldots$ are consecutive integers
$n, n + 2, n + 4, \ldots$ are consecutive even or consecutive odd integers

MODEL PROBLEMS

1 Find two consecutive integers whose product is 72.

SOLUTION

Let n and $n + 1$ = the two integers

Then $n(n + 1) = 72$

Solve:

Simplify.	$n^2 + n = 72$
Write in standard form.	$n^2 + n - 72 = 0$
Factor.	$(n + 9)(n - 8) = 0$
Set each factor equal to 0.	$n + 9 = 0$ or $n - 8 = 0$
Find n.	$n = -9$ or $n = 8$
Find $n + 1$.	$n + 1 = -8$ or $n + 1 = 9$

Check:

$n = -9$	$n = 8$
$n(n + 1) = 72$	$n(n + 1) = 72$
$-9(-9 + 1) = 72$	$8(8 + 1) = 72$
$-9(-8) = 72$	$8(9) = 72$
$72 = 72$ ✓	$72 = 72$ ✓

Answer: The integers are -9 and -8 or 8 and 9.

2 The sum of two numbers is 13. The sum of their squares is 89. Find the numbers.

SOLUTION

Let x = one number

$13 - x$ = the other number

Then $x^2 + (13 - x)^2 = 89$

Solve:

Simplify and write in standard form.	$x^2 + 169 - 26x + x^2 = 89$
	$2x^2 - 26x + 169 - 89 = 0$
	$2x^2 - 26x + 80 = 0$
Divide by 2.	$x^2 - 13x + 40 = 0$
Factor.	$(x - 8)(x - 5) = 0$
Find x.	$x - 8 = 0$ so $x = 8$
	$x - 5 = 0$ so $x = 5$
Find $13 - x$.	$13 - x = 5$ or $13 - x = 8$

Answer: Both factors yield the solution 5 and 8.

Practice

1. Twice the square of an integer is 3 less than 7 times the integer. Find the integer.
   - (1) −3
   - (2) −1
   - (3) 1
   - (4) 3

2. The square of a number increased by 3 times the number equals 4. Find all possible solutions.
   - (1) {1, −4}
   - (2) {−1, −5}
   - (3) {1}
   - (4) {0, −3}

3. The sum of the squares of two consecutive odd integers is 202. Find the integers.
   - (1) {11, 9} or {−9, −11}
   - (2) {11, 9}
   - (3) {5, 7} or {−7, −5}
   - (4) {−11, −13} or {9, 11}

4. When a number is decreased by its reciprocal, the result is $2\frac{1}{10}$. Find the number.
   - (1) $2\frac{1}{2}$
   - (2) $\frac{2}{5}$
   - (3) $2\frac{1}{2}$ or $-\frac{2}{5}$
   - (4) $2\frac{1}{2}$ or $-2\frac{1}{2}$

5. Six times the square of a number decreased by 5 times the number equals 1. Find the negative solution.

6. Find two pairs of consecutive odd integers whose product is 63.

7. The sum of the squares of two consecutive even integers is 164. Find the integers.

8. The square of a number is 12 more than the number. Find the number.

9. The square of a number decreased by 3 times the number is 18. Find the number.

10. The product of a number and 5 less than the number is 24. Find the number.

11. The product of two consecutive integers is 42. Find the integers.

12. The square of 1 more than a number is equal to 4 more than 4 times the number. Find the number.

13. The sum of two numbers is 14 and their product is 48. Find the numbers.

14. One number is 3 less than another number. Their product is 40. Find the numbers.

15. The sum of 3 times the square of a number and 6 times the number is equal to twice the square of that number, decreased by 8. Find the number.

16. If the square of a number is increased by 4 times the number, the result is 12. Find the number.

17. The difference of the squares of a number and one-half the number is 27. Find the number.

18. Two numbers are consecutive integers. The square of the lesser added to twice the greater is 37. Find the numbers.

19. The sum of a number and its reciprocal is $2\frac{1}{6}$. Find all solutions.

20. If 4 is added to 7 times a number, the result is twice the square of the number. Find the number.

Real-World Applications

In a real-world problem, not all values that satisfy a given equation are necessarily the answer. We need to consider the conditions of the problem and reject answers that are not reasonable. Some typical problems, like the models below, involve geometry. See Chapters 6 and 8 for a review of geometric concepts.

MODEL PROBLEMS

1 The length of a rectangle is twice the width. The area is 32 square units. Find the length and width. (Area = length times width.)

SOLUTION

Let w = width

$2w$ = length

Then:

$A = lw$

$32 = (2w)(w) = 2w^2$

$w^2 = 16$

Here, the value of w is obvious by inspection, but when that is not the case, we can solve by taking a square root or by factoring.

Method 1: Take the square root of both sides of the equation.

$\sqrt{w^2} = \sqrt{16}$

$w = 4$ or -4

We reject -4, since a length cannot be negative. The solution is the principal square root.

Answer: The width is 4 units. Therefore, the length is $2 \cdot 4 = 8$ units.

Method 2: Factor. First write the equation in standard form.

$w^2 = 16$

$w^2 - 16 = 0$

$(w - 4)(w + 4) = 0$

Then:

$w - 4 = 0 \qquad w + 4 = 0$

$w = 4 \qquad\quad w = -4$

Reject the negative root, since a length must be positive.

$w = 4$

$l = 2w = 2(4) = 8$

Answer: As above, width is 4 units and length is 8 units.

2 The area of Mr. Lamb's rectangular garden is 100 square feet. One side is a stone wall. He encloses the other 3 sides with 30 feet of fencing. What are the dimensions of his garden? Perimeter = (2 times length) + (2 times width). Area = length times width.

SOLUTION

Let w = width in feet

Now use $P = 2l + 2w$. For 3 sides, $l + 2w = 30$. Therefore:

$l = 30 - 2w$

Then:

$A = lw$

$100 = w(30 - 2w)$

$100 = 30w - 2w^2$

$2w^2 - 30w + 100 = 0$

$w^2 - 15w + 50 = 0$

Factor:

$(w - 10)(w - 5) = 0$

Solve:

$w - 10 = 0 \qquad w - 5 = 0$

$\quad w = 10 \qquad\qquad w = 5$

Both solutions are postive, so both are possible.

Answers: If $w = 10$ feet, then length = $30 - 2w = 30 - 20 = 10$ feet.

If $w = 5$ feet, then length = $30 - 2w = 30 - 10 = 20$ feet.

ALTERNATIVE SOLUTION

Assume that the wall is width w.

Then for 3 sides $2l + w = 30$, so $w = 30 - 2l$.

Therefore $100 = l(30 - 2l) = 30l - 2l^2$.

Proceed as above, solving for l.

Answer: If $l = 10$ feet, then $w = 10$ feet.
If $l = 5$ feet, then $w = 20$ feet.

Practice

Linear dimensions are units. Areas are square units. Formulas to use are:

Perimeter of rectangle	Perimeter = (2 • length) + (2 • width) $P = 2l + 2w$
Area of rectangle	Area = length • width $A = lw$
Area of triangle	Area = $\frac{1}{2}$ • base • height $A = \frac{1}{2}bh$

1. The length l of a rectangle is twice the width w. The area is 18. Find the width.
   (1) $w = 9$
   (2) $w = 8$
   (3) $w = 4$
   (4) $w = 3$

2. The width w of a rectangle is 7 less than the length l. The area is 60. Find the length.
   (1) $l = 10$
   (2) $l = 11$
   (3) $l = 12$
   (4) $l = 15$

3. The length l of a rectangle is 3 more than twice the width w. The area is 90. Find the width.
   (1) $w = 3$
   (2) $w = 6$
   (3) $w = 7.5$
   (4) $w = 9$

4. Base b and height h of a triangle have the same measure. The area of the triangle is 8. Find the base.
   (1) $b = 8$
   (2) $b = 4$
   (3) $b = 2\sqrt{2}$
   (4) $b = 2$

5. The area of a rectangle is 99 square inches. The length is is 2 feet longer than the width. Find the dimensions.

6. The length of a rectangle is 8 feet longer than the width. The area is 105 square feet. Find the dimensions of the rectangle.

7. A pool is in the shape of a triangle. The sum of the base and height is 19 yards. The area is 42 square yards. What are the dimensions of the pool?

8. The altitude (height) of a triangle is 8 feet more than twice its base. The area is 45 square feet. Find base and altitude.

9. A rectangle is 10 feet longer than it is wide. Its area is 39 square feet. Find the dimensions.

10. The length of a rectangle is 3 inches less than twice the width. The area is 65 square inches. Find length and width.

11. The length of a rectangular garage is 2 yards more than its width. The area is 80 square yards. What are the dimensions of the garage?

12. The area of Janina's rectangular garden is 45 square meters. The length of the garden is 4 meters more than the width. What are the dimensions of the garden?

13. The area of a rectangle is 70 square inches. The width is 3 inches less than the length. What are the dimensions?

14. The perimeter of a rectangle is 22 feet, and the area is 24 feet. Find length and width.

15. A rectangular mini-park is 16 yards long and 40 yards wide. The town adds the same amount to the length and the width, increasing the area by 305 square yards. How much is added to each dimension?

16. Two opposite sides of a square are each increased by 6 inches. The other two opposite sides are each decreased by 1 inch. The result is a rectangle that is twice the area of the

144   Chapter 4: Factoring and Quadratic Equations

square. What is the length of a side of the square?

17 The perimeter of a rectangular walk-in closet is 26 feet. If the length is increased by 4 feet and the width is increased by 3 feet, the area of the new room will be 96 square feet. Find the dimensions of the new room.

18 One side of a rectangular garden plot is the bank of a stream. The other 3 sides are enclosed by 12 yards of fencing. The area is 16 square yards. Find the possible dimensions of the garden.

19 A rectangle is 3 times as long as it is wide. If the width is increased by 6 feet and the length is decreased by 3 feet, the area is doubled. Find the dimensions of the original rectangle.

20 The length of a rectangle exceeds its width by 7 inches. The length of the rectangle is decreased by 2 inches, and the width is increased by 3 inches. The resulting new rectangle has an area 1 square inch less than twice the area of the original rectangle. Find the dimensions of the original rectangle.

CHAPTER REVIEW

1 Solve for x: $64x^2 = 9m^2$

(1) $\{24m, -24m\}$

(2) $\{2\frac{2}{3}m\}$

(3) $\{\frac{3}{8}m\}$

(4) $\{-\frac{3}{8}m, \frac{3}{8}m\}$

2 One root of $x^2 - 10x + k = 0$ is 5. Find k and the other root.

(1) $k = 25$ and the other root is 5
(2) $k = -25$ and the other root is 5
(3) $k = 25$ and the other root is -5
(4) $k = -25$ and the other root is -5

3 Solve: $\frac{x+7}{9} = \frac{3}{x+1}$

(1) $\{-2, 10\}$
(2) $\{-10, 2\}$
(3) $\{2, 10\}$
(4) $\{10, 2\}$

4 Three numbers are consecutive integers. The square of the second number is 8 more than the sum of the other two numbers. Which of the following is a solution?

(1) 0, 1, 2
(2) 1, 2, 3
(3) 2, 3, 4
(4) 3, 4, 5

Exercises 5–14: Solve each equation and check your solution.

5 $x(x - 1) = 56$

6 $x^2 + 15x + 44 = 0$

7 $x^2 - 13x + 22 = 0$

8 $25x^2 - 4 = 32$

9 $x^2 - 12 = 11x$

10 $4x^2 + 28x = 0$

11 $5x^2 + 4x - 1 = 0$

12 $2x^2 - 7x - 15 = 0$

13 $(x + 8)^2 = 25$

14 $\frac{x+3}{5} = \frac{6}{x+4}$

15 Solve this system and check your solution:

$y = x^2 + 5$

$3x + 2y = 10$

16 Find a quadratic equation that has roots -1 and 9.

17 Write a quadratic equation for which both roots are -6.

18 One root of $x^2 - 8x + k = 0$ is 3. Find:

a k

b the other root

19 The length of a rectangle is 3 units more than its width. The area is 28 square units. Find length and width. (Area = length • width)

20 In a triangle, the measures of the base and height are in a ratio of 3 : 4. The area is 54 square centimeters. Find base and height.

(Area of a triangle = $\frac{1}{2}$ • base • height)

Note: Now turn to page 410 and take the Cumulative Review test for Chapters 1–4. This will help you monitor your progress and keep all your skills sharp.

Chapter Review 145

Chapter 5:
Functions and the Coordinate Plane

The **Cartesian coordinate system** provides an easy way to locate points on a plane, which is a two-dimensional surface. It consists of a coordinate grid separated by a horizontal number line called the *x*-axis and a vertical number line called the *y*-axis. The intersection of these number lines is called the **origin**. The two axes divide the grid into four **quadrants**, numbered I to IV counterclockwise.

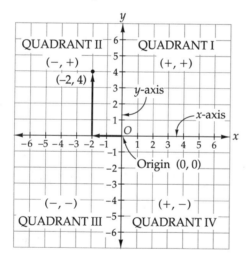

Every point on the grid can be identified with two numbers called **coordinates**:

- The first number is the ***x*-coordinate** (or **abscissa**). This is the distance from the origin horizontally to the point on the *x*-axis above or below the point.
- The second number is the ***y*-coordinate** (or **ordinate**). This is the distance from the origin vertically to the point on the *y*-axis to the right or left of the point.
- For any point, the coordinates are written (x, y). When in this form, they are called an **ordered pair**. The coordinates of the origin are $(0, 0)$.
- Point $(-2, 4)$ on the graph has an *x*-coordinate of -2 and a *y*-coordinate of 4.
- The correct signs for coordinates in each quadrant are shown in parentheses.

Note: All graphs you draw should be labeled. The *x*-axis and *y*-axis must be identified and the scale $(1, 2, 3, \ldots; 2, 4, 6, \ldots)$ must be indicated.

Relations and Functions

A **relation** is any set of ordered pairs. The **domain** of the relation is the first element of the ordered pair. The **range** of the relation is the second element of the ordered pair. The **rule** for a relation describes the relationship between values of the domain and the range. A rule assigns or **maps** one or more values in the range to each value in the domain.

Relations with a limited number of ordered pairs can be expressed in many ways. Consider the following relation: The domain is the set of integers greater than 0 and less than 4, {1, 2, 3}. The range is the same set {1, 2, 3}. The rule for this relation maps odds to 1, evens to 2, and prime numbers to 3.

This relation can be expressed in several ways, including:

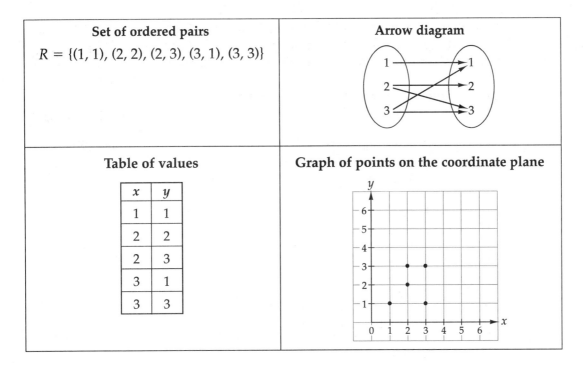

If no set is specified for the domain, assume that the domain is the set of real numbers, or the entire x-axis. A relation with an infinite domain would have an infinite number of ordered pairs. Relations with infinite numbers of ordered pairs are usually expressed as algebraic rules or drawn as graphs.

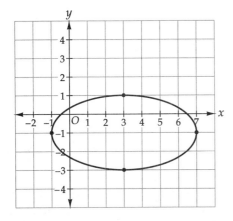

For the relation shown here, the graph indicates all values of x between -1 and 7. Therefore, the domain is the infinite set of real numbers such that $-1 \leq x \leq 7$. All values of y between -3 and 1 are shown in the graph, so the range is the infinite set of real numbers such that $-3 \leq y \leq 1$. The domain and range are infinite sets because a line has an infinite number of points.

A **function** is a relation in which each element of the domain corresponds with *one and only one* element of the range. In other words, no two ordered pairs have the same *x* value.

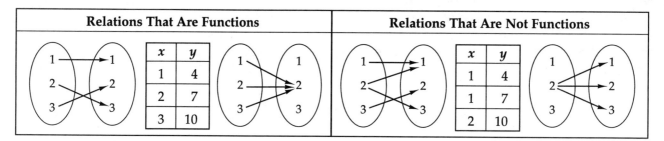

Vertical Line Test for Functions A simple way to check whether a relation is a function is to look at its graph. If you draw a vertical line through any part of the domain, the line should intersect the graph of a function at one and only one point. If you can draw a vertical line that touches two or more points of the graph, the relation is *not* a function.

An example of a relation that is a function is $y = x$. Notice its graph.

> The notation $f(x)$ is sometimes used in the place of y, since y is a function of x.
>
> (You will need to know this notation for Math B.)

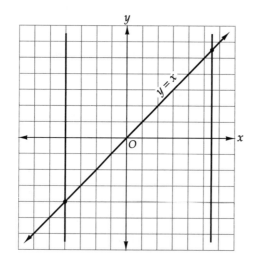

Compare this graph with the graph of a circle. A vertical line can be drawn that intersects the circle at more than one point. The circle is not a function.

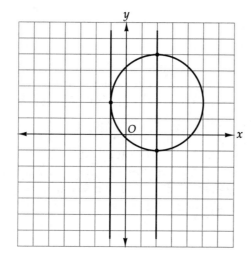

148 Chapter 5: Functions and the Coordinate Plane

Practice

1. If a relation has a rule $y = 2x$ and a domain of {2, 4, 6, 8}, identify the range.
   (1) {1, 2, 3, 4}
   (2) {2, 4, 6, 8}
   (3) {4, 8, 12, 16}
   (4) {(2, 4), (4, 8), (6, 12), (8, 16)}

2. Choose the rule for a relation that describes this table of values:

x	y
1	1
1	−1
2	2
2	−2

   (1) The first value is the absolute value of the second value.
   (2) The second value is the absolute value of the first value.
   (3) The first value is the opposite of the second value.
   (4) The second value is the opposite of the first value.

3. Choose the rule for a function that describes this table of values:

x	y
0	2
1	6
2	10
3	14

   (1) $y = x + 2$
   (2) $y = 4x + 2$
   (3) $y = 4x - 2$
   (4) $y = 6x - 2$

4. Which letter, when drawn on the coordinate plane, could represent a function?
   (1) W
   (2) X
   (3) Y
   (4) Z

5. Which of the following relations represents a function?
   (1) {(1, 1), (1, 2), (1, 3)}
   (2) {(2, 1), (1, 2), (2, 3)}
   (3) {(1, 1), (2, 1), (3, 1)}
   (4) {(1, 1), (2, 1), (2, 2)}

Exercises 6–10: State the domain and range for each relation and tell whether the relation is a function.

6. {(1,1), (2,4), (3,9)}

7. {(5,1), (6,2), (7,1)}

8.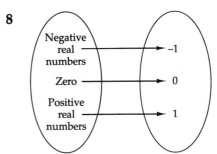

9.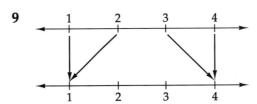

10.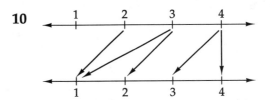

Exercises 11–15: Use the vertical line test to determine whether or not the graph represents a function.

11.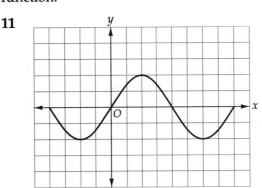

12.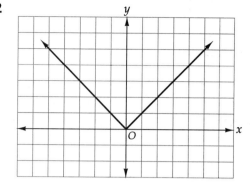
14.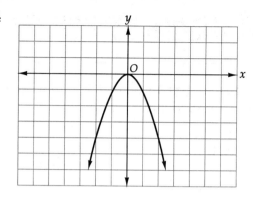
13.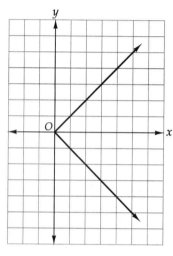
15.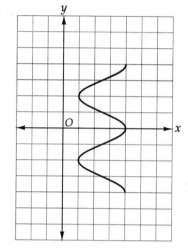

Lines on the Coordinate Plane

Graphing a Linear Equation

The standard form of a linear equation in two variables is $ax + by = c$, where a and b are both not 0. For example, $2x = 4$, $2x + 3y = 8$, and $-x + 4y = -5$ are linear equations in standard form. The solution set of a linear equation is the set of all the ordered pairs (x, y) that make the algebraic sentence true. The graph of a linear equation is the graph of its solution set, which is always a straight line. Since a linear equation assigns only one value of y to each value of x, it is a function.

To Graph a Linear Equation

- Solve the equation for y in terms of x.
- Find at least three points in the solution set by picking three convenient values for x and computing the corresponding values of y. (Although two points determine a straight line, we use a third point as a check.)
- Graph the three points determined by the ordered pairs (x, y).
- Draw a line through the three points. (If the three points are not in a straight line, there has been a mistake.)

You can also use a graphing calculator to graph a line. See the appendix.

150   Chapter 5: Functions and the Coordinate Plane

MODEL PROBLEMS

1 Graph $x - y + 4 = 0$

SOLUTION

Solve the equation for y: $y = x + 4$

Make a table using three values of x:

x	y
0	4
1	5
2	6

Plot these points and draw the line between them.

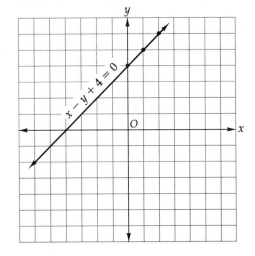

2 Explain why the point (2, 3) does or does not lie on the line $y = 4x - 3$.

SOLUTION

Substitute the coordinates of the point in the equation of the line.

$y = 4x - 3$
$(3) = 4(2) - 3$
$3 = 8 - 3$
$3 \neq 5$

If the coordinates of the point make the equation true, the point lies on the line.

Answer: Since the coordinates of the point are not a solution of the equation, the point does not lie on the line.

The graph of any linear equation of the form $ax + by + c = 0$, where a and b are both not 0, eventually crosses both the x-axis and y-axis. The point at which the graph crosses an axis is called an **intercept**.

To Find the Intercepts of the Graph of an Equation

- For the x-intercept, substitute 0 for y in the given equation and solve for x.
- For the y-intercept, substitute 0 for x in the given equation and solve for y.

 MODEL PROBLEM

Find the x- and y-intercepts of $x - y + 4 = 0$.

SOLUTION

To find the x-intercept, replace y with 0. $x - 0 + 4 = 0$
Solve for x. $x = -4$
To find the y-intercept, replace x with 0. $0 - y + 4 = 0$
Solve for y. $y = 4$

Answer: The x-intercept is -4 and the y-intercept is 4.

Practice

1. The abscissa for an ordered pair in the solution set of $x + 2y = 11$ is 3. Find the ordinate.
   (1) 3
   (2) 4
   (3) 5
   (4) 9

2. Which point lies on the graph of $y = x + 3$?
   (1) $(-3, 0)$
   (2) $(1, 3)$
   (3) $(3, 9)$
   (4) $(5, 9)$

3. Which line passes through the point $(2, -1)$?
   (1) $2y - x = 0$
   (2) $2x - y = 0$
   (3) $3x - 2y = 8$
   (4) $3x + 2y = 8$

4. Which line has a y-intercept of 2?
   (1) $2 = x$
   (2) $4 = x + y$
   (3) $5 = x + 2y$
   (4) $6 = x + 3y$

5. If the ordinate is -1 in the equation $4x - y = 9$, what is the abscissa?

6. If the abscissa is -2 in the equation $2x + 3y = -10$, what is the ordinate?

Exercises 7–9: Which of the three given points, if any, lie on the given line?

7. $x + y = 7$ $(2, 5)$ $(-2, -5)$ $(-3, 10)$
8. $2x = y + 1$ $(-1, -3)$ $(0, 1)$ $(2, 3)$
9. $2x - y = 8$ $(4, 0)$ $(1, -7)$ $(-2, -12)$

Exercises 10–13: Find the unknown coordinate of the point on the given line.

10. $x + y = 5$ $(x, 4)$
11. $x - 2y = 8$ $(6, y)$
12. $3y - 1 = x$ $(-3, y)$
13. $3x + 7 = 5y$ $(x, 5)$

Exercises 14–17: State whether the given line passes through the given point. Show your work.

14. $2y + x = 7$ $(1, 3)$
15. $4x + y = 10$ $(2, -2)$
16. $2y = 3x - 5$ $(-1, -4)$
17. $y = -2x + 4$ $(3, 10)$

Exercises 18–20: Use each equation to create a table of ordered pairs. Graph the line using the ordered pairs in the table. State the y-intercept and the x-intercept.

18. $2x - y = 2$
19. $x - 2y = -6$
20. $2x + y = 0$

Horizontal and Vertical Lines

When we graph a line such as $y = 4$, we find that the ordinate (y-value) of each member of the solution set is 4 regardless of the value of the abscissa (x-value). Thus, $(-2, 4)$, $(0, 4)$, and $(2, 4)$ are all members of the solution set. The graph is a line parallel to the x-axis. The y-intercept is 4.

- In general, an equation of the form $y = b$ is parallel to the x-axis, with y-intercept equal to b.

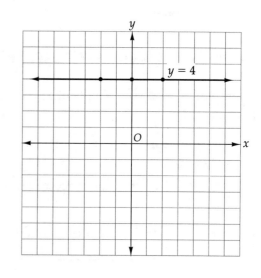

When we graph a line such as $x = 1$, we find that the abscissa (x-value) of each member of the solution set is 1 regardless of the value of the ordinate (y-value). Thus, $(1, -1)$, $(1,1)$, and $(1,3)$ are all members of the solution set. The graph is a line parallel to the y-axis. The x-intercept is 1.

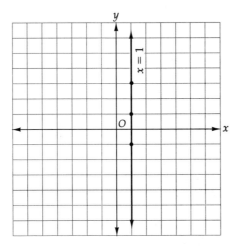

- In general, an equation of the form $x = a$ is parallel to the y-axis, with x-intercept equal to a.

Practice

1 The equation of the y-axis is

   (1) $x = 0$
   (2) $x = y$
   (3) $y = 0$
   (4) $y = y$

2 Which equation describes the line in this graph?

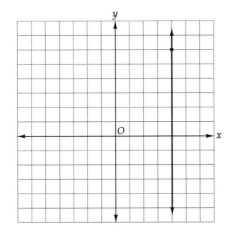

   (1) $x = 4$
   (2) $x = 6$
   (3) $y = 4$
   (4) $y = 6$

3 What is the x-intercept of $x = 7$?

   (1) 0
   (2) 1
   (3) 7
   (4) $x = 7$ has no x-intercept

4 Which equation has a graph that is parallel to the x-axis?

   (1) $x = 0$
   (2) $x = 25$
   (3) $x = 25 + y$
   (4) $x = 25 + y + x$

Exercises 5–10: Graph each line.

5 $y = 2$

6 $x = 3$

7 $y = -1$

8 $x = -3$

9 $y = 0$

10 $x = -1$

Slope of a Line

The **slope** of a line is the ratio of the difference in *y*-values to the difference in *x*-values between any two points on the line. Thus,

$$\text{slope} = \frac{\text{difference in } y\text{-values}}{\text{difference in } x\text{-values}} \text{ or } \frac{\text{vertical change}}{\text{horizontal change}}$$

Sometimes the slope is shown as

$$\frac{\text{rise}}{\text{run}}$$

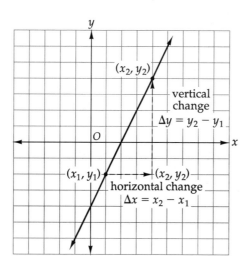

Mathematicians use the Greek letter Δ (delta) as a symbol for difference. Slope is usually symbolized *m*. Therefore, the formula for the slope of the line connecting points (x_1, y_1) and (x_2, y_2) is

$$m = \frac{\Delta y}{\Delta x} \text{ or } m = \frac{y_2 - y_1}{x_2 - x_1}$$

It makes no difference which points are labeled (x_1, y_1) and (x_2, y_2), the value of *m* will always be the same.

To find the Slope of a Line

- Choose any two points (x_1, y_1) and (x_2, y_2) on the line.
- Find the difference in *y*-values by subtracting y_1 from y_2.
- Find the difference in *x*-values by subtracting x_1 from x_2.
- Write the ratio: $\text{slope} = \frac{\text{difference in } y\text{-values}}{\text{difference in } x\text{-values}}$ as $m = \frac{y_2 - y_1}{x_2 - x_1}$ or $m = \frac{\Delta y}{\Delta x}$.

When graphing, express the slope as a fraction. This makes it easier to determine points on the graph. Thus, a slope of 7 is really:

$$\frac{7}{1}$$

Possible values of *m*

Value of *m*	Values of Δ*y* and Δ*x*	Appearance of Graph
positive	Δ*y* and Δ*x* have the same sign.	Line rises from left to right. /
negative	Δ*y* and Δ*x* have opposite signs.	Line falls from left to right. \
zero	Δ*y* = 0	Line is horizontal. —
undefined	Δ*x* = 0	Line is vertical. |

Chapter 5: Functions and the Coordinate Plane

MODEL PROBLEMS

1 Draw the graph of $2x - y = 4$ and find the slope.

SOLUTION

Solve the equation for y.
$$-y = -2x + 4$$
$$y = 2x - 4$$

Create a table of values.

x	0	2	3
y	-4	0	2

Plot the points and draw the graph.

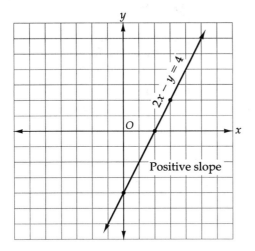

Choose points $(0, -4)$ as (x_1, y_1) and $(3, 2)$ as (x_2, y_2).

Then $m = \dfrac{\Delta y}{\Delta x} = \dfrac{y_2 - y_1}{x_2 - x_1} = \dfrac{2 - (-4)}{3 - 0} = \dfrac{6}{3} = \dfrac{2}{1} = 2$

Note: Slope can also be found in reverse order:
$$\dfrac{y_1 - y_2}{x_1 - x_2} = \dfrac{-4 - 2}{0 - 3} = \dfrac{-6}{-3} = \dfrac{2}{1} = 2$$

2 Draw the graph of $x + 2y = -4$ and find the slope.

SOLUTION

Solve the equation for y.
$$2y = -x - 4$$
$$y = -\dfrac{x}{2} - 2$$

Create a table of values.

x	-4	-2	0
y	0	-1	-2

Plot these points and draw the graph.

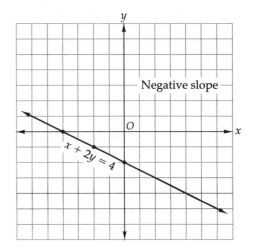

Choose points $(-4, 0)$ as (x_1, y_1) and $(0, -2)$ as (x_2, y_2). Then

$m = \dfrac{\Delta y}{\Delta x} = \dfrac{y_2 - y_1}{x_2 - x_1} = \dfrac{-2 - 0}{0 - (-4)} = \dfrac{-2}{4} = \dfrac{-1}{2}$ or $-\dfrac{1}{2}$

3 Draw the graph of $y = 3$ and find the slope.

SOLUTION

We have already discussed the idea that graphs of this type are parallel to the x-axis with y-intercept 3.

Draw the graph.

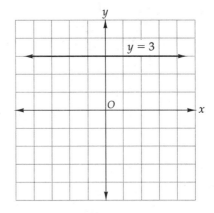

Notice that, since all y-values are the same, the change in y or Δy is 0. Thus, slope $m = \dfrac{\Delta y}{\Delta x} = \dfrac{0}{\Delta x} = 0$.

Lines on the Coordinate Plane **155**

4 Draw the graph of $x = 1$ and find the slope.

SOLUTION

We have already discussed the idea that graphs of this type are parallel to the y-axis with x-intercept 1.

Draw the graph.

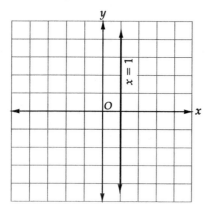

Notice that, since all x-values are the same, the change in x, or Δx, is 0, which means the denominator in the ratio $\frac{\Delta y}{\Delta x}$ is 0. Thus $\frac{\Delta y}{\Delta x} = \frac{\Delta y}{0}$. However, since division by zero is undefined, the slope is *undefined*.

All graphs of the form $y = b$ have zero slope.

The slope of all graphs of the form $x = a$ is undefined.

Practice

1 What is the slope of the line represented by this table of values?

x	y
-3	1
-2	3
-1	5

(1) -2
(2) -0.5
(3) 0.5
(4) 2

2 What is the slope of the line passing through $(0, 0)$ and $(-2, -5)$?

(1) -2.5
(2) -0.4
(3) 0.4
(4) 2.5

3 What is the slope of the line passing through $(5, -4)$ and $(-2, -1)$?

(1) -1
(2) $-\dfrac{3}{7}$
(3) $\dfrac{5}{3}$
(4) $\dfrac{5}{7}$

4 Which line has a slope of 5? The line passing through:

(1) $(0, -5)$ and $(-5, 0)$
(2) $(0, 10)$ and $(2, 0)$
(3) $(0, -10)$ and $(2, 0)$
(4) $(0, 5)$ and $(5, 0)$

156 Chapter 5: Functions and the Coordinate Plane

Exercises 5–8: Find the slope of the line represented by each table of values.

5.
x	y
0	0
1	−1
2	−2

6.
x	y
−5	1
−5	3
−5	5

7.
x	y
1	4
2	4
3	4

8.
x	y
−2	0
0	−2
2	−4

Exercises 9–15: Find the slope of the line passing through the following pairs of points.

9. (1, 1), (6, 7)
10. (2, −5), (4, 0)
11. (4, −3), (6, −5)
12. (−3, −5), (−2, −4)
13. (4, −7), (−7, 4)
14. (−1, 2), (−3, 6)
15. (6, −2), (7, −2)

Direct Variation

A teacher hands out a 25-question quiz with a note that each correct answer is worth 4 points. A table can be created to show the relation between the number of correct answers (x) and points earned (y).

x	1	2	3	...	25
y	4	8	12	...	100

Notice that in comparing the values of x and the corresponding values of y, the ratios, $\frac{y}{x}$, are all equivalent: $\frac{y}{x} = \frac{4}{1} = \frac{8}{2} = \frac{12}{3} = 4$. This relationship between the variables y and x is a **direct variation**. We say that y varies directly as x or that y is directly proportional to x. This constant ratio of $\frac{y}{x}$ is called the **constant of variation** or k, and we can write the relation as an equation $y = kx$. The ratio of any set of (x_1, y_1) values is equal to the ratio of any other corresponding set of (x_2, y_2) values in the table. Thus, $\frac{y_1}{x_1} = \frac{y_2}{x_2}$.

It is necessary to indicate the order in which the variables are being compared. The constant of variation of y with respect to x is $\frac{y}{x} = \frac{4}{1}$, since y equals $4x$. The constant of variation of x with respect to y is $\frac{x}{y} = \frac{1}{4}$, since x equals $\frac{1}{4}y$. Note that the two constants are reciprocals.

The graph of a direct variation is always a straight line passing through the origin, with a slope equal to the constant of variation of y with respect to x. In this case, since $y = 4x$, the line has a slope of 4. See the graph on the following page.

$$m = \frac{\Delta y}{\Delta x} = \frac{y_2 - y_1}{x_2 - x_1} = \frac{8 - 4}{2 - 1} = \frac{4}{1} = 4$$

Unless a problem states otherwise, assume that the constant of variation is the constant of the equation for y in terms of x.

Lines on the Coordinate Plane **157**

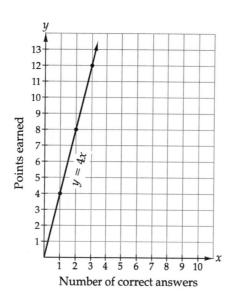

Direct variation problems may be conveniently solved by proportion when we are not required to find the constant of variation. In many problems, however, finding the constant k is very useful information.

Note: In direct variation...

... if x increases, y increases.
... if x decreases, y decreases.
... if x is multiplied by a number, y is multiplied by the same number.
... if x is divided by a number, y is divided by the same number.

MODEL PROBLEMS

1 If x varies directly as y and $x = 1.5$ when $y = 3.75$,

   a Find the constant of variation for x in terms of y.
   b Find the constant of variation for y in terms of x.
   c Graph this situation

SOLUTION

a Constant of variation $= \dfrac{x}{y} = \dfrac{1.5}{3.75} = 0.4 = \dfrac{2}{5}$

b Constant of variation $= \dfrac{y}{x} = \dfrac{3.75}{1.5} = 2.5 = \dfrac{5}{2}$

c The graph of the direct variation passes through the origin (0, 0) and the given point (1.5, 3.75)

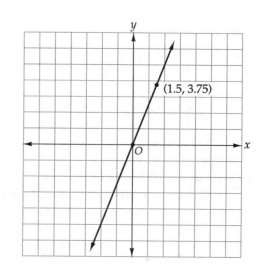

158   Chapter 5: Functions and the Coordinate Plane

2 The cost (C) of ground beef varies directly with the weight (W) of the package. Find the cost of 5 pounds of ground beef if 2 pounds cost $3.90.

SOLUTION

The constant of variation is $= \dfrac{\text{cost}}{\text{weight}} = \dfrac{3.90}{2}$.

METHOD 1

Using the proportion method:

Thus, $\dfrac{3.90}{2} = \dfrac{C}{5}$.

Cross multiply.

$2C = 5(3.90)$

$2C = 19.50$

$\dfrac{2C}{2} = \dfrac{19.50}{2}$

$C = \$9.75$

METHOD 2

Using constant of variation:

$k = \dfrac{3.90}{2}$ (constant of variation)

$C = kW$

$C = \dfrac{3.90}{2} \cdot 5$

$C = 1.95(5)$

$C = \$9.75$

Answer: Five pounds of ground beef cost $9.75.

3 The table gives pairs of values for s and P, where s is the side of an equilateral triangle and P is the perimeter. The units of measure are the same.

s	1	2	3	4
P	3	6	9	12

a Express the relationship between s and P as a graph and as a formula.
b Find the value of P when $s = 9$.
c Find the value of s when $P = 21$.

SOLUTION

a Plot the points $(0, 0)$ and $(1, 3)$, and draw a straight line through them.

$\dfrac{P}{s} = \dfrac{3}{1}$ and $P = 3s$

Answer: $P = 3s$

b $P = 3s$
$P = 3 \times 9$
$P = 27$

Answer: 27 units

c $P = 3s$
$21 = 3s$
$s = \dfrac{21}{3} = 7$

Answer: 7 units

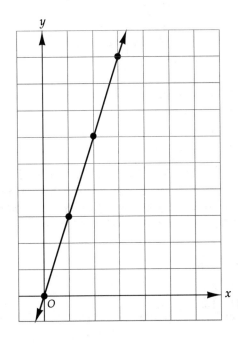

4 Express each of the following relations as an equation using k as the constant of variation.

SOLUTIONS

a The cost (c) of a railroad ticket varies directly as the distance in miles (m) of the trip.

$c = km$

b The surface area (A) of a sphere varies directly as the square of its radius (r).

$A = kr^2$

c The volume of sphere (V) varies directly as the cube of its radius (r).

$V = kr^3$

d The velocity of sound (V) varies directly as the square root of the absolute temperature (t) of the air.

$V = k\sqrt{t}$

Practice

1 The graph shows the number of gallons of water that leaked from a rusty water pipe. Assume the leakage occurred at a constant rate. How many gallons leaked in $5\frac{1}{2}$ hours?

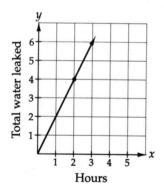

(1) 10 gallons
(2) 11 gallons
(3) 11.5 gallons
(4) 12 gallons

2 Two quantities, A and h, vary directly. When A = 18.75, h = 7.25. Find the value of h rounded to the nearest hundredth when A = 8.75.

(1) 3.38
(2) 2.59
(3) 2.14
(4) 1.21

3 A car travels 4 miles in 5 minutes. What distance will the car cover in 33 minutes?

(1) 6.6 miles
(2) 8.25 miles
(3) 26.4 miles
(4) 41.25 miles

4 Paul set up a car wash to raise money for a school fund-raiser. He found that he needed 5 buckets of water for every 2 cars. If 35 cars were washed, how many buckets of water were used?

(1) 14 buckets
(2) 62.5 buckets
(3) 75 buckets
(4) 87.5 buckets

5 Andrea is making cookies from a recipe that calls for $2\frac{1}{2}$ cups of flour for a yield of 2 dozen cookies. How much flour will she need to make 5 dozen cookies?

(1) $5\frac{1}{5}$ cups
(2) $5\frac{1}{2}$ cups
(3) 6 cups
(4) $6\frac{1}{4}$ cups

Exercises 6–10: The table shows a relationship between two variables.

  a Sketch the graph to determine whether or not the variation is direct.

  b If the variation is direct, find the constant of variation and express the relationship between the variables as a formula.

6. | X | 2 | 3 | 4 | 5 | 6 |
    |---|---|---|---|---|---|
    | Y | 4 | 8 | 12 | 16 | 20 |

7. | R | 0.5 | 1 | 1.5 | 2 | 2.5 |
    |---|---|---|---|---|---|
    | D | 2 | 4 | 6 | 8 | 10 |

8. | S | 2 | 3 | 4 | 5 | 6 |
    |---|---|---|---|---|---|
    | A | 4 | 9 | 16 | 25 | 36 |

9. | H | 6.0 | 6.2 | 6.4 | 6.6 | 6.8 |
    |---|---|---|---|---|---|
    | W | 10 | 11 | 12 | 13 | 14 |

10. | X | 0 | 1 | 2 | 3 | 4 |
    |---|---|---|---|---|---|
    | Y | 2 | 3 | 4 | 5 | 6 |

11 In the following table, one variable varies directly as the other.

  a Find the missing numbers.

  b Write the formula that connects the variables.

H	1	2	?	7
d	5	?	25	35

12 In the following table, one variable varies directly as the other.

  a Find the missing numbers.

  b Write the formula that connects the variables.

D	4	8	?	22
S	6	?	15	33

13 Two quantities, x and y, vary directly. When $x = 4$, $y = 10$. Find the value of y when $x = 6$.

14 If x varies directly as y when $x_1 = 13$, $x_2 = 3$, and $y_1 = 7$ then what is the value of y_2?

15 If x varies directly with y and $x = 6$ when $y = 10$, what is the value of x when $y = 25$?

16 This graph shows the cost of a round-trip railroad ticket compared with distance traveled in miles. What is the distance in miles for a trip that cost $90?

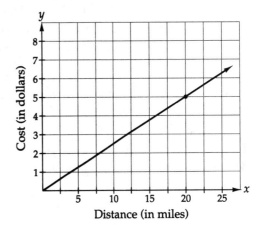

Slope-Intercept and Point-Slope Forms

Certain forms of a linear equation can provide enough information to easily graph the corresponding line.

Slope-Intercept Form In several of the examples discussed in previous sections, the first step in graphing the equation was to solve the equation for y. You may have noticed that in each of these examples, the value of the slope happened to equal the coefficient of x, and the y-intercept equaled the numeric constant. These equalities are always true.

> If a linear equation is expressed in the form $y = mx + b$, m represents the slope of the line and b represents the y-intercept. Thus, $y = mx + b$ is called the **slope-intercept form** of the equation.

For example, if $y = \dfrac{-2}{3}x + 4$, then the slope is $\dfrac{-2}{3}$ and the y-intercept is 4. Conversely, if we are told that the slope of a line is $\dfrac{3}{4}$ and the y-intercept is -2, then the equation of the line is $y = \dfrac{3}{4}x - 2$.

The slope-intercept form is the easiest form to use when a graph of the line is required. Using the last example, $y = \dfrac{3}{4}x - 2$, we can graph the y-intercept (-2), at the point $(0, -2)$. *Remember that b is the value of y where the graph crosses the y-axis.* To find a second point, use the slope, $m = \dfrac{3}{4}$. Remember that $\dfrac{3}{4} = \dfrac{\Delta y}{\Delta x}$. This means that when y changes $+3$, x changes $+4$. So starting at the y-intercept $(0, -2)$, the next point would be $(0 + 4, -2 + 3)$ or $(4, 1)$. Plot that point. We can repeat the procedure by adding 3 to y and 4 to x. The next point would then be $(8, 4)$. Plot that point and draw a line connecting the three points.

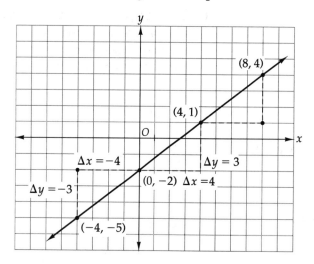

The slope-intercept form is the appropriate form for use with a graphing calculator. Use Y=. Make sure the y-intercept is within your graphing window.

MODEL PROBLEM

Find the slope and y-intercept of $4x + 3y = 3$.

SOLUTION

Solve for y.
$$4x + 3y = 3$$
$$3y = 3 - 4x$$
$$y = 1 - \dfrac{4}{3}x$$

Rewrite in slope-intercept form: $y = mx + b$.
$$y = -\dfrac{4}{3}x + 1$$

Answer: Slope $m = -\dfrac{4}{3}$ and y-intercept $b = 1$

Point-Slope Form Given the slope m of line l and the coordinates (x_1, y_1) of any point on the line, it is possible to write an equation for the line. Since the slope is the same for any two points on the line, we can let (x, y) represent another point on the line and write the slope as $m = \dfrac{y - y_1}{x - x_1}$. By cross multiplying we get $y - y_1 = m(x - x_1)$.

> If a line passes through the given point (x_1, y_1) and has slope m, the **point-slope form** of the equation of the line is $y - y_1 = m(x - x_1)$.

MODEL PROBLEM

Given a line with point $P(4, 6)$ on the line and slope $m = 3$, write the correct linear equation in slope-intercept form.

SOLUTION

Substitute $(4, 6)$ in the slope equation $m = \dfrac{y - y_1}{x - x_1}$, and substitute 3 for m. $\qquad 3 = \dfrac{y - 6}{x - 4}$

Multiply both sides by $x - 4$. $\qquad 3(x - 4) = y - 6$

Remove parentheses. $\qquad 3x - 12 = y - 6$

Rewrite the equation in slope-intercept form. $\qquad 3x - 6 = y$

$\qquad$ or $\quad y = 3x - 6$

Practice

1 Which is the graph of $x + y = 5$?

(1)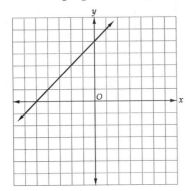

(2)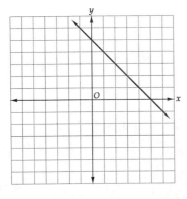

(3)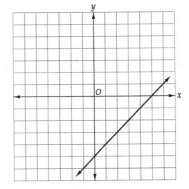

(4)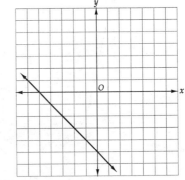

2 Which is the graph of $3y + 12x = 2$?

(1)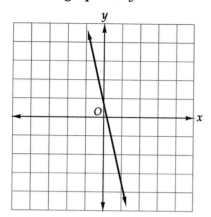

(2)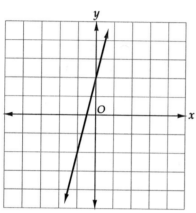

(3)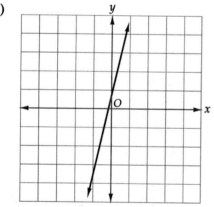

(4)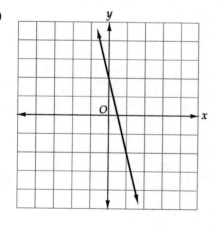

3 Which is the graph of $y - 3 = x - 5$?

(1)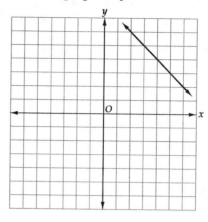

(2)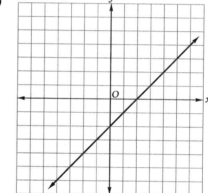

(3)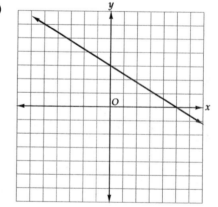

(4)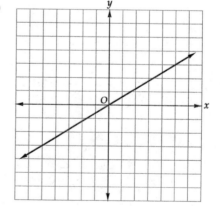

4 Which is the graph of $y - 2 = \frac{1}{3}(x - 3)$?

(1)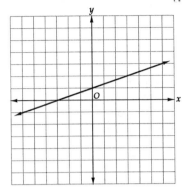

(3)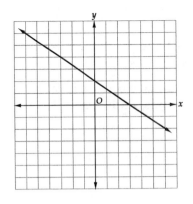

(2)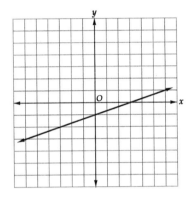

(4)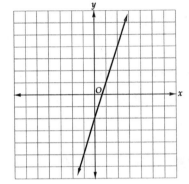

Exercises 5–12: Fill in the table.

Equation	Solve for y	Slope	y-intercept
5 $2x + 2y = 6$			
6 $2x + 8y = 8$			
7 $3x - 3y = 12$			
8 $4x + 2y = 0$			
9 $10x + 7y = 7$			
10 $3x - y + 7 = 0$			
11 $-x + 3y = -9$			
12 $3y - 7 = 0$			

Exercises 13–16: Write an equation of the line whose slope and y-intercept are given, and graph the line.

13 Slope is 4 and y-intercept is 1.

14 Slope is $-\frac{1}{2}$ and y-intercept is 0.

15 Slope is -1 and y-intercept is 1.

16 Slope is 0 and y-intercept is 6.

Exercises 17–20: Write an equation of the line with the given slope, passing through the given point, and graph the line.

17 $m = -2$, $(4, 2)$

18 $m = -\frac{1}{2}$, $(-4, -2)$

19 $m = \frac{2}{3}$, $(-3, 1)$

20 $m = $ undefined, $(2, -2)$

Graphing a Line From Partial Information

We can determine whether two lines are perpendicular or parallel from observing their slopes.

- **If two lines have the same slope, then they are parallel.**
  For example, $y = 4x + 7$ and $y = 4x - 105$ are parallel.
- **If two lines have slopes that would equal −1 when multiplied together, then the lines are perpendicular.**
  For example, $y = 2x + 6$ and $y = -\frac{1}{2}x - 9$ are perpendicular, since $2 \cdot \left(-\frac{1}{2}\right) = -1$. Number pairs like these are called **negative reciprocals**.

To find the graph of a line, given the slope and one point on the line without an equation:

- Plot the known point, (x_1, y_1).
- Use the slope in fraction form, $m = \frac{\Delta y}{\Delta x}$, to find points $(x_1 + \Delta x, y_1 + \Delta y)$ and $(x_1 - \Delta x, y_1 - \Delta y)$. Plot these two points.
- Draw a line through the three points.

 MODEL PROBLEM

Given a line passing through point $P(4, 6)$ and perpendicular to $y - 5 = -\frac{1}{3}(x + 3)$, graph the line.

SOLUTION

Plot the known point $(4, 6)$. Since $y - 5 = -\frac{1}{3}(x + 3)$ is in point-slope form, the slope of that line is $-\frac{1}{3}$.

Since our line is perpendicular, it has a slope of $\frac{3}{1}$.

Find the point on the graph with a y-value that is 3 more than 6 and an x-value that is 1 more than 4. That is point $(5, 9)$. Find a y-value that is 3 less than 6 and an x-value that is 1 less than 4. That point is $(3, 3)$. Plot these two points.

Plot the second point and draw a line through both points.

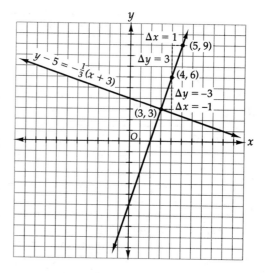

To find the graph and equation of a line from two points, you can use either of two methods.

Method 1

- Use the two points to find slope m of the line.
- Use slope m and either one of the given points to substitute in the slope-intercept formula: $y = mx + b$.
- Solve for b, the y-intercept.
- Substitute the values for m and b in $y = mx + b$.

Method 2

- Use the two points to find slope m of the line.
- Then use slope m and either one of the given points to substitute in the point-slope formula: $y - y_1 = m(x - x_1)$
- Simplify and change into $y = mx + b$ form.
- Use the slope and the y-intercept to graph the equation.

MODEL PROBLEM

Find the equation of the line that joins the points (1, 3) and (2, 5).

SOLUTION

METHOD 1

Substitute and find slope m:

$$m = \frac{y_2 - y_1}{x_2 - x_1} = \frac{5 - 3}{2 - 1} = \frac{2}{1} \text{ or } 2$$

Now substitute 2 for slope m, and point (1, 3) for (x, y) in the slope-intercept formula $y = mx + b$. Solve for b:

$3 = 2(1) + b$, and $b = 1$

Once again, substitute 2 for m, and 1 for b in $y = mx + b$:

The equation of the line is $y = 2x + 1$.

Check by substituting the coordinates of one of the points for (x, y) in the equation of the line.

METHOD 2

Substitute and find the slope m.

$$m = \frac{y_2 - y_1}{x_2 - x_1} = \frac{5 - 3}{2 - 1} = \frac{2}{1} \text{ or } 2$$

Now substitute 2 for slope m and point (1, 3) for (x_1, y_1) in the point-slope formula:

$y - y_1 = m(x - x_1)$
$y - 3 = 2(x - 1)$

If we had chosen the point (2, 5), then

$y - y_1 = m(x - x_1)$
$y - 5 = 2(x - 2)$

In fact, these are dependent lines; that is, they have the same graph. If we convert each equation to slope-intercept form, the result is the same equation.

$y - 3 = 2(x - 1)$ $y - 5 = 2(x - 2)$
$y - 3 = 2x - 2$ $y - 5 = 2x - 4$
$y = 2x + 1$ $y = 2x + 1$

Practice

1. Which pair of lines are parallel?
   (1) $y = 4x - 9$ and $y - 4x = 3$
   (2) $y = 3x + 7$ and $y + 2 = 3(2x - 5)$
   (3) $y = 2x$ and $2y = x - 9$
   (4) $y + x = 0$ and $y = x$

2. Which pair of lines are perpendicular?
   (1) $y = 4x - 9$ and $y - 4x = 3$
   (2) $y = 3x + 7$ and $y + 2 = 3(x - 5)$
   (3) $y = 2x$ and $2y = x - 9$
   (4) $y + x = 0$ and $y = x$

3. Which equation passes through (3, −5) and (−1, 3)?
   (1) $y + 5 = x - 3$
   (2) $y = 2x + 1$
   (3) $y - 3 = -2(x + 1)$
   (4) $y = x + 2$

4 Which line passes through (2, 3) with a slope of −2?

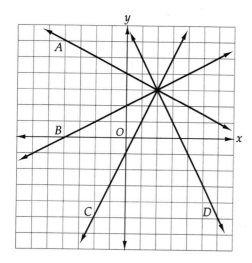

(1) line A
(2) line B
(3) line C
(4) line D

Exercises 5–8: Write an equation of the line that passes through the given points.

5 (1, 3) and (−2, −6)

6 (6, 1) and (−4, −4)

7 (4, 0) and (2, −2)

8 (7, 4) and (1, 1)

Exercises 9–12: Write an equation of the line that satisfies the given conditions.

9 parallel to $x + 2y = 6$; passes through (1, 1)

10 parallel to $2x - 4y + 4 = 0$; passes through (−2, 3)

11 perpendicular to $x + y = 0$; passes through (0, 5)

12 perpendicular to $3x + y - 1 = 0$; passes through (3, −1)

Exercises 13–15: Select the one equation whose graph would *not* be parallel to the other two.

13 $\{y - 6x = 3, y = 6x, x = 5y\}$

14 $\{2x - y = 9, 2y - x = 7, 2x - y = 2\}$

15 $\{x + 3y = 7, 3x + y = 7, y = 7 - 3x\}$

Quadratic Equations on the Coordinate Plane

Graphing a Parabola

In Chapter 4, we learned to factor and solve quadratic equations of the form $0 = ax^2 + bx + c$, where $a \neq 0$. The graph of a quadratic equation in the form $y = ax^2 + bx + c$, where $a \neq 0$, is a special curve called a **parabola**.

To Graph a Parabola, Given a Domain

- Express the equation in the form $ax^2 + bx + c = 0$.
- Create a table of coordinates by using values from the given domain.
- Plot the points from the table of values and join them to make a smooth curve.

MODEL PROBLEM

Graph $y = x^2 + 4x + 3$ using integral values from -5 to 1.

SOLUTION

Make a table of values.

x	$x^2 + 4x + 3$	y
-5	$(-5)^2 + 4(-5) + 3$	8
-4	$(-4)^2 + 4(-4) + 3$	3
-3	$(-3)^2 + 4(-3) + 3$	0
-2	$(-2)^2 + 4(-2) + 3$	-1
-1	$(-1)^2 + 4(-1) + 3$	0
0	$(0)^2 + 4(0) + 3$	3
1	$(1)^2 + 4(1) + 3$	8

You can use the graphing calculator to find this table of values quickly. See the appendix.

When you plot the ordered pairs (x, y) you get the smooth curve shown.

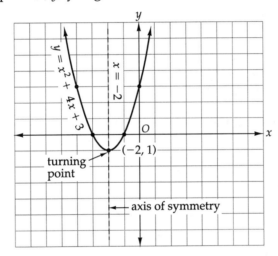

The graph of this parabola has certain interesting features. The graph is symmetric with respect to the line $x = -2$. That line has a special name, the **axis of symmetry**. Every parabola $y = ax^2 + bx + c$ has an axis of symmetry. The equation for the axis of symmetry is $x = \dfrac{-b}{2a}$. For the parabola $y = x^2 + 4x + 3$, $a = 1$ and $b = 4$. The equation of the axis of symmetry can be found by using the formula: $x = \dfrac{-b}{2a}$.

$x = \dfrac{-4}{2(1)}$ so $x = -2$

The curve of this parabola has a lowest point on the graph where it crosses the axis of symmetry. This minimum point is called the **turning point** of the parabola. The turning point is on the axis of symmetry. Therefore, its abscissa, or x-value, is $\dfrac{-b}{2a}$. In this example, the abscissa is -2. The ordinate, or y-value, is found by substituting the x-value (-2) back into the quadratic equation. In the table, when $x = -2$, $y = -1$, so the turning point is $(-2, -1)$.

Quadratic Equations on the Coordinate Plane

This parabola happens to turn upward, like a bowl, but not all parabolas must point that way. The direction of the parabola can be determined by the value of the coefficient a. If a is positive, then the graph of the equation $y = ax^2 + bx + c$ opens upward and the turning point is a **minimum point**. If a is negative, then the graph of the equation $y = ax^2 + bx + c$ opens downward and the turning point is a **maximum point**.

By applying the vertical line test, we find that this parabola is a function. In fact, *every parabola of the form $y = ax^2 + bx + c$ is a function.*

To Graph a Parabola, Not Given a Domain

- Express the equation in the form $ax^2 + bx + c = 0$.
- Obtain a table of values by finding the axis of symmetry, $x = \dfrac{-b}{2a}$, and then choose integer values of x both greater and smaller than $\dfrac{-b}{2a}$.
- Plot the points from the table of values and join them to make a smooth curve.

Note: (1) The turning point of the parabola must occur on the axis of symmetry. (2) The axis of symmetry is always an equation. It is never just a number.

MODEL PROBLEM

Graph $y = -x^2 + 6x - 5$ and label the turning point.

SOLUTION

Substitute -1 for a and 6 for b in the equation for the axis of symmetry.

$$x = \frac{-b}{2a} = \frac{-6}{2(-1)} = \frac{6}{2} = 3$$

The equation of the axis of symmetry is $x = 3$, so choose values for x near 3 from the domain, {0, 1, 2, 3, 4, 5, 6}.

Make a table of values.

x	$-x^2 + 6x - 5$	y
0	$-(0)^2 + 6(0) - 5$	-5
1	$-(1)^2 + 6(1) - 5$	0
2	$-(2)^2 + 6(2) - 5$	3
3	$-(3)^2 + 6(3) - 5$	4
4	$-(4)^2 + 6(4) - 5$	3
5	$-(5)^2 + 6(5) - 5$	0
6	$-(6)^2 + 6(6) - 5$	-5

Plot and graph the ordered pairs.

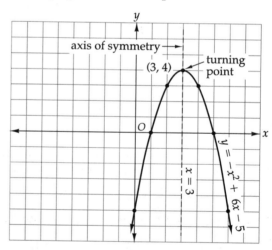

The turning point of the parabola is on the axis of symmetry where the x-value is 3. The ordinate, or y-value, is found by substituting 3 for x in the quadratic equation. Looking at our table, when $x = 3$, $y = 4$. The turning point is (3, 4).

Practice

1. What is the equation of the axis of symmetry for $y = x^2 - 4x$?

   (1) $x = 2$
   (2) $x = -\dfrac{1}{8}$
   (3) $y = \dfrac{1}{8}$
   (4) $y = -2$

2. Identify the turning point of $y = -x^2 - 2x$.

   (1) $(4, 8)$
   (2) $(1, -3)$
   (3) $(-1, 1)$
   (4) $(-4, -8)$

3. Which equation does the parabola in the graph represent?

   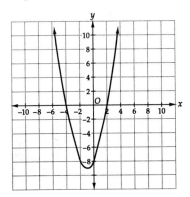

   (1) $y = x^2 + 2x + 8$
   (2) $y = x^2 + 2x - 8$
   (3) $y = x^2 - 2x - 8$
   (4) $y = x^2 - 2x + 8$

4. Which parabola is the graph of the equation $y = -x^2 + 4x - 3$?

   (1)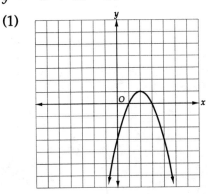

   (2)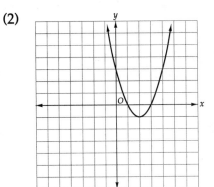

   (3)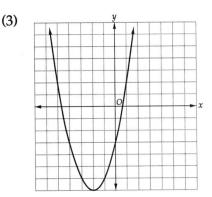

   (4)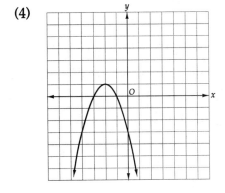

Quadratic Equations on the Coordinate Plane

Exercises 5–10: Using the domain indicated in the parentheses, prepare a table of values and graph the quadratic equation. Find the equation of the axis of symmetry and the turning point.

5. $y = x^2 - 6x - 7$ $(0 \leq x \leq 6)$
6. $y = -2x^2$ $(-3 \leq x \leq 3)$
7. $y = x^2 + 2$ $(-3 \leq x \leq 3)$
8. $y = x^2 - 3x - 4$ $(-1 \leq x \leq 5)$
9. $y = 2(x^2 - 1)$ $(-3 \leq x \leq 3)$
10. $y = -x^2 + 4x - 1$ $(1 \leq x \leq 5)$

Exercises 11–15:

a. Find the axis of symmetry and turning point.

b. Create a table to get 7 points for the graph, centered on the turning point.

c. Graph the equation.

11. $y = x^2 + 2x + 1$
12. $y = -2x^2 + 1$
13. $y = 2x^2 + 4x + 1$
14. $y = -2x^2 + 8x$
15. $y = x^2 - 6x$

Solving Quadratic Equations Graphically

In Chapter 4, we found solutions for quadratic equations by factoring and setting each factored term equal to 0. The solutions, or **roots**, of a quadratic equation can also be found by graphing the equation and looking at the x-intercepts of the graph.

To Solve a Quadratic Equation Graphically

- Express the equation in the form $ax^2 + bx + c = 0$.
- Graph the parabola $y = ax^2 + bx + c$ using a suitable table of values.
- Identify where the parabola crosses the x-axis (where $y = 0$). These are the roots or solutions to the quadratic equation.

Using a graphing calculator, we can find noninteger roots to a specified degree of accuracy. See the appendix.

As before, a quadratic equation can have two, one, or zero solutions.

Case I: Parabola Intersects the x-Axis at Two Distinct Points An example is the graph of $x^2 - 4 = 0$. We can solve this equation algebraically.

$$x^2 - 4 = 0$$
$$(x + 2)(x - 2) = 0$$

$x + 2 = 0 \qquad x - 2 = 0$

$x = -2 \qquad x = 2$ The solution set is $\{2, -2\}$.

Now we solve the equation graphically.

Find the axis of symmetry, $x = \dfrac{-b}{2a} = \dfrac{0}{2} = 0$.

172 Chapter 5: Functions and the Coordinate Plane

Make a table of values.

x	x² − 4	y
−3	(−3)² − 4	5
−2	(−2)² − 4	0
−1	(−1)² − 4	−3
0	(0)² − 4	−4
1	(1)² − 4	−3
2	(2)² − 4	0
3	(3)² − 4	5

Using the table, graph $y = x^2 - 4$.

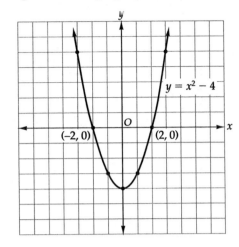

The x-intercepts of the graph are (−2, 0) and (2, 0). The roots of the equation are −2 and 2. *If the graph crosses the x-axis at two distinct points, the roots are real and unequal.*

Case II: Parabola Intersects the x-Axis at Exactly One Point An example is the graph of $x^2 + 4x + 4 = 0$. We can solve this equation algebraically.

$$x^2 + 4x + 4 = 0$$
$$(x + 2)(x + 2) = 0$$
$$x + 2 = 0 \quad x + 2 = 0$$
$$x = -2 \quad x = -2 \quad \text{The roots of the equation are } both -2.$$

Now we solve the equation graphically.

Find the axis of symmetry, $x = \dfrac{-b}{2a} = \dfrac{-4}{2} = -2$.

Make a table of values.

x	x² + 4x + 4	y
−5	(−5)² + 4(−5) + 4	9
−4	(−4)² + 4(−4) + 4	4
−3	(−3)² + 4(−3) + 4	1
−2	(−2)² + 4(−2) + 4	0
−1	(−1)² + 4(−1) + 4	1
0	(0)² + 4(0) + 4	4
1	(1)² + 4(1) + 4	9

Using the table, graph $y = x^2 + 4x + 4$ around the axis of symmetry.

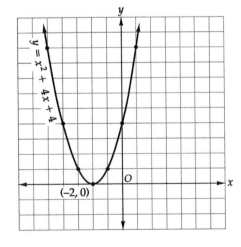

The x-intercept of the graph is the point (−2,0), which is also the turning point of the parabola. The solution set is {−2}. *Since the graph crosses the x-axis at exactly one point, the roots are real and equal.*

Quadratic Equations on the Coordinate Plane

Case III. Parabola Does Not Intersect the x-Axis An example is the graph of $x^2 - 2x + 2 = 0$.

x	$x^2 - 2x + 2$	y
-2	$(-2)^2 - 2(-2) + 2$	10
-1	$(-1)^2 - 2(-1) + 2$	5
0	$(0)^2 - 2(0) + 2$	2
1	$(1)^2 - 2(1) + 2$	1
2	$(2)^2 - 2(2) + 2$	2
3	$(3)^2 - 2(3) + 2$	5
4	$(4)^2 - 2(4) + 2$	10

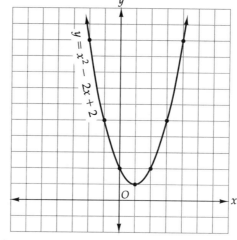

Since the graph does not intersect the x-axis, the equation has no real roots.

 Practice

1. How many x-intercepts does the graph of $x^2 + x + 5 = 0$ have?
   - (1) 0
   - (2) 1
   - (3) 2
   - (4) an infinite number of intercepts

2. What is the largest root of $x^2 + 5x - 6 = 0$?
   - (1) 6
   - (2) 5
   - (3) 3
   - (4) 1

3.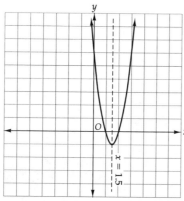

   From this graph of $3x^2 - 9x + 6 = 0$, we can conclude that the equation has
   - (1) no roots
   - (2) one root at 6
   - (3) one root at 1.5
   - (4) two roots, one at 1 and one at 2

4. The single solution of $x^2 + 7x + 12.25 = 0$ lies between what two integer values of x?
   - (1) -6 and -5
   - (2) -4 and -3
   - (3) 3 and 4
   - (4) 5 and 6

Exercises 5–12: Solve the quadratic equations by graphing and label the roots. You may find equations with no real roots.

5. $x^2 - x - 6 = 0$
6. $x^2 + 2x = 0$
7. $x^2 - 1 = 0$
8. $x^2 - 6x + 9 = 0$
9. $x^2 + 8x + 16 = 0$
10. $-x^2 + 6x - 27 = 0$
11. $x^2 + 3x = 0$
12. $x^2 - 4x - 21 = 0$

174 Chapter 5: Functions and the Coordinate Plane

Graphing Systems of Equations and Inequalities

Systems of Linear Equations

Often, problems on qualifying examinations will include information relating two variables without providing a value for either variable. To find a solution, we need two sets of information in the form of equations.

Suppose that the perimeter of a rectangular vegetable patch is 12 feet and the length is twice the width. If we let x represent the width and y represent the length, we know that $2x + 2y = 12$, or $x + y = 6$. We also know that $y = 2x$. The two equations, $x + y = 6$, and $y = 2x$, are called a **system of simultaneous linear equations**. While neither equation, alone, is enough to find the dimensions of the rectangle, they can be solved together algebraically, as discussed in Chapter 3, or by graphing the lines on the same coordinate plane.

To Solve Simultaneous Linear Equations Graphically

- Draw the graph of each equation on the same coordinate plane.
- The coordinates of the point of intersection of the graphs are the required values of x and y.
- Check the solution in both original equations.

To find the dimensions of the vegetable patch, we graph the two equations on the same coordinate plane.

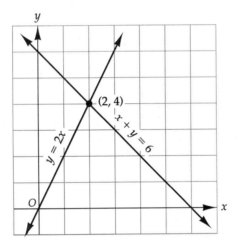

Check the solution (2, 4) in both original equations:

$x + y = 6$ $y = 2x$
$2 + 4 = 6$ $4 = 2(2)$
$6 = 6$ ✓ $4 = 4$ ✓

Since this system has one solution, it is a system of consistent equations. As discussed in Chapter 3, a system of equations can be consistent, inconsistent, or dependent. An example of each of these is illustrated below.

Consistent Equations

- Lines have different slopes.
- Lines intersect in exactly one point.

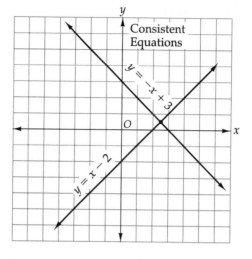

Inconsistent Equations

- Lines have the same slope but different y-intercepts.
- Lines are parallel and do not intersect.

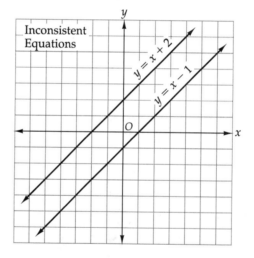

Dependent Equations

- Lines have the same slope and the same y-intercept.
- Lines are identical and share all points in common.

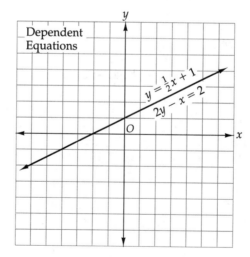

1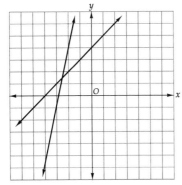

The system of equations represented by this graph is:

(1) consistent
(2) inconsistent
(3) dependent
(4) codependent

2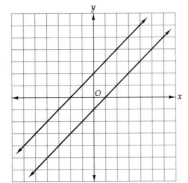

The system of equations represented by this graph is:

(1) consistent
(2) inconsistent
(3) dependent
(4) codependent

3 The simultaneous solution to $y = 5$ and $y = 4x - 3$ is:

(1) (5, 17)
(2) (2, 5)
(3) (0, 5)
(4) ∅, since the lines have different y-intercepts

4 The lines $y = 2x + 2$ and $y = 3x - 1$ intersect at point:

(1) (−3, −8)
(2) (−1, −4)
(3) (1, 4)
(4) (3, 8)

Exercises 5–7: Identify each system of linear equations as consistent or inconsistent.

5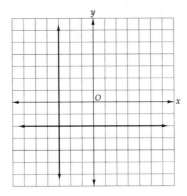

6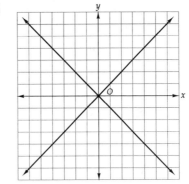

7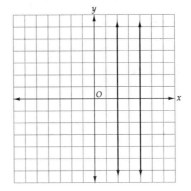

Exercises 8–15: Solve each system of equations graphically and check. If there is no common solution, write "Inconsistent system."

8 $x + y = 4$
$y = 3x$

9 $y = -x + 7$
$y = 2x + 1$

10 $x - y = 4$
$y = \dfrac{1}{2}x$

11 $2x + y = 2$
$y - x = 5$

Graphing Systems of Equations and Inequalities **177**

12 $y - x = 2$
    $y - 2x - 2 = 0$

13 $x = 0$
    $y = -3$

14 $y = -3x$
    $2x + y + 2 = 0$

15 $y = -2x - 1$
    $x + y + 4 = 0$

Linear Inequalities

A line graphed in the coordinate plane divides the plane into two regions, called **half-planes**. When the equation of the line is written in slope-intercept form or $y = mx + b$ form, the half-plane *above* the line is the graph of $y > mx + b$ and the half-plane *below* the line is the graph of $y < mx + b$. If a half-plane is to be included in a solution set, we show this by shading its entire region. The line itself, considered a **boundary line**, is drawn as a solid line when it is part of the solution set, and as a dashed line if it is not.

To Graph a Linear Inequality

- Graph the boundary line for the inequality by expressing the inequality as an equation in slope-intercept form or by creating a table of values.
    If the sign is > or <, the boundary line will be broken or dashed.
    If the sign is ≥ or ≤, the boundary line will be solid.

- Shade the half-plane of the inequality.

    Method 1: Select two points, one on each side of the boundary line, and substitute them into the inequality. Shade the half-plane with the point that makes the inequality true.

    Method 2: Solve the inequality for y or for x.

    If the inequality begins with $y >$ or $y \geq$, shade the half-plane *above* the boundary line. If the inequality begins with $y <$ or $y \leq$, shade the half-plane *below* the boundary line.

    If the inequality begins with $x >$ or $x \geq$, shade the half-plane to the *left* of the boundary line. If the inequality begins with $x <$ or $x \leq$, shade the half-plane to the *right* of the boundary line.

If you solve without selecting points, check your answer by substituting the coordinates of a point in the given equation.

MODEL PROBLEMS

1 Graph $y > 2x + 1$ and $y < 2x + 1$.

SOLUTION

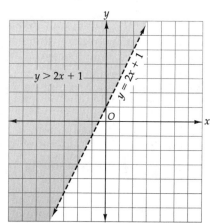
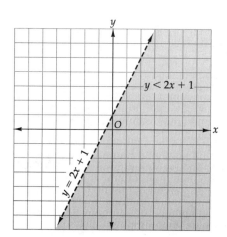

2 Graph $y \geq 2x + 1$ and check.

SOLUTION

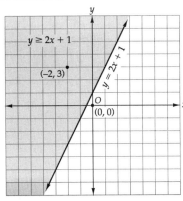

Check $(-2, 3)$ by substituting.

$y \geq 2x + 1$
$3 \geq 2(-2) + 1$
$3 \geq -4 + 1$
$3 \geq -3$ ✓

Check $(0, 0)$ from outside the solution set.

$y \geq 2x + 1$
$0 \geq 2(0) + 1$
$0 \geq 0 + 1$
$0 \not\geq 1$ (not in solution set) ✓

3 Graph the inequality $x - 3y > 6$.

SOLUTION

Rewrite the inequality in $y = mx + b$ form:
$x - 3y > 6$
$x - 6 > 3y$ or $3y < x - 6$
$y < \dfrac{x}{3} - 2$

Create a table of values for $y = \dfrac{x}{3} - 2$.

x	−3	0	3
y	−3	−2	−1

Graph the inequality.

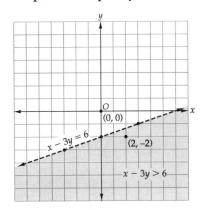

Check $(2, -2)$ by substituting.

$x - 3y > 6$
$(2) - 3(-2) > 6$
$2 + 6 > 6$
$8 > 6$ ✓

Check $(0, 0)$ from outside the solution set.

$x - 3y > 6$
$(0) - 3(0) > 6$
$0 \not> 6$ (Not in solution set) ✓

4 Graph $x > 3$ and $x \leq 3$ on separate coordinate planes.

SOLUTION

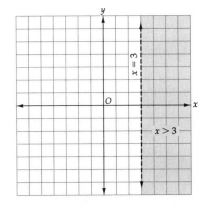 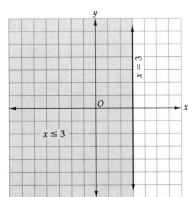

Graphing Systems of Equations and Inequalities

Practice

1 Which graph shows $y < 3x$?

(1)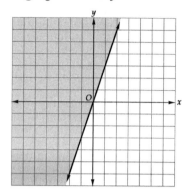

(2)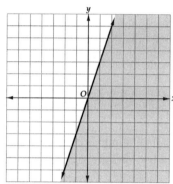

(3)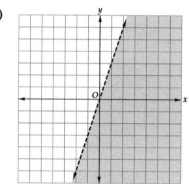

(4)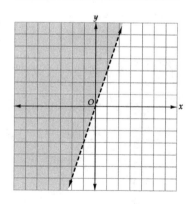

2 Which graph shows $y > -x - 1$?

(1)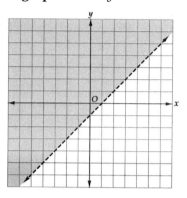

(2)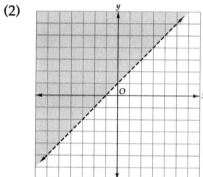

(3)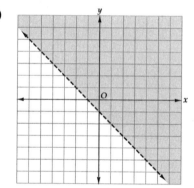

(4)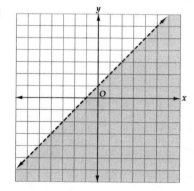

3 Which equation describes the shaded area of the graph?

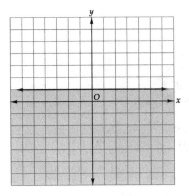

(1) $y > 1$ (3) $y \leq 1$
(2) $y < 1$ (4) $y \geq 1$

4 Which equation describes the shaded area of the graph?

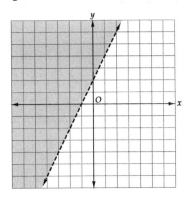

(1) $y \leq \frac{1}{2}x + 2$

(2) $y \geq -\frac{1}{2}x + 2$

(3) $y < -2x + 2$

(4) $y > 2x + 2$

For Exercises 5–10, graph each inequality.

5 $y \leq -2x$

6 $x + y \geq 1$

7 $2x + y > -4$

8 $2x - y < 3$

9 $x \leq 0$

10 $x < 1 + y + x$

For Exercises 11–15, write each verbal sentence as an open sentence and graph the inequality.

11 The ordinate of a point is greater than 2 less than the abscissa.

12 The sum of the abscissa and the ordinate of a point is greater than or equal to 4.

13 Twice the ordinate of a point decreased by the abscissa is less than or equal to -1.

14 The ordinate of a point decreased by half the abscissa is greater than 0.

15 The abscissa of a point decreased by the ordinate is less than or equal to 3.

Systems of Linear Inequalities

The graph of the solution of a system of linear inequalities is the shaded region of the plane containing all the points that are common solutions of all the inequalities given.

To Find the Solution Set of a System of Linear Inequalities

- Solve each inequality for y.
- Identify slope and y-intercept and graph the boundary lines for each inequality.
- Shade the correct region for each graph.
- Label the common region shaded by both graphs with a capital letter, such as S for *solution*.
- If the two regions do not overlap, indicate that there is no solution.

MODEL PROBLEM

Graph the following system of inequalities and label the solution set S:

$y > x + 4$

$x + y \leq 0$

SOLUTION

The first graph is a dashed line, with y-intercept or $b = 4$ and a slope of $m = 1$. The shading is above the line. The second inequality must be rewritten as $y \leq -x$. The second graph is a solid line, with y-intercept at the origin or $b = 0$ and a slope of $m = -1$. The shading is below the line. The solution to the system is the common region, marked S.

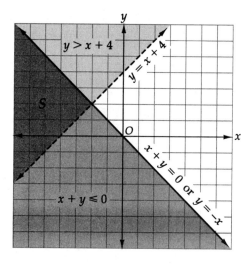

Practice

1 Which graph shows the solution to the system $y \geq -3x + 2$ and $x < 0$?

(1)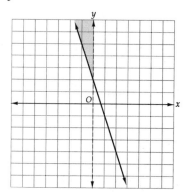

(2)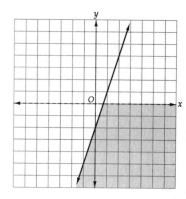

(3)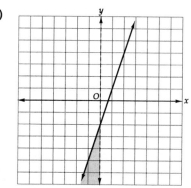

(4)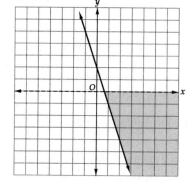

182 Chapter 5: Functions and the Coordinate Plane

2 Which graph shows the solution set of $y \leq 1$ and $x + y \geq 2$?

(1)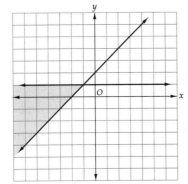

(2)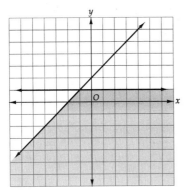

(3)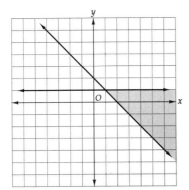

(4)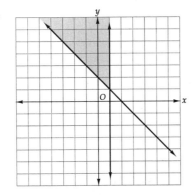

3 Which set of inequalities describes the shaded area of the graph?

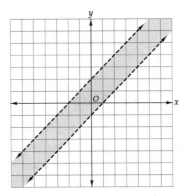

(1) $y > x - 1$ and $y < x + 2$
(2) $y < x - 1$ and $y > x + 2$
(3) $y < x - 1$ or $y > x + 2$
(4) $y > -1$ and $y < 2$

4 Which set of inequalities describes the shaded area of the graph?

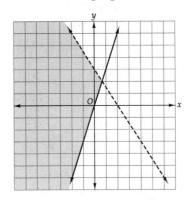

(1) $y < 3x$ and $y \geq -2x + 3$
(2) $y > \frac{1}{3}x$ and $y \geq -2x - 3$
(3) $y \leq 3x$ and $y < -2x + 3$
(4) $y \geq 3x$ and $y < -2x + 3$

Exercises 5–12: Graph each system of inequalities and label the solution set S.

5 $y > x - 3$
$y < x + 5$

6 $y > -2x - 8$
$x - y > 1$

7 $y < x$
$y > 0$

8 $y \geq x + 5$
$x > -5$

Graphing Systems of Equations and Inequalities

9 $x + y < 7$
      $x - y > 3$

10 $y > -x$
      $x + 2y < 4$

11 $x + y \geq -2$
      $3x + 2y > 1$

12 $y > 2x + 1$
      $y < x - 2$

Quadratic-Linear Pairs

The graph of a quadratic-linear system consists of the graph of a quadratic equation (a parabola) and a linear equation (a line). As before, the intersections of the graphs reveal the solutions common to both equations.

To Solve Quadratic-Linear Systems Graphically

- Draw the graphs of both equations on the same coordinate plane.
- Find the common solutions by reading the points of intersection of the two graphs.
- Write the solutions as ordered pairs.
- Check the solutions in both original equations.

The solution set to a quadratic-linear pair may consist of two points, one point, or no points. Some possibilities for these systems are illustrated below.

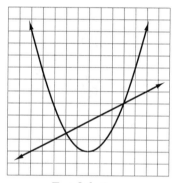
Two Solutions

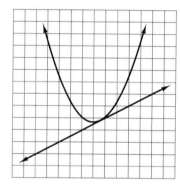
One Solution

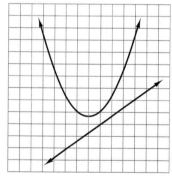
No Solution

Chapter 5: Functions and the Coordinate Plane

The following model problems, all using the same quadratic equation, demonstrate each case.

MODEL PROBLEMS

1 Find the solution set for
$y = x^2 - 2x + 1$
$y = -x + 3$

SOLUTION

For $y = x^2 - 2x + 1$, the equation of the axis of symmetry is $x = \dfrac{-b}{2a} = \dfrac{-(-2)}{2(1)} = 1$.

Create a table centered on $x = 1$.

x	$x^2 - 2x + 1$	y
−2	$(-2)^2 - 2(-2) + 1$	9
−1	$(-1)^2 - 2(-1) + 1$	4
0	$(0)^2 - 2(0) + 1$	1
1	$(1)^2 - 2(1) + 1$	0
2	$(2)^2 - 2(2) + 1$	1
3	$(3)^2 - 2(3) + 1$	4
4	$(4)^2 - 2(4) + 1$	9

Plot the points and draw the parabola. The line $y = -x + 3$ is already in slope-intercept form. The y-intercept is 3 and the slope is −1. Graph the line.

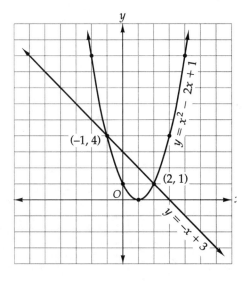

The graphs intersect at the two points (−1, 4) and (2, 1). The solution set is {(−1, 4), (2, 1)}.

Check for (−1, 4):
$y = x^2 - 2x + 1$ $y = -x + 3$
$4 = (-1)^2 - 2(-1) + 1$ $4 = -(-1) + 3$
$4 = 1 + 2 + 1$ $4 = 1 + 3$
$4 = 4$ ✓ $4 = 4$ ✓

Check for (2, 1):
$y = x^2 - 2x + 1$ $y = -x + 3$
$1 = (2)^2 - 2(2) + 1$ $1 = -(2) + 3$
$1 = 4 - 4 + 1$ $1 = 1$ ✓
$1 = 1$ ✓

2 Find the solution set for
$y = x^2 - 2x + 1$
$y = -2x + 1$

SOLUTION

The parabola is copied from the example above. For the line $y = -2x + 1$, the slope is −2 and the y-intercept is 1.

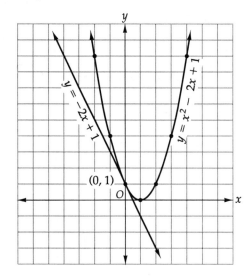

The graphs intersect at only one point (0, 1). The solution set is {(0,1)}.

Check for (0, 1):
$y = x^2 - 2x + 1$ $y = -2x + 1$
$1 = (0)^2 - 2(0) + 1$ $1 = -(0) + 1$
$1 = 1$ ✓ $1 = 1$ ✓

Graphing Systems of Equations and Inequalities

3 Find the solution set for
$y = x^2 - 2x + 1$
$y = -x - 3$

SOLUTION

Again, the parabola is copied from the example above. For the line $y = -x - 3$, the slope is -1 and the y-intercept is -3.

The graphs do not intersect. The solution is the empty set, $\varnothing$.

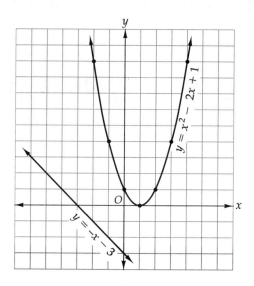

Practice

1 Which graph shows the solution to the system $y = x - 2$ and $y = x^2 - 6x + 4$?

(1)

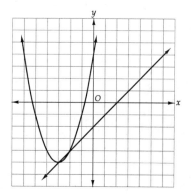

(2)

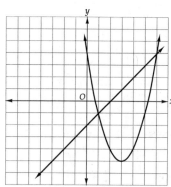

(3)

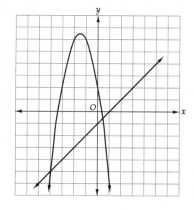

(4)

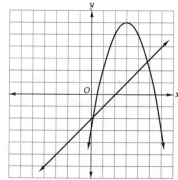

2 Which graph shows the solution set of $y = -\frac{1}{2}x - 4$ and $y = x^2 - 4$?

(1)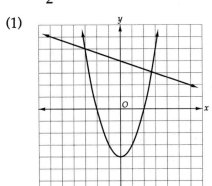

(2)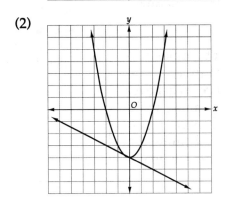

(3)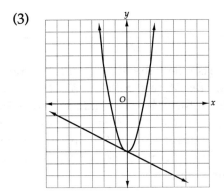

(4)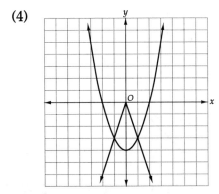

3 Which set of equations describes the graph?

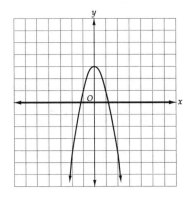

(1) $x = 0$ and $y = -2x^2 + 3$
(2) $x = 0$ and $y = 2x^2 + 3$
(3) $y = 0$ and $y = 2x^2 + 3$
(4) $y = 0$ and $y = -2x^2 + 3$

4 Which set of equations describes the graph?

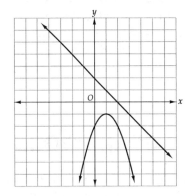

(1) $y = x + 2$ and $(x - 1)^2 + (y - 1) = 0$
(2) $y = x + 2$ and $(x - 1)^2 - (y - 1) = 0$
(3) $x + y = 2$ and $-(x - 1)^2 + (y - 1) = 0$
(4) $x + y = 2$ and $(x - 1)^2 + (y + 1) = 0$

Exercises 5–13: Solve each system of equations graphically and check. If there is no solution, write "no solution."

5 $y = x - 6$
$y = x^2 - 6$

6 $y = x^2 - 2x$
$y = x + 4$

7 $y = -x^2 + 4x + 1$
$y = x - 3$

8 $(x - 3)^2 + (y + 2) = 4$
$y = -2$

9 $y = x^2 + 2$
$y = x + 4$

Graphing Systems of Equations and Inequalities **187**

10 $y = x^2 + 4x + 4$
    $y = -2x - 5$

11 $y = -x^2 + 6x - 4$
    $y = 2x$

12 $y = 2x^2$
    $y = -2x + 4$

13 $2x^2 + 2y = 18$
    $x = 3$

Graphic Solutions of Real Situations

Graphs are used to represent relationships between two variables. They help us understand the relationships and often help us generalize beyond the data supplied. In functions, a value of y is assigned to each value of x. The primary relationship in a function is that the y-value depends on the x-value. For this reason, the x-axis is called the **independent axis** and the y-axis is called the **dependent axis**.

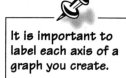

It is important to label each axis of a graph you create.

Drawing the Graph of a Situation In dealing with real situations, it becomes necessary to think of the x-axis as some independent variable, such as time, and to think of the y-axis as a dependent variable, such as height or distance. Whenever it is clear that one variable depends on the other, the dependent variable should be the y and the independent variable should be the x. If a graph is comparing air pressure at different altitudes, then the pressure (y) depends on the altitude (x). But if a graph is charting a mountain climber's progress, then the altitude (y) depends on the time spent climbing (x).

MODEL PROBLEM

Deluxe Limousine Service charges $2 for the initial pickup of a passenger and then $1 per mile. Make a graph of this situation.

SOLUTION

Note that the fee for the ride *depends* upon the number of miles traveled. Therefore, the fee will be represented by the variable y and distance will be represented by the variable x. The equation will then be $y = \$2 + \$1x$.

We can make a table of miles and fees for Deluxe Limousine Service. The label for the x-axis is *miles traveled* and the label for the y-axis is *taxi fee in dollars*.

Miles Traveled x	Taxi Fee in Dollars y
0	2
1	3
2	4

The graph starts where the number of miles traveled is 0. The fee at zero miles is $2. Thus the point (0, 2) is on the graph. A 1-mile trip costs $3 and a 2-mile trip costs $4. When we plot these points, we see a linear pattern. We can extend the graph in the first quadrant and we can read the fees for longer trips.

Answer:

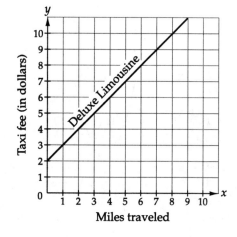

Note: The graph is in the first quadrant only. Why does the graph stop at the y-axis?

Reading the Graph of a Situation On an examination, you may be given the graph of a situation and then asked various questions about the situation. When answering these questions, remember:

- Questions that involve rates (time per room painted, daily charge) or speeds (miles per hour, births per year) will usually require information about the slope.
- Questions that involve starting times, opening deposits, beginning locations, or similar **initial conditions** are usually asking questions about the y-intercept.
- Questions that start "When will…" or "How many…" usually provide a value for one variable and expect you to find the value for the other variable from the corresponding point on the graph.

MODEL PROBLEM

This graph represents the billing structure for Friendly Taxi. What is the initial pickup fee and charge per mile for this service?

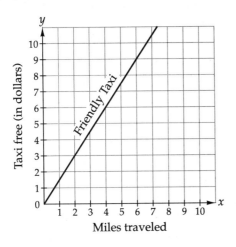

SOLUTION

The initial pickup fee is the charge before any driving is done, in other words when x is 0. The point on the graph with an x value of zero is the origin, (0, 0). Therefore, the pickup fee is $0.

The charge per mile is the change in price per change in miles, or $\frac{\Delta y}{\Delta x}$, the slope. Two points on the line include (0, 0) and (2, 3), so
$$\frac{\Delta y}{\Delta x} = \frac{3-0}{2-0} = \frac{3 \text{ dollars}}{2 \text{ miles}} = 1.5 \text{ dollars per mile}.$$

Answer: There is no pickup fee, and the charge per mile is $1.50.

Parabolas as Graphs of Situations The parabolic shape appears often in real situations. When answering questions about parabolas, remember:

- Questions that involve minimum or maximum values of the dependent variable will usually require you to find the coordinates of the turning point.
- Questions that provide a value for the dependent variable and ask for the corresponding value of the independent variable may have more than one answer.

Situations Involving Graphs of Systems When two equations are graphed on the same coordinate plane, many types of questions involve comparing the graphs.

- Questions that ask "When is situation A better/lower than situation B?" want you to find an inequality involving x that describes what part of the graph satisfies the situation. For example, the answer might be "When the time is longer than 5 days."
- Questions that ask "When are the values the same?" want you to find the coordinates of the point of intersection.
- Questions that ask about differences at specific values of x require you to subtract the y-value of one line from the other.

Graphing Solutions of Real Situations

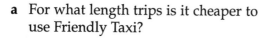

MODEL PROBLEM

Deluxe Limousine Service charges $2 for the initial pickup of a passenger and then $1 per mile. By comparison, Friendly Taxi does not have an initial pickup charge but charges $1.50 per mile. Graph the fee schedule for Deluxe Limousine on the same coordinate plane as the graph for Friendly Taxi. Then answer the following questions:

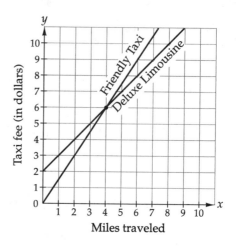

   a For what length trips is it cheaper to use Friendly Taxi?

   b At what mileage is the cost the same for both services? What is the cost?

   c What is the difference in cost at the 6-mile mark?

SOLUTION

The graphs of the costs of each service were discussed earlier in this section. By graphing them on the same coordinate plane, you can answer the three questions.

Answers:

a The graph for Friendly Taxi is below the graph for Deluxe Limousine up to the 4-mile mark. So for trips under 4 miles long, it is cheaper to use Friendly Taxi.

b The lines intersect at (4, 6). So at 4 miles, the cost for both services is $6.

c At 6 miles, the graph shows the cost for Deluxe Limousine is $8 and the cost for Friendly Taxi is $9. The difference is $9 − $8 = $1.

 Practice

1 Terry and Claire had a list of eight books they each had to read over the summer. They decided to race each other to see who would finish first. According to the results in the figure, who reads faster and by how much?

(1) Claire reads 8 books more per week than Terry.
(2) Claire reads 2 books more per week than Terry.
(3) Claire reads 1 book more per week than Terry.
(4) Terry reads twice as many books per week as Claire.

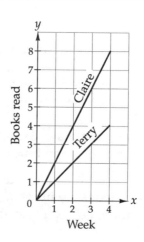

190 Chapter 5: Functions and the Coordinate Plane

2 Alyssa bicycled for 2 hours at 4 miles per hour. She stopped for one hour to visit a friend. She bicycled for another hour at 5 miles per hour. Which graph best represents Alyssa's trip if the horizontal axis is time and the vertical axis is distance?

(1)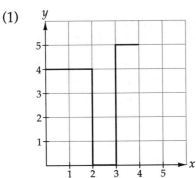

(2)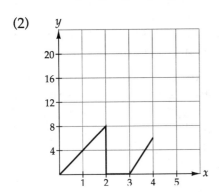

(3)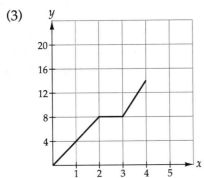

(4)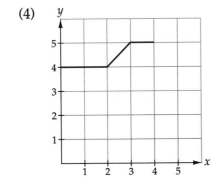

3 In a certain high school, the percentage of seniors who failed to graduate was monitored over a 4-year period. The data can be modeled by the quadratic equation $y = x^2 + 8$, where x is the number of years since the start of the study (the start of the study is at $x = 0$) and y represents the percent of seniors who fail to graduate. If the 4-year study began in 1990, in what year was the percent smallest?

(1) 1990
(2) 1991
(3) 1992
(4) 1993

4 Juan drops a penny from the top of a building at the same time that Carlo releases a balloon from the ground. The equation describing the height above ground of the penny in feet is $y = 161 - 16x^2$ where x is seconds. The equation describing the elevation of the balloon in feet is $y = 4x + 5$ where x is seconds. After how many seconds will the balloon and penny pass each other?

(1) 2 seconds
(2) 3 seconds
(3) 4 seconds
(4) 5 seconds

5 Both Janine and Fran used exercise equipment at their gym. Fran walked on a treadmill and Janine rode a stationary bicycle. Their times and calories burned are shown in the graph below. In calories burned per hour, how much faster was Janine than Fran?

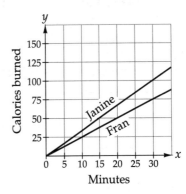

Graphing Solutions of Real Situations

6. The figure below shows the cost of membership for two recreational clubs. For each club, the cost includes an initial fee to join the club plus a monthly charge.

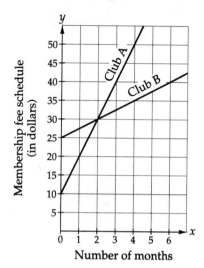

a. What is the initial fee for Club A? For Club B?
b. For which month will the total expenses be the same for both clubs? What is that total cost?
c. What is the monthly charge for Club A? For Club B?

7. Two different health clubs have the following rates. Sammy's Spa charges a flat fee of $350 a year for the use of the club, machines, the pool, and the classes. Shape Up! charges $150 a year for the use of the club, machines, and pool, plus $20 per exercise class.

a. Write an equation to represent each health club's yearly fees.
b. Graph each equation with appropriate labels for the axes.
c. Under what circumstances is it more economical to join Shape Up?

8. An artist drew a rainbow mural on a school wall. The mural is 8 feet wide at the base. Its shape can be represented by a parabola with equation $y = -\frac{1}{2}x^2 + 4x$ where y is the height of the rainbow.

a. Graph the parabola from $x = 0$ to $x = 8$.
b. Determine height y of the rainbow.

9. The height (in feet) of a golf ball hit into the air is given by $h = 64t - 16t^2$, where t is the number of seconds elapsed since the ball was hit.

a. Graph the height of the ball versus time for the first 4 seconds.
b. What is the maximum height of the ball during the first 4 seconds?
c. How long will it take for the ball to reach its maximum height?

10. During a tropical storm, an antenna broke loose from the roof of a building 144 feet high. Its height h above the ground after t seconds is given by $h = -16t^2 + 144$. Graph the height of the antenna with respect to time until it hits the ground.

11. When an arrow is shot into the air, its height h (in feet) above the ground is given by $h = -16t^2 + 32t + 5$ where t is the time elapsed in seconds. If the arrow hit a building after 0.75 of a second, what was the maximum height of the arrow?

CHAPTER REVIEW

1 Which graph does *not* represent a function?

(1)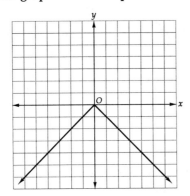

(2)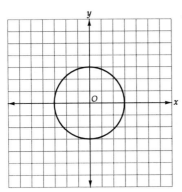

(3)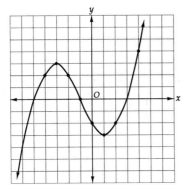

(4)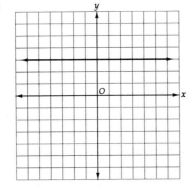

2 What is the slope of the line in the graph?

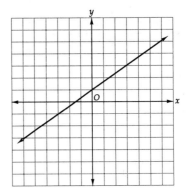

(1) $\dfrac{3}{2}$

(2) 1

(3) $\dfrac{2}{3}$

(4) $-\dfrac{2}{3}$

3 The line in this graph passes through which point?

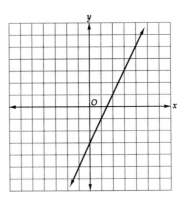

(1) (12, 21)
(2) (12, 22)
(3) (14, 23)
(4) (14, 24)

4 What is the equation of this line?

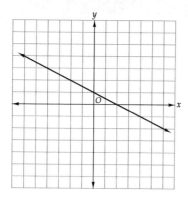

(1) $y + 2 = -2(x - 2)$

(2) $y - 2 = -\frac{1}{2}(x + 2)$

(3) $y + 2 = -\frac{1}{2}(x - 2)$

(4) $y - 2 = -2(x + 2)$

5 Which parabola is the graph of $y + x^2 - 2 = 0$?

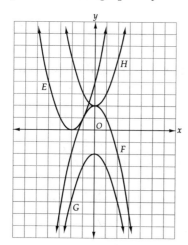

(1) parabola E
(2) parabola F
(3) parabola G
(4) parabola H

6 If x varies directly as y and $x = 2.4$ when $y = 6$, which is a possible ordered pair for (x, y)?

(1) (24, 6)
(2) (0.4, 1)
(3) (0.4, 2.5)
(4) (1, 2.4)

7 Categorize the seven lines ($a - g$) in the figure by slope as positive, negative, zero, or undefined.

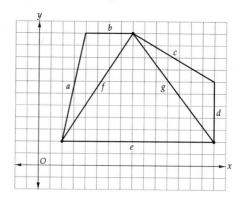

8 What is the slope of the line representing the equation $x + 3y + 6 = 0$?

9 What is the solution of the following system of equations?

$x - y = -5$
$x + y = -3$

10 On the same coordinate plane, graph the following system of inequalities:

$y < 3x + 1$
$y \leq x - 1$

11 A rocket carrying fireworks is launched from a hill 80 feet above a lake. The rocket will fall into the lake after exploding at its maximum height. The rocket's height h above the surface of the lake is given by $h = -16t^2 + 64t + 80$, where t is time in seconds.

   a What is the maximum height of the rocket?
   b How long will it take for the rocket to hit the lake?

12 Solve the following system of equations graphically and check.

$x - y = 5$
$y = -2x + 4$

13 Given the equation $y = -\frac{1}{2}x^2 + x + 4$:

   a Draw the axis of symmetry and write its equation.
   b Find the turning point and label its coordinates.
   c Graph the quadratic equation.
   d Using the graph, determine the roots of the equation.

14. Graph $y > 0$ and $x \leq 0$ on the same coordinate plane and label the solution set of both equations S. Are points $(0, 0)$ and $(4, -3)$ in the solution set?

15. Find the equation for the line passing through points $(2, -3)$ and $(-2, 5)$. Show whether or not the line will pass through point $(8, -13)$.

16. Find the y-intercept of the line passing through the point $(9, 2)$ with a slope of $-\frac{1}{3}$.

17. Use the graphing method to find the solution set for this quadratic-linear pair.
    $y = x^2 - 6x + 5$
    $y = 2x - 10$

18. Solve the system of equations graphically and check.
    $3x + 2y = -5$
    $2x + 4y = 2$

19. A regional telephone service provider offered two different plans for unlimited monthly local service. The graph below represents the total monthly cost of plan A and plan B.

    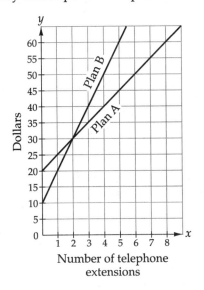

    a. If the cost per month includes a base fee and a charge for each extension to the primary phone, what is the base fee for plan B?
    b. What is the charge per extension for plan A?
    c. For what number of extensions is the cost the same for both plans?

20. A pirate ship shot a cannonball onto an island. The path of the cannonball can be represented with the equation $y = \frac{1}{3}x(10 - x)$ where the horizontal distance x and the height y are in yards and the x-axis is sea level. The slope of the island up from the water can be represented by $y = \frac{2}{3}x - 3$.

    a. Graph both equations on the same axes. (The quadratic equation is already factored for you.)
    b. What will be the highest elevation the cannonball reaches?
    c. How far will the cannonball have traveled horizontally when it strikes the island?
    d. What will be the elevation above sea level when the cannonball strikes the island?
    e. Check your answers for c and d by substituting in the original equations.

Now turn to page 412 and take the Cumulative Review test for Chapters 1–5. This will help you monitor your progress and keep all your skills sharp.

Chapter 6:

Foundations of Geometry

Basic Elements

Geometry is a branch of mathematics concerned with the properties, position, measurement, and relationships of points, lines, angles, planes, and solids.

Points, Lines, and Planes

Most terms in mathematics are **defined** by referring to more basic terms that have already been defined. But as we go back to concepts that are more and more basic, we eventually come to a starting place where terms are **undefined**. In geometry, the point, the line, and the plane are undefined. The characteristics of these three concepts are described below.

- **Point**: A location in space. A point has *no* dimensions—no length, no width, no height. It has only position. We represent a point as a dot and usually name it with a capital letter, such as point P.

  • P

- **Line**: An infinite set of points, with no endpoints. A **straight line**, which could be represented by a string stretched tight, extends endlessly in two opposite directions. (Unless otherwise described, lines are assumed to be straight.) A line has only one dimension, length. We can name a line by any two points on it. For example, if the two points are A and B, we have line AB or $\overleftrightarrow{AB}$. We can also name a line by a single lowercase letter: line ℓ.

  A **curved line** is a line of which no portion is straight.

- **Plane**: A set of points forming a flat surface that extends infinitely in all directions. A plane has two dimensions, length and width. It has *no* thickness.

All other terms in geometry are defined in terms of points, lines, and planes. Here are some key definitions.

- **Figure**: A point or a set of points.
- **Collinear points**: A set of points all lying on the same line. (Noncollinear points do not all lie on the same line.) Any two points are collinear.

- **Coplanar points**. A set of points all lying in the same plane. (Noncoplanar points do not all lie in the same plane.) Any three points are coplanar.

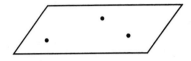

- **Coplanar lines**: Lines lying in the same plane. (Lines that do not lie in the same plane are called **skew** lines.)

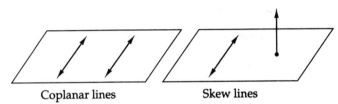

- **Ray**: Part of a line, consisting of one **endpoint** and all the points on one side of that endpoint. A ray is named by the endpoint and any other point. If the endpoint is A and the ray passes through B, we can refer to the figure as ray AB or $\overrightarrow{AB}$. (The endpoint is the first letter.)

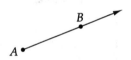

$\overleftrightarrow{AB}$: line
$\overrightarrow{AB}$: ray
$\overline{AB}$: line segment
AB: measure of a segment

- **Line segment**: Part of a line, consisting of two endpoints and all the points on the line between them. A line segment is named by its endpoints. If the endpoints are A and B, we have line segment AB or $\overline{AB}$, which we could also call $\overline{BA}$. (Either endpoint can be the first letter.) The measure—or measured length—of the segment is the distance between A and B and is written AB or BA.

- **Congruent line segments** have the same measure. The symbol for congruence is $\cong$. $\overline{AB} \cong \overline{CD}$ compares segments and indicates that they have the same length. $AB = CD$ compares the numerical measures of the segments and states that they are equal.

Congruence is reviewed further at the end of this chapter.

 **Practice**

1. Which statement is true of ray KL?
   (1) It extends infinitely in both directions.
   (2) Its endpoint is L.
   (3) It contains exactly two points, K and L.
   (4) It is a portion of a line beginning with point K.

2. What does FG represent?
   (1) length of a line
   (2) length of a line segment
   (3) length of a ray
   (4) a plane

3. Which *must* be collinear?
   (1) points A, B, and C
   (2) points A and C
   (3) lines LM and NO when $\overline{LM} \cong \overline{NO}$
   (4) skew lines ST and UV

4. We can measure the length of:
   (1) point X
   (2) $\overleftrightarrow{XY}$
   (3) $\overline{XY}$
   (4) none of the above

Exercises 5–10: Draw each of the following.

5. Line segment JK
6. Ray ZX
7. Noncollinear points Q, R, S, T
8. Coplanar lines ℓ, m, and n
9. Skew lines s and t
10. Curved line segment OP

Drawing reasonable sketches of geometric figures is a useful skill that gets easier with practice.

Angles

An **angle** (∠) is formed by two rays that share an endpoint. The rays are called its **sides**, and the endpoint is called the **vertex**. We can also say that angles are formed by **interesecting lines**—lines that meet or cross. In this case the vertex is the point of intersection.

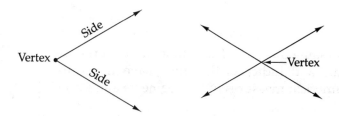

There are several ways to **name** an angle:

- By a capital letter that names its vertex, such as ∠B.
- By a lowercase letter or a number placed inside the angle, such as ∠x or ∠1.

198 Chapter 6: Foundations of Geometry

- By three capital letters, such as ∠ABC. The middle letter is the vertex, and the other two letters name points on different rays (or lines), in either order: ∠ABC can also be named ∠CBA.

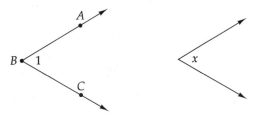

Angles are measured by **degrees** (°) and are classified by their measure. *The measure of angle ABC is 50 degrees* can be written m∠ABC = 50. **Congruent angles** are equal in measure. Above, ∠1 ≅ ∠x.

To review measurement of angles, see Chapter 8.

Classification of Angles

Angle	Defintion
Right angle	An angle of exactly 90°. The symbol for a right angle is a square at the vertex.
Acute angle	An angle less than 90°.
Obtuse angle	An angle greater than 90° but less than 180°.
Straight angle	An angle of exactly 180°. A straight angle is a line.
Reflex angle	An angle greater than 180° but less than 360°.

 MODEL PROBLEMS

1 Refer to the following angle:

a Name this angle in four different ways.
b Name its vertex and its sides.

SOLUTIONS

a Four names for the angle:
∠1, ∠S, ∠RST, ∠TSR
b Vertex: S. Sides: $\overrightarrow{SR}$ and $\overrightarrow{ST}$.

Basic Elements **199**

2 Refer to the illustration below:

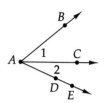

a Why is ∠A an incorrect name for any of the angles shown?

b Why is ∠EAD an incorrect name for ∠2?

c Give four correct names for ∠2.

SOLUTIONS

a All three angles have the same vertex, A. Therefore, the name ∠A does not identify any one of them.

b ∠EAD is an incorrect name for ∠2 because D and E are on the same side of the angle.

c Four names for ∠2 are: ∠CAD, ∠CAE, ∠DAC, ∠EAC.

 Practice

1 Choose the correct name or names for ∠1.

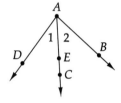

(1) ∠DAE
(2) ∠A and ∠DAE
(3) ∠CEA and ∠DAE
(4) ∠A, ∠CEA, and ∠DAE

2 In which case are points X, Y, and Z collinear?

(1) ∠XYZ is acute.
(2) ∠XYZ is a 90° angle.
(3) ∠XYZ is obtuse.
(4) ∠XYZ is a 180° angle.

3 The measure of a right angle is:

(1) less than 90°
(2) at least 90°
(3) equal to 90°
(4) approximately 90°

4 An obtuse angle could be made up of:

(1) a straight angle plus an acute angle
(2) three acute angles
(3) two smaller obtuse angles
(4) a smaller obtuse angle plus a right angle

Exercises 5 and 6: Refer to the illustration below.

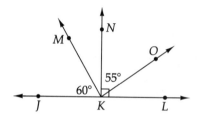

5 Name: 5 acute angles, 2 obtuse angles, 2 right angles, 1 straight angle.

6 How many degrees does each of the following have: ∠LKO, ∠MKN, ∠MKO?

Exercises 7–10: Draw and label an angle to fit each description.

7 Acute angle DEF

8 Obtuse angle with sides $\overrightarrow{BR}$ and $\overrightarrow{BT}$

9 Right angle V

10 Reflex angle GHI

Line and Angle Relationships

The following table reviews important relationships of angles and lines.

Relationship	Description	Examples
Adjacent angles	Two angles having the same vertex and sharing one side.	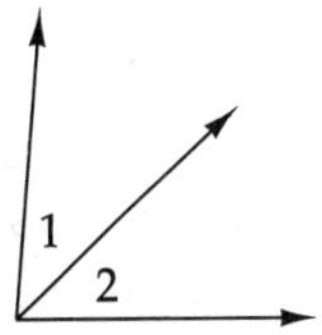 ∠1 and ∠2 are adjacent.
Complementary angles	Two angles, adjacent or nonadjacent, whose sum is 90°.	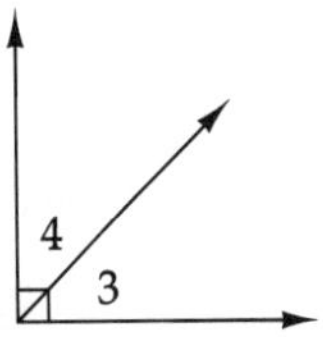 ∠3 and ∠4 are complementary.
Supplementary angles	Two angles, adjacent or nonadjacent, whose sum is 180°.	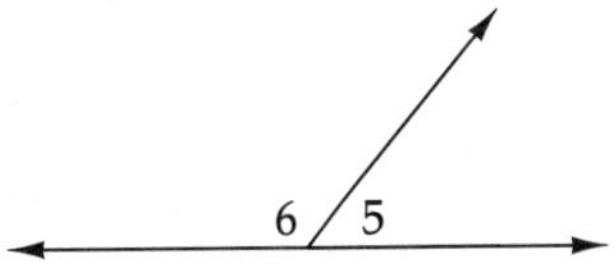 ∠5 and ∠6 are supplementary.
Congruent angles	Angles having the same measure.	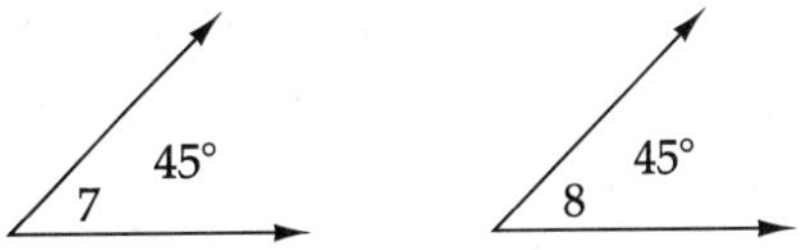 ∠7 and ∠8 are congruent; ∠7 ≅ ∠8
Vertical angles	Opposite angles formed by intersecting lines. Vertical angles are congruent: their measures are equal.	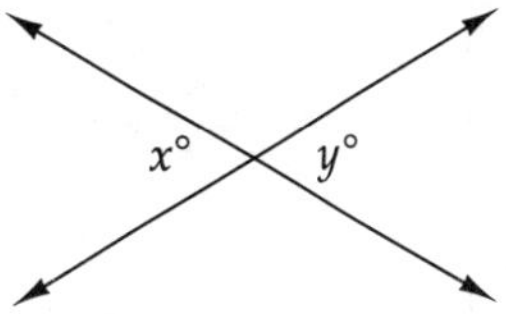 $x = y$. ∠x and ∠y are vertical angles.
Bisector	A line that divides a line segment or an angle into two equal (congruent) parts.	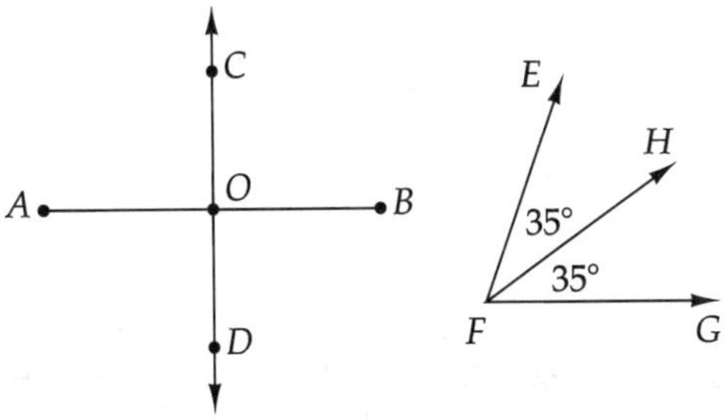 If line CD bisects $\overline{AB}$, then $\overline{AO} \cong \overline{OB}$. If line FH bisects ∠EFG, then ∠EFH ≅ ∠HFG.

(continues)

Basic Elements **201**

Relationship	Description	Examples
Perpendicular (⊥) lines	Lines that form right angles.	If line BC ⊥ line AC, then ∠C measures 90°.
Parallel (‖) lines	Lines that are the same perpendicular distance apart and do not intersect.	$AB = CD$ If $AB = CD$, then line ℓ_1 ‖ line ℓ_2.
Transversal	A line that intersects two parallel lines.	Transversal t crosses lines ℓ and m.
Corresponding angles	Angles in the same relative position when a transversal intersects two parallel lines.	Corresponding angles: ∠1 and ∠5 ∠3 and ∠7 ∠2 and ∠6 ∠4 and ∠8
Alternate interior angles	Angles on opposite sides of a transversal, inside the two parallel lines.	Alternate interior angles: ∠4 and ∠6 ∠3 and ∠5
Alternate exterior angles	Angles on opposite sides of a transversal, outside the two parallel lines.	Alternate exterior angles: ∠1 and ∠7 ∠2 and ∠8

Relationship	Description	Examples
Consecutive interior angles	Inside angles on the same side of a transversal.	(diagram showing two lines cut by a transversal with angles 1,2,3,4 at upper intersection and 5,6,7,8 at lower intersection) Consecutive interior angles: ∠4 and ∠5 ∠3 and ∠6
Parallel lines cut by a transversal	Any pair of angles formed by the intersection of a transversal and two parallel lines are either congruent or supplementary. When parallel lines are cut by a perpendicular transversal, all angles formed are right angles.	(diagram showing parallel lines ℓ and m cut by a transversal with angles 1,2,3,4 and 5,6,7,8) If line ℓ ∥ m, then: • Corresponding angles are congruent (for example, ∠1 ≅ ∠5). • Pairs of alternate interior and exterior angles are congruent (for example, ∠4 ≅ ∠6 and ∠1 ≅ ∠7). • Consecutive interior angles are supplementary (for example, ∠4 + ∠5 = 180°).

Note: In working with intersecting lines and transversals, remember that pairs of vertical angles are congruent and pairs of adjacent angles are supplementary.

MODEL PROBLEMS

1 If $\overleftrightarrow{AB} \parallel \overleftrightarrow{CD}$ and t is not perpendicular to them, which pair of angles are NOT congruent?

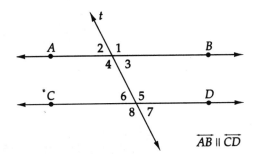

$\overleftrightarrow{AB} \parallel \overleftrightarrow{CD}$

(1) ∠1 and ∠4
(2) ∠2 and ∠7
(3) ∠6 and ∠3
(4) ∠3 and ∠5

SOLUTION

Consider each pair.

Choice (1): ∠1 ≅ ∠4 because they are vertical angles.

Choice (2): ∠2 ≅ ∠7 because they are alternate exterior angles.

Choice (3): ∠6 ≅ ∠3 because they are alternate interior angles.

Choice (4): ∠3 and ∠5 are same-side interior angles, so they are supplementary. Two supplementary angles are congruent only if each measures 90°, and that is impossible here, because t is not perpendicular to lines AB and CD.

Answer: (4)

Basic Elements **203**

2 Which are adjacent angles?
  (1) ∠2 and ∠3
  (2) ∠7 and ∠8
  (3) ∠4 and ∠5
  (4) ∠2 and ∠7

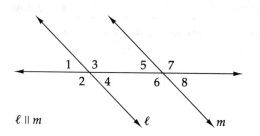
ℓ ∥ m

SOLUTION

Choice (2) is the answer, because ∠7 and ∠8 have the same vertex and a common side.

 Practice

1 Two parallel lines are cut by a nonperpendicular transversal. Which are *not* congruent?

  (1) same-side exterior angles
  (2) corresponding angles
  (3) alternate interior angles
  (4) alternate exterior angles

2 If m∠3 = 50, what is m∠8?

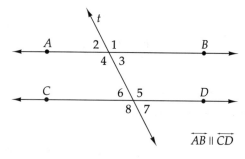

  (1) 40
  (2) 50
  (3) 90
  (4) 130

3 In the figure below, if m∠8 = 120, what is m∠1?

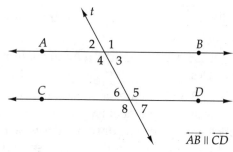

  (1) 60
  (2) 120
  (3) 180
  (4) 300

4 In the figure below, if m∠6 = 60, what is m∠4?

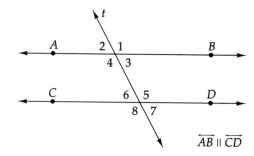

  (1) 30
  (2) 60
  (3) 120
  (4) 180

5 Which choice lists all the angles that are supplementary to ∠13 in the figure below?

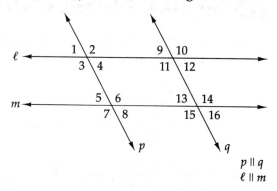
p ∥ q
ℓ ∥ m

  (1) ∠1, ∠4, ∠5, ∠8, ∠16
  (2) ∠2, ∠3, ∠6, ∠7, ∠10, ∠11, ∠14, ∠15
  (3) ∠1, ∠4, ∠5, ∠8, ∠9, ∠12, ∠16
  (4) ∠1, ∠2, ∠3, ∠4, ∠5, ∠6, ∠7, ∠8

204   Chapter 6: Foundations of Geometry

Exercises 6–10: Find the value of x.

6
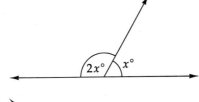

7
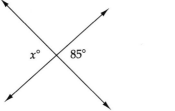

8
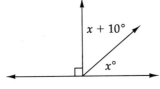

9
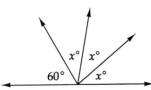

10
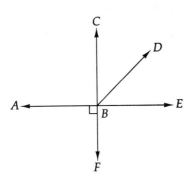

11 In the figure below, $\overline{BD}$ bisects $\angle CBE$. What is the measure of $\angle ABD$?

12 Draw lines ℓ_1, ℓ_2, and ℓ_3 if $\ell_1 \parallel \ell_2$ and $\ell_1 \perp \ell_3$.

13 If $\angle x$ and $\angle y$ are complementary, $\angle y$ and $\angle z$ are supplementary, and $m\angle x = 50$, what is $m\angle z$?

14 In the figure below, $\ell_1 \parallel \ell_2$. Fill in the missing angle values.

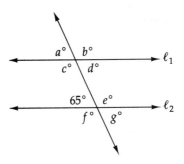

15 In the figure below, $\ell_1 \parallel \ell_2$ and $\ell_3 \parallel \ell_4$. What are the values of w, x, y, and z?

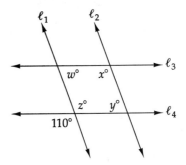

16 Find the indicated measures.

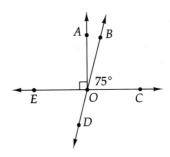

a $m\angle DOE$
b $m\angle AOB$
c $m\angle DOC$

Plane Figures

A **plane figure** is a two-dimensional figure. All its points lie in the same plane. Two-dimensional geometric figures include circles and polygons. Typical polygons are triangles and quadrilaterals.

Note: Measurement of plane figures is reviewed in Chapter 8.

Plane Figures   **205**

Circles

A **circle** (⊙) is the set of all points in a plane that are the same distance from a point called the **center**. A circle is named by its center: for example, ⊙C. Two or more circles with the same center are called **concentric circles**.

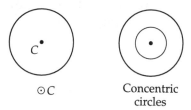

⊙C   Concentric circles

Here are some important definitions and relationships.

- **Chord**—A line segment with both endpoints on the circle.
- **Radius**—A line segment with one endpoint at the center of the circle and the other on the circle. The plural of *radius* is **radii**. Radius $r = \frac{1}{2}$ diameter d.
- **Diameter**—A chord that passes through the center of the circle. All diameters are chords, but only those chords that pass through the center are diameters. Diameter $d = 2r$.
- **Semicircle**—Half a circle. A diameter divides a circle into two semicircles.
- **Arc**—Part of a circle. An arc is identifed with the symbol ⌢.
- **Degrees of arc**—An arc (like an angle) can be measured in degrees. For example: $m\widehat{AB} = 90$. A full circle has 360°.
- **Central angle**—An angle formed by two radii. Its sides (the radii) **intercept** an arc. The vertex of a central angle is the center of the circle. The measure of a central angle equals the measure (degrees) of the intercepted arc.
- **Inscribed angle**—An angle whose vertex is on the circle and whose sides are chords. The sides intercept an arc. The measure of an inscribed angle equals $\frac{1}{2}$ the measure (degrees) of the intercepted arc.
- **Circumference**—The distance (length) around a circle.
- **Arc length**—A part of the circumference, measured in linear units.

  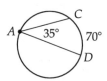

Radii: $\overline{OC}, \overline{OA}, \overline{OB}$
Diameter: $\overline{AB}$
Chords: $\overline{AB}, \overline{DE}$

Central angle: $\angle AOB$
Intercepted arc: $\widehat{AB}$
$m\angle O = m\widehat{AB}$

Inscribed angle: $\angle DAC$
Intercepted arc: $\widehat{DC}$
$m\angle A = \frac{1}{2}m\widehat{CD}$

Chapter 6: Foundations of Geometry

- **Tangent**—A line coplanar with a circle and intersecting the circle at exactly one point. A line tangent to two or more circles is a **common internal tangent** or a **common external tangent**.

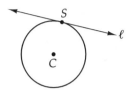

ℓ is tangent to $\odot C$.

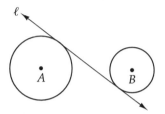 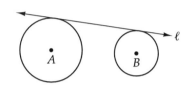

Common internal tangent   Common external tangent

Circles are reviewed further in Chapters 8 and 9.

MODEL PROBLEMS

1 Draw a circle with center C. Then draw and label: radius CR; diameter ST; and chord $\overline{XY}$ that is not a diameter.

SOLUTION

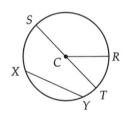

2 Circle D has a radius of 6 centimeters. Locate each of these points inside, outside, or on the circle:

Point A is 7 centimeters from D.

Point B is 5 centimeters from D.

Point C is 6 centimeters from D.

SOLUTIONS

Point A is outside $\odot D$.

Point B is inside $\odot D$.

Point C is on $\odot D$.

Practice

1 Which statement is *false*?

(1) Every diameter of a circle is a chord.
(2) The longest chord of any circle is a diameter.
(3) Every radius of a circle is a chord.
(4) A circle with a radius of 6 centimeters can have a chord of 10 centimeters.

2 Which statement is *not always* true?

(1) If the radius of a circle is 6 centimeters, its diameter is 12 centimeters.
(2) If point A is on $\odot C$, only one radius of $\odot C$ contains point A.
(3) If points A and B are outside a circle, $\overline{AB}$ intersects the circle at two points.
(4) Every diameter of $\odot C$ contains point C.

Plane Figures

3 In ⊙P, radius = 5 centimeters, PQ = 6 centimeters, and PR = 7 centimeters. Which statement is true?

   (1) Points Q and R are inside the circle.
   (2) Points Q and R are on the circle.
   (3) Points Q and R are outside the circle.
   (4) The distance between Q and R is 1 centimeter.

4 Points A and B are both inside ⊙C, and its radius = 6 centimeters. Which statement can *not* be true?

   (1) AB = 14 centimeters
   (2) AB = 8 centimeters
   (3) AB = 6 centimeters
   (4) AB = 4 centimeters

5 Concentric circles *always* have:

   (1) equal radii
   (2) the same center
   (3) inscribed angles
   (4) all of the above

Exercises 6–11: Refer to the illustration (⊙C) and name the following.

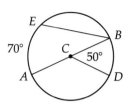

6   3 radii
7   1 diameter
8   2 chords
9   3 arcs
10 1 central angle
11 1 inscribed angle

12 Explain why any inscribed angle that intercepts a semicircle is a right angle.

Polygons: General Properties

A **polygon** is a closed figure formed by three or more coplanar line segments joined at their endpoints. The line segments are the **sides** of the polygon, and the endpoints are its **vertices**. A polygon is named by the letters of its vertices, starting with any vertex and going in order in either direction: for instance, square ABCD.

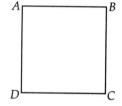

Polygons are classifed by number of sides. The table lists some common polygons:

Polygons

Number of sides	Name	Example
3	Triangle	
4	Quadrilateral	
5	Pentagon	

(continues)

Number of sides	Name	Example
6	Hexagon	
8	Octagon	
10	Decagon	
n	n-gon	16-gon

A **regular** polygon is both equilateral and equiangular: all its sides are congruent, and all its angles are congruent.

Remember: The symbols used in geometric drawings give information about lines and angles:

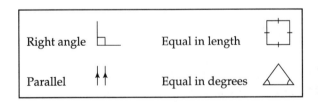

A **diagonal** of a polygon is a line segment connecting one vertex to any other nonconsecutive vertex.

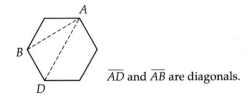

$\overline{AD}$ and $\overline{AB}$ are diagonals.

Interior angles of a polygon are inside it. In **convex polygons**, such as those in the table on pages 208–209, each interior angle measures less than 180°.

Exterior angles of polygons are formed by extending one of two adjacent sides. An exterior angle is supplementary to its adjacent interior angle. Shown below are interior and exterior angles of a regular pentagon, a regular hexagon, and a regular octagon.

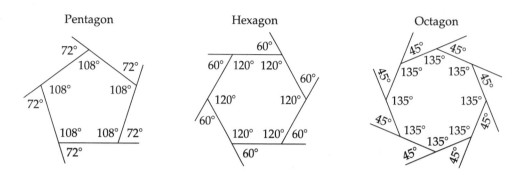

All convex polygons have the following angle properties:

- The sum of the measures of the exterior angles is 360°.
- The sum of the measures of the interior angles is $180(n - 2)$, where n is the number of sides.

Regular polygons have additional properties. Where n is the number of sides:

- The measure of an exterior angle is $\dfrac{360}{n}$.
- The measure of an interior angle is $\dfrac{180(n - 2)}{n}$.

Remember: Pairs of exterior and interior angles are supplementary.

If every vertex of a polygon is on the same circle, we say that the polygon is **inscribed** in the circle, or that the circle is **circumscribed** about the polygon. Each angle of an inscribed polygon is an inscribed angle of the circle. Below, $ABCD$ is inscribed in $\odot O$, and $\odot O$ is circumscribed around $ABCD$.

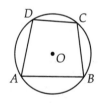

If every side of a polygon is tangent to the same circle, we say that the polygon is circumscribed around the circle or the circle is inscribed in the polygon. Here, $WXYZ$ is circumscribed around $\odot C$, and $\odot C$ is inscribed in $WXYZ$:

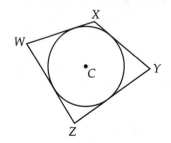

MODEL PROBLEMS

1. Given a regular pentagon, find the following degree measures.

SOLUTIONS

a Sum of the exterior angles.   360°. This is the sum of the exterior angles in any convex polygon.

b Sum of the interior angles. $180(n - 2) = 180(5 - 2) = 180 \cdot 3 = 540°$

c An exterior angle. $\dfrac{360}{n} = \dfrac{360}{5} = 72°$

d An interior angle.   Use the formula:

$$\dfrac{180(n - 2)}{n} = \dfrac{180(5 - 2)}{5} = \dfrac{180 \cdot 3}{5} = 108°$$

Or find the supplement of the exterior angle found in c:

$180° - 72° = 108°$

2. How many sides does a regular polygon have if the measure of an exterior angle is 15°?

SOLUTION

Use the formula: measure of an exterior angle $= \dfrac{360}{n}$ where $n =$ number of sides. Then:

$15 = \dfrac{360}{n}$

$15n = 360$

$n = 360 \div 15 = 24$

Answer: 24 sides

3. If the sum of the measures of the interior angles of a regular polygon is 1,080°, what is the measure of one exterior angle?

SOLUTION

First, find the number n of sides by using the formula for the sum of the angles.

$180(n - 2) = 1,080$

$\dfrac{180(n - 2)}{180} = \dfrac{1,080}{180}$

$n - 2 = 6$

$n = 8$

Next, use the formula for an exterior angle:

$\dfrac{360}{n} = \dfrac{360}{8} = 45$

Answer: 45°

Plane Figures

Practice

1. A polygon is classified by its number of sides. This is the same as the number of:
   (1) interior angles
   (2) vertices
   (3) both of the above
   (4) neither of the above

2. In a convex polygon, the measure of each interior angle is:
   (1) greater than 180°
   (2) less than 180°
   (3) approximately 180°
   (4) at most 180°

3. If every side of a polygon is tangent to the same circle:
   (1) the polygon is inscribed in the circle
   (2) the circle is inscribed in the polygon
   (3) every vertex of the polygon is on the circle
   (4) each angle of the polygon is an inscribed angle of the circle

4. A regular polygon is:
   (1) equilateral but not necessarily equiangular
   (2) equiangular but not necessarily equilateral
   (3) both equilateral and equiangular
   (4) not necessarily equilateral or equiangular

5. In a polygon, a vertex is connected to any other nonadjacent vertex by:
   (1) a side
   (2) an interior angle
   (3) an exterior angle
   (4) a diagonal

6. Pairs of interior and exterior angles in a polygon are:
   (1) always equal
   (2) never equal
   (3) complementary
   (4) supplementary

7. How many sides does a regular polygon have if the measure of an exterior angle is 36°?

8. What is the measure of an exterior angle of a square?

9. A STOP sign is a regular octagon. What is the measure of an interior angle?

10. The sum of the measures of the interior angles of a polygon is 1,440°. Name the polygon.

11. The Pentagon in Washington, D.C., is built in the shape of a regular pentagon. What is the sum of the measures of the interior angles?

12. A swimming pool has a hexagonal shape. However, it is not a regular hexagon. The deep end of the pool has three congruent interior angles that are twice as large as the three congruent angles at the shallow end. Find the measure of an interior angle at the shallow end.

13. What is the sum of the interior angles of a 20-gon?

14. If the sum of the interior angles of a regular polygon is 720°, what is the measure of one exterior angle?

15. Suppose that you draw a regular hexagon and draw diagonals connecting each pair of opposite vertices. Name the figures formed in the interior of the hexagon.

Triangles

Following are key definitions and facts about triangles.

- **Triangle** (△): A closed figure formed by three line segments.

- **Side**: One of the line segments making up the triangle.
- **Vertex**: A point where two sides of the triangle meet.
- **Interior angle**: An angle within the triangle. A triangle has three interior angles.
- **Exterior angle**: An angle formed outside the triangle by one side and the extension of the adjacent side.
- **Altitude, or height**: A line segment with one endpoint at any vertex of the triangle, extending to the line containing the opposite side, and perpendicular to that side.

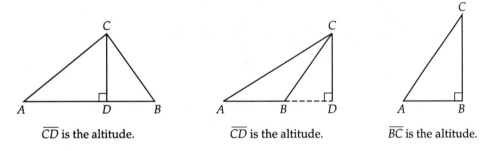

- **Median**: A line segment with one endpoint at any vertex of the triangle, extending to the **midpoint** (middle) of the opposite side.

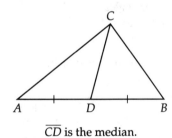

$\overline{CD}$ is the median.

- **Angle bisector**: A line segment with one endpoint at any vertex of the triangle, extending to the opposite side so that it bisects (evenly divides) the vertex angle.

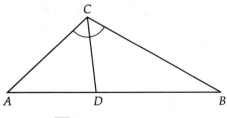

$\overline{CD}$ is the angle bisector.

For all triangles, the following statements are true. Note the examples and applications.

- *The sum of the measures of the interior angles is 180°.* This fact can be used to find the measure of an unknown angle.

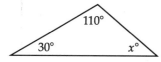

Add up all the angles. $x + 30 + 110 = 180$
Solve for x. $x + 140 = 180$
                                      $x = 40$

- *The longest side is opposite the largest angle.* (Also, the largest angle is opposite the longest side.)

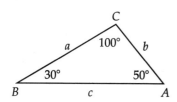

Since $\angle C$ is the largest angle, c is the longest side. The size order of the sides is:
$$b < a < c$$

- *The smallest angle is opposite the shortest side.* (Also, the shortest side is opposite the smallest angle.)

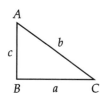

Since c is the shortest side, $\angle C$ is the smallest angle. The size order of the angles is:
$$m\angle C < m\angle A < m\angle B$$

- *An exterior angle of a triangle equals the sum of the measures of the two nonadjacent interior angles.*

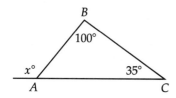

$\angle x$ is the exterior angle. $\angle B$ and $\angle C$ are the nonadjacent interior angles.
$$x = 100 + 35 = 135$$

- *The sum of any two sides of a triangle must be greater than the third side.*

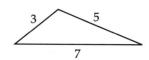 $3 + 5 > 7$ 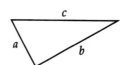 $a + b > c$
                 $5 + 7 > 3$ $b + c > a$
                 $3 + 7 > 5$ $a + c > b$

Also: *Any side of a triangle is greater than the difference of the other two sides.*

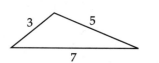 $3 > 7 - 5$ 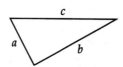 $a > c - b$
                 $5 > 7 - 3$ $b > c - a$
                 $7 > 5 - 3$ $c > b - a$

Therefore, where a, b, and c are sides:

$(a - b) < c < (a + b)$

A triangle can be classified by the number of its congruent (equal) sides or by its angles.

Classification of Triangles by Sides

Name	Description	Example
Scalene triangle	No two sides are equal. No two angles are equal.	Triangle with angles $C = 60°$, $A = 80°$, $B = 40°$
Isosceles triangle	Two sides, called the **legs**, are equal. The third side is the **base**. Two angles, called the **base angles**, are equal. The third angle, called the **vertex angle**, is opposite the base. An altitude drawn from the vertex angle bisects the angle and the base.	Triangle ABC with vertex angle at B, base AC, base angles at A and C. $AB = BC$, $\angle A \cong \angle C$
Equilateral triangle	All three sides are equal. Equilateral triangles are **equiangular**: all angles are equal and each angle $= 60°$.	Triangle with all angles $60°$. $AB = BC = AC$

Classification of Triangles by Angles

Name	Description	Example
Acute triangle	All angles are acute. That is, each angle is less than $90°$.	Triangle with angles $B = 70°$, $A = 50°$, $C = 60°$
Obtuse triangle	One angle is obtuse. That is, one angle is between $90°$ and $180°$.	Triangle with angles $B = 100°$, $A = 50°$, $C = 30°$
Right triangle	One angle is a right ($90°$) angle. The side opposite the right angle is the **hypotenuse**. The other two sides are the **legs**. The **Pythagorean theorem** is true of all right triangles: $a^2 + b^2 = c^2$.	Right triangle with right angle at C, hypotenuse c, legs a and b. $a^2 + b^2 = c^2$

Plane Figures

The Pythagorean theorem is an important geometric relationship. It applies to every right triangle.

Pythagorean Theorem

- In a right triangle, the square of the hypotenuse c equals the sum of the squares of the legs a and b:

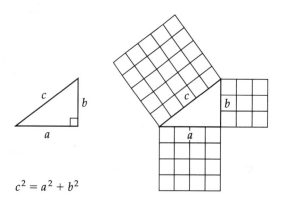

$$c^2 = a^2 + b^2$$

- Conversely, if the lengths of the sides of any triangle satisfy the Pythagorean theorem, that triangle is a right triangle.

- Any three whole numbers that satisfy the equation $c^2 = a^2 + b^2$ form a **Pythagorean triple**. For example:

  3, 4, 5   5, 12, 13   7, 24, 25   8, 15, 17

- Multiples of any set of triples are also triples. For example:

  3, 4, 5 → 6, 8, 10 → 15, 20, 25 …

> The Pythagorean theorem is reviewed further in Chapter 8.

MODEL PROBLEMS

1 Classify each triangle by its sides and by its angles.

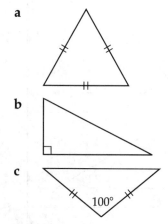

SOLUTIONS

  a Equilateral, acute
  b Scalene, right
  c Isosceles, obtuse

2 For any triangle, the sum of the measures of the interior angles is 180°. In △ABC, m∠A = 52 and m∠B = 16. Classify △ABC as acute, right, or obtuse.

SOLUTION

To classify △ABC, find the measure of the third angle by subtracting the two given angles from 180 degrees (the total):

   m∠C = 180 − (52 + 16) = 180 − 68 = 112

The third angle measures 112°, and an angle of 112° is obtuse. Therefore, △ABC is an obtuse triangle.

216 Chapter 6: Foundations of Geometry

3 Determine whether each set of lengths represents a right triangle:

SOLUTIONS

a 5, 6, 8

$8^2 = 64$ and $5^2 + 6^2 = 25 + 36 = 61$

$8^2 \neq 5^2 + 6^2$

Answer: No

b 30, 24, 18

$30^2 = 900$ and $24^2 + 18^2 = 576 + 324 = 900$

$30^2 = 24^2 + 18^2$

Answer: Yes

Practice

1 Which statement about triangles is true?

(1) If a triangle has two sides of equal length, it has two angles of equal measure.
(2) A triangle can have more than one right angle.
(3) A triangle always has one angle of at least 90°.
(4) The sum of the measures of the interior angles of a triangle is greater than 180°.

2 Which statement about the triangle below is *wrong*?

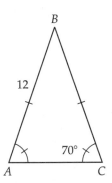

(1) $m\angle B = 40$
(2) $m\angle C + m\angle B = 100$
(3) $AB + BC = 24$
(4) $m\angle A = 70$

3 In the triangle below, if $m\angle A = 40$, what is the degree measure of the vertex angle?

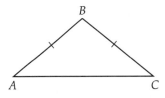

(1) 40
(2) 50
(3) 90
(4) 100

4 Which statement must be *false*?

(1) A scalene triangle has angles measuring 40°, 60°, and 80°.
(2) An acute triangle has angles measuring 60°, 60°, and 60°.
(3) An obtuse triangle has angles measuring 45°, 45°, and 90°.
(4) An isosceles triangle has angles measuring 45°, 45°, and 90°.

5 If the lengths of two sides of a triangle are 2 and 4, which is a possible length of the third side?

(1) 1
(2) 2
(3) 4
(4) 6

Plane Figures

6. The direct distance between city A and city B is 400 miles. The direct distance between city B and city C is 500 miles. Which could be the direct distance between city C and city A?

   (1) 50 miles
   (2) 150 miles
   (3) 950 miles
   (4) 1,050 miles

7. If the ratio of the degree measures of a triangle is 3 : 4 : 11, what is the degree measure of the smallest angle?

8. In △ABC below, $\overline{AB}$ is extended to D, ∠CBD is an exterior angle, m∠CBD = 130, and m∠C = 90. What is m∠CAB?

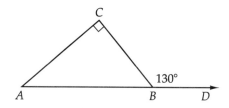

9. In △ABC below, $\overline{AB}$ is extended to point D, m∠A = 45, and m∠C = 85. What is m∠CBD?

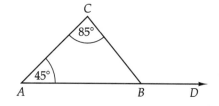

10. ABC and ABD below are isosceles triangles with m∠CAB = 45 and m∠BDA = 50. If AB = AC and AB = BD, what is m∠CBD?

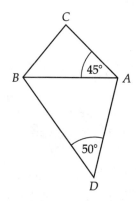

11. In the figure below, what is the value of x?

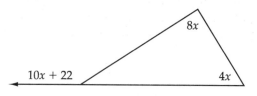

12. Find the length of $\overline{BC}$ in the figure below if BD = 4.

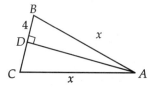

13. Explain why the altitude drawn from the vertex angle to the base of an isosceles triangle also bisects the vertex angle.

14. Richard says that he has measured a triangle and found two obtuse angles. Explain why he is wrong.

15. Find the values of x and y.

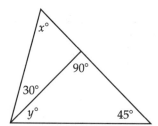

Exercises 16–20: Determine whether the lengths represent the sides of a right triangle:

16. $\sqrt{3}, \sqrt{5}, \sqrt{8}$

17. $4, 4, 4\sqrt{2}$

18. $4\sqrt{3}, 4, 8$

19. 5, 6, 8

20. 6, 12, 17

Quadrilaterals

A **quadrilateral** is a polygon with four sides. Figure *ABCD* is an example of a quadrilateral. Refer to *ABCD* as the parts of a quadrilateral are defined.

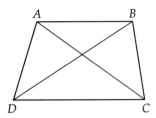

Parts of a Quadrilateral

- **Opposite sides** are sides that do not touch. $\overline{AB}$ and $\overline{CD}$ are opposite sides. $\overline{AD}$ and $\overline{BC}$ are opposite sides.
- **Opposite vertices** are vertices not connected by a side. *A* and *C* are opposite vertices, and *B* and *D* are opposite vertices.
- **Consecutive angles** are angles whose vertices are consecutive. That is, their vertices are next to each other, either clockwise or counterclockwise. In our example, consecutive angles are:

  $\angle ABC$ and $\angle BCD$ $\angle BCD$ and $\angle CDA$ $\angle CDA$ and $\angle DAB$ $\angle DAB$ and $\angle ABC$

- **Opposite angles** are angles whose vertices are not consecutive. $\angle ABC$ and $\angle CDA$ are opposite angles. $\angle BCD$ and $\angle DAB$ are opposite angles.
- **Diagonals** of a quadrilateral are line segments whose endpoints are pairs of opposite vertices. In our example, $\overline{AC}$ and $\overline{BD}$ are the diagonals.
- The **altitude**, or **height**, of a quadrilateral is a line segment extending from a vertex to the line containing the opposite side, and perpendicular to that side. In a rectangle or a square, the height is a side.

Some quadrilaterals have special names and properties. These include parallelograms, rectangles, rhombuses, squares, and trapezoids.

Special Quadrilaterals

Name	Properties
Parallelogram	Opposite sides are parallel: $AB \parallel CD$ and $BC \parallel AD$
	Opposite sides are congruent:
	$AB \cong CD$ $BC \cong AD$
	Opposite angles are congruent:
	$\angle ABC \cong \angle CDA$ $\angle BCD \cong \angle DAB$
	Consecutive angles are supplementary. For example:
	$m\angle ABC + m\angle DAB = 180$ $m\angle ABC + m\angle BCD = 180$
	Diagonals bisect each other:
	$AE \cong EC$ $BE \cong ED$
	A diagonal forms two congruent triangles (triangles of the same size and shape).

(continues)

Name	Properties
Rectangle (figure with vertices G, H, I, F)	Parallelogram in which all angles are right angles: $m\angle FGH = m\angle GHI = m\angle HIF = m\angle IFG = 90$ Diagonals are congruent: $FH \cong GI$
Rhombus (figure with vertices K, L, M, J)	Parallelogram in which all sides are congruent: $JK \cong KL \cong LM \cong MJ$ Diagonals are perpendicular: $JL \perp KM$ Diagonals bisect the angles of the rhombus: $\angle KJL \cong \angle MJL \quad \angle JKM \cong \angle LKM$ $\angle KLJ \cong \angle MLJ \quad \angle LMK \cong \angle JMK$
Square (figure with vertices O, P, Q, N)	Rhombus in which all angles are right angles: $m\angle NOP = m\angle OPQ = m\angle PQN = m\angle QNO = 90$ Also—Rectangle in which all sides are congruent: $NO \cong OP \cong PQ \cong QN$ Diagonals are equal and perpendicular: $NP \cong OQ \quad NP \perp OQ$ Diagonals bisect the angles of the square.
Trapezoid (figure with vertices S, T, U, R)	Has only one pair of parallel sides, called the **bases**: $ST \parallel RU$ The nonparallel sides are the **legs**. If one leg is perpendicular to the bases ($m\angle R = m\angle S = 90$), then the figure is a **right trapezoid**.
Isosceles trapezoid (figure with vertices W, X, Y, V and angles 1, 2, 3, 4)	Trapezoid with congruent legs: $VW \cong YX$ Diagonals are congruent: $VX \cong YW$ **Base angles** are congruent: $\angle 1 \cong \angle 2 \quad \angle 3 \cong \angle 4$

The following chart categorizes special quadrilaterals. The correct or most appropriate name for a quadrilateral is usually the one that *gives the most information about it*—in other words, the most *specific* name.

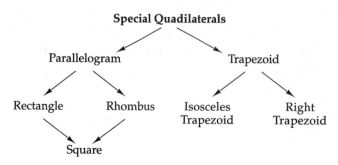

Special Quadrilaterals

220   Chapter 6: Foundations of Geometry

MODEL PROBLEMS

1 BEST is a parallelogram. Find m∠ETS, m∠BTE, and m∠TEB.

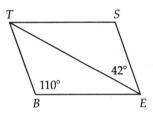

SOLUTION

Since ∠S and ∠B are opposite angles, m∠S = m∠B = 110.

The sum of the angles in △ETS = 180°, so:

   m∠ETS = 180 − (110 + 42) = 180 − 152 = 28

Both ∠BTS and ∠BES are supplementary to both ∠S and ∠B. Thus:

   m∠BTS = m∠BES = 180 − 110 = 70

Parts of each of these angles are already known:

m∠BTS = m∠BTE + m∠ETS	m∠BES = m∠TEB + m∠TES
70 = m∠BTE + 28	70 = m∠TEB + 42
m∠BTE = 70 − 28 = 42	m∠TEB = 70 − 42 = 28

Answer: m∠ETS = 28, m∠BTE = 42, m∠TEB = 28

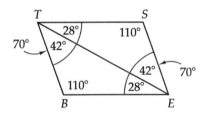

2 In rectangle GROW, how long is RW?

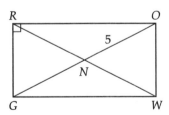

SOLUTION

A rectangle is a parallelogram, and the diagonals of a parallelogram bisect each other. Thus ON is half of GO, so length of GO is 10.

The diagonals of a rectangle are congruent. Therefore, length of RW is also 10.

Answer: 10 units

3 In rhombus *PINK*, the diagonals measure 6 units and 8 units. What is the length of a side?

SOLUTION

Diagonals of a rhombus are perpendicular to each other. Thus, they form 4 right triangles within the rhombus, and for each of these triangles the hypotenuse is a side of the rhombus:

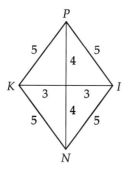

Diagonals of a rhombus bisect each other, so for each right triangle the legs are 3 units and 4 units.

Substitute these values in the Pythagorean theorem, where c is the hypotenuse and a and b are the legs:

$c^2 = a^2 + b^2$

$c^2 = 3^2 + 4^2 = 9 + 16 = 25$

$c = \sqrt{25} = 5$

Answer: 5 units

Shortcut: Recognize the Pythagorean triple 3, 4, 5.

4 The bases of an isosceles trapezoid measure 10 cm and 20 cm. The height (altitude) is 12 cm. How long are the legs?

SOLUTION

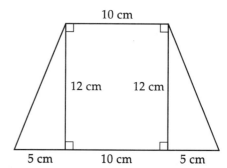

Because the legs are congruent, the shorter base must be centered over the larger base. Thus altitudes drawn from the upper vertices form a rectangle with sides of 10 cm and 12 cm.

The lower base is 10 cm longer than the upper base. This difference must be evenly divided to the left and right of the rectangle, so each leftover segment is 5 cm.

Thus two right triangles are formed with legs 5 cm and 12 cm, and hypotenuse c is a leg of the trapezoid.

Substitute these values in the Pythagorean theorem:

$c^2 = a^2 + b^2$

$c^2 = 5^2 + 12^2 = 25 + 144 = 169$

$c = \sqrt{169} = 13$

Answer: 13 cm

Shortcut: Recognize the Pythagorean triple 5, 12, 13.

 Practice

---

1 Which is *not* a correct name for this figure?

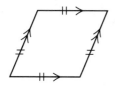

(1) polygon
(2) regular polygon
(3) parallelogram
(4) rhombus

2 Which statement is *false*?

   (1) Every parallelogram is a quadrilateral.
   (2) Every parallelogram is a rectangle.
   (3) Every rhombus is a parallelogram.
   (4) Every rectangle is a parallelogram.

3 Which polygons are always regular?

(A) equilateral triangle	(C) rectangle
(B) square	(D) rhombus

   (1) A and B only
   (2) A, B, and C only
   (3) A, B, and D only
   (4) A, B, C, and D

4 Identify the necessarily true statement or statements.

   (A) A rhombus with right angles is a square.
   (B) A rectangle is a square.
   (C) A parallelogram with right angles is a square.

   (1) A only
   (2) B only
   (3) C only
   (4) A and C only

5 The bases of a trapezoid are:

   (1) the parallel sides
   (2) the nonparallel sides
   (3) always congruent
   (4) always perpendicular to the legs

Exercises 6–9: Name the special quadrilateral that corresponds to each set of diagonals.

6

8.

7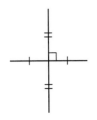

9.

10 The shorter base of an isosceles trapezoid is 8 units, the longer base is 20 units, and each nonparallel side is 10 units. Find the altitude of the trapezoid.

11 Two congruent isosceles triangles are joined at their bases. Which special parallelogram is formed? Explain.

Exercises 12–15: Solve for x.

12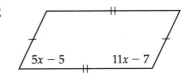

13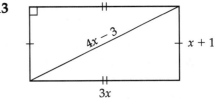

14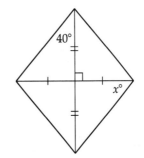

15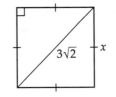

Solid Figures

A **solid figure** is a three-dimensional figure. It has length, width, and thickness. Here are some important definitions for solid figures.

A **polyhedron** is a three-dimensional figure in which each **surface**, called a **face**, is a polygon. The intersections of the faces are the **edges** of the polyhedron. The intersections of the edges are its **vertices**.

A **prism** is a polyhedron formed by connecting the vertices of any two parallel congruent faces.

- The parallel congruent faces of a prism are called the **bases**.
- The surfaces that are not the bases are called **lateral faces**.
- A **right prism** has all its lateral edges perpendicular to the bases.
- A prism is named by the shape of its bases: for example, a triangular prism, a rectangular prism, or a pentagonal prism.
- In a **regular** prism, the bases are regular polygons. In a **rectangular prism**, also called a **rectangular solid**, every face is a rectangle.

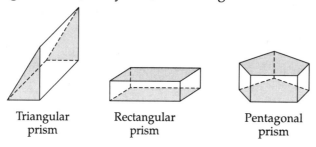

Triangular prism   Rectangular prism   Pentagonal prism

Note: In the illustrations of solids here, bases are shaded and dashed line segments indicate "hidden" edges that cannot be seen.

A **pyramid** is a polyhedron with only one base. All the vertices of the base are connected to a single point called the **vertex of the pyramid**.

- All faces of a pyramid that are not the base are the **lateral faces**.
- A pyramid is named by its base: for instance, a triangular pyramid (**tetrahedron**), a square pyramid, or a pentagonal pyramid.
- **Slant height** of a pyramid, often symbolized ℓ, is the length of the altitude, or height, of a lateral (triangular) face.
- **Height** h of a pyramid is the perpendicular distance from the vertex to the base.
- In a **regular** or **right pyramid**, the base is a regular polygon and the lateral edges are congruent to each other. The line representing the height of a regular pyramid intersects the center of the base.

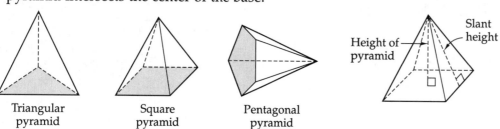

Triangular pyramid   Square pyramid   Pentagonal pyramid

Remember: A polygon is a closed plane figure formed by 3 or more line segments.

224   Chapter 6: Foundations of Geometry

Some geometric solids are related to the circle. (These solids are *not* polyhedrons.)

- **Cylinder**: A circular cylinder has two congruent parallel circles (planes) as its bases. Its cylindrical (curved) surface is called the **lateral surface**. In a **right circular cylinder**, a line connecting the centers of the bases is perpendicular to the plane of each circle.

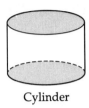
Cylinder

- **Cone**: A circular cone has only one base, which is a circle. Its conical (curved) surface is the lateral surface. The lateral surface has one vertex. In a **right circular cone**, a line connecting the vertex to the center of the circular base is perpendicular to the plane of the base.

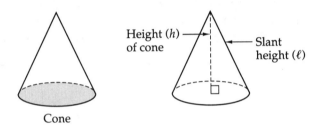
Cone

Slant height ℓ of a cone is the length of a line segment between the vertex and any point on the circumference of the base.

Height h of a cone is the distance from the vertex to the plane of the base.

- **Sphere**: The surface of a sphere is formed by all points in space at a given distance from a point called its **center**.

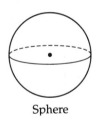
Sphere

Practice drawing sketches of solids. This is a useful problem-solving skill.

MODEL PROBLEMS

  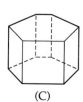

(A) (B) (C)

For each prism (A, B, and C):

1. State the number of faces, edges, and vertices.
2. Describe the shape of the base.
3. Classify (name) the figure.

SOLUTIONS

Prism A
1. 6 faces, 12 edges, 8 vertices
2. Rectangular (or square)
3. Rectangular solid (or cube)

Prism B
1. 5 faces, 9 edges, 6 vertices
2. Bases are triangles.
3. Triangular prism

Prism C
1. 8 faces, 18 edges, 12 vertices
2. Bases are hexagons.
3. Hexagonal prism

Practice

1. Which statement is true?
   (1) A regular (right) pyramid is a prism.
   (2) Every right prism has rectangular bases.
   (3) A cylinder is a prism.
   (4) A cube is a rectangular prism.

2. Which statement is *false*?
   (1) All lateral faces of a right prism are rectangular.
   (2) All lateral faces of a right pyramid are congruent triangles.
   (3) A triangular pyramid has four triangular faces.
   (4) A lateral edge of a right pyramid and the height of the pyramid can be congruent.

3. For which cube or cubes does the drawing indicate that you are seeing the figure from the bottom?

   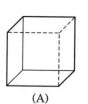

   (A) (B) (C)

   (1) A only
   (2) A and B only
   (3) A and C only
   (4) B and C only

4 Which pattern will *not* fold into a cube?

(1)

(2)

(3)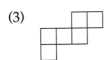

(4)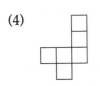

5 Which has more than one base?

(1) cone
(2) cylinder
(3) regular pyramid
(4) tetrahedron

6 Which has *no* vertex?

(1) cylinder
(2) cone
(3) obtuse angle
(4) hexagonal pyramid

7 Name the figure formed by folding each two-dimensional pattern:

a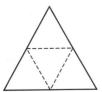

c.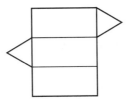

b

d.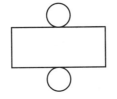

8 This cube is formed by 27 equal smaller cubes.

If the surface of the large cube is painted blue, state how many smaller cubes will have:

a 3 blue faces
b 2 blue faces
c 1 blue face
d no blue faces

9 Sketch each of the following:

a cone
b rectangular prism that is *not* a cube
c triangular prism
d triangular pyramid
e cylinder

10 The French mathematician René Descartes first stated a formula for the relationship between the number of faces, vertices, and edges of any polyhedron. This formula was later proved by a Swiss mathematician, Leonhard Euler. Count the faces, vertices, and edges of the polyhedrons reviewed here. Compare the numbers and write a formula. Let F = number of faces, V = number of vertices, and E = number of edges.

Congruence and Similarity

Congruent Figures

We have already reviewed **congruence** (≅) with regard to lines and angles: congruent lines are equal in length, and congruent angles are equal in measure.

Congruent figures have the same size and the same shape. When plane figures are congruent, one would fit exactly on top of the other. The parts that fit together, or match, are called **corresponding vertices**, **corresponding sides**, and **corresponding angles**.

Here are two important **properties of congruent figures**:

- Corresponding sides are congruent (equal in length).
- Corresponding angles are congruent (equal in measure).

We can also define congruency using a biconditional sentence. For example:

- Two polygons are congruent if and only if their corresponding sides are congruent and their corresponding angles are congruent.

Biconditional sentences are reviewed in Chapter 7.

- Two circles are congruent if and only if their radii are congruent.

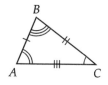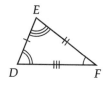

△ABC ≅ △DEF

In a correct **congruence statement**, such as △ABC ≅ △DEF above, corresponding vertices are named in the same order. If △ABC ≅ △DEF, then:

∠A ≅ ∠D $\overline{AB} \cong \overline{DE}$

∠B ≅ ∠E $\overline{BC} \cong \overline{EF}$

∠C ≅ ∠F $\overline{AC} \cong \overline{DF}$

An important application of congruence is showing that two triangles are congruent, given certain information about corresponding parts. This can be done in five different ways.

Proving Triangles Congruent

- **Side-angle-side: SAS ≅ SAS.** Two triangles are congruent if two sides and the included angle of one triangle are congruent, respectively, to two sides and the included angle of the other.

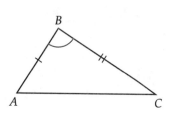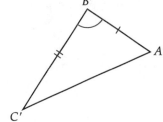

- **Angle-side-angle: ASA ≅ ASA.** Two triangles are congruent if two angles and the included side of one triangle are congruent, respectively, to two angles and the included side of the other.

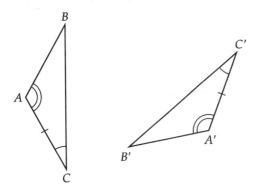

- **Angle-angle-side: AAS ≅ AAS.** Two triangles are congruent if two angles and a non-included side of one triangle are congruent, respectively, to two angles and the corresponding side of the other.

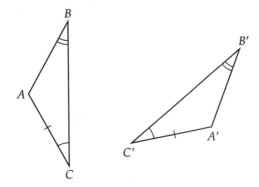

- **Side-side-side: SSS ≅ SSS.** Two triangles are congruent if three sides of one triangle are congruent to three sides of the other.

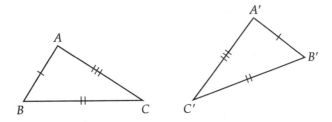

- **Hypotenuse-leg: HL ≅ HL.** This rule applies to right triangles only. Two right triangles are congruent if the hypotenuse and a leg of one triangle are congruent to the hypotenuse and a leg of the other triangle. (Since the Pythagorean theorem allows us to find the third side of a right triangle, HL implies SSS.)

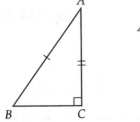 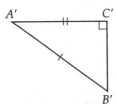

Congruence and Similarity

Note: If any two pairs of corresponding angles of a triangle are congruent, their measures have the same sum. The third angles must then also be congruent, so that the sum of all of the interior angles is 180°. Thus if two pairs of corresponding angles and a nonincluded side are given, that side actually is included between a pair of corresponding congruent angles.

MODEL PROBLEMS

1 △RST ≅ △XYZ. Complete each statement.

SOLUTIONS

$\overline{ST} \cong ?$ $\overline{ST} \cong \overline{YZ}$

$\overline{TR} \cong ?$ $\overline{TR} \cong \overline{ZX}$

$\angle T \cong ?$ $\angle T \cong \angle Z$

$\triangle SRT \cong ?$ $\triangle SRT \cong \triangle YXZ$

$\triangle TSR \cong ?$ $\triangle TSR \cong \triangle ZYX$

2 For each pair of triangles, determine whether △I and △II are congruent.

SOLUTIONS

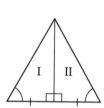

△I ≅ △II by SAS, since they share a side. They are also congruent by ASA and AAS.

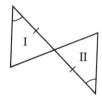

△I ≅ △II by ASA, since vertical angles are congruent.

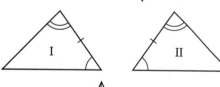

△I ≅ △II by ASA.

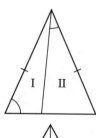

△I and △II are not necessarily congruent, because the congruent angles are not corresponding angles.

△I ≅ △II by hypotenuse-leg (HL), since these two right triangles share a leg.

Chapter 6: Foundations of Geometry

1 △GHI ≅ △LMN. Which is *not* a congruence statement for these triangles?
  (1) △HGI ≅ △MLN
  (2) △HIG ≅ △MNL
  (3) △IHG ≅ △MNL
  (4) △GIH ≅ △LNM

2 Two squares are congruent if and only if:
  (1) any two corresponding angles are congruent
  (2) two pairs of corresponding angles are congruent
  (3) all four pairs of corresponding angles are congruent
  (4) a side of one is congruent to a side of the other

Exercises 3–5: Refer to pairs I to V of triangles.

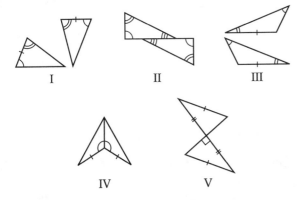

3 Which pair of triangles are congruent by SAS?
  (1) I
  (2) II
  (3) III
  (4) IV

4 Which pair of triangles are congruent by ASA?
  (1) I
  (2) III
  (3) IV
  (4) V

5 Which pair of triangles can *not* be shown to be congruent?
  (1) II
  (2) III
  (3) IV
  (4) V

6 △ADB ≅ △BCA. Complete each statement.

  a $\overline{BD}$ ≅ ?
  b △ABD ≅ ?
  c $\overline{AD}$ ≅ ?
  d △DAB ≅ ?

7 Write a congruence statement for each pair of triangles.
  a
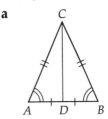

  b
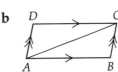

8 △ABC ≅ △EFG

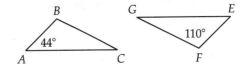

Find the measure of:
  a ∠E
  b ∠B
  c ∠C

9 Provide an example that shows this statement is *false*: "If the angles of one triangle are congruent to the angles of another triangle, then the two triangles are congruent."

Exercises 10 and 11: Draw and label a figure to fit the description.

10 ABCD ≅ MNOP

11 ABCDE ≅ FGHIJ

Congruence and Similarity   **231**

12 In the diagram below, $\overline{AE} \cong \overline{BE}$ and $\overline{CE} \cong \overline{DE}$.

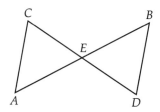

   a Explain why △AEC ≅ △BED.
   b Given the following, find the lengths of the three sides of each triangle:
      AC = 2x
      BD = x + 5
      AE = x + 6
      EC = x + 7

13 In the diagram below, $\overline{AE} \cong \overline{BE}$, △CAE ≅ △BDE, m∠ACE = 5x, and m∠BDE = 7x − 18. Show that △BDE is a right triangle.

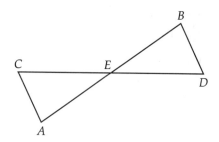

14 Find the value of AB, given the following:
    △ABC ≅ △A'B'C'
    $\overline{AB} \cong \overline{A'B'}$
    $AB = x^2 - 4x$
    $A'B' = x + 14$

15 Alyssa says that two triangles are congruent if two pairs of sides are congruent and one pair of angles are congruent. Is she right? Explain.

Similar Figures

Similar figures have the same shape, but not necessarily the same size. Thus congruent figures are similar, but not all similar figures are congruent. *Same shape* means that similar figures are the same type of figure, and also that one may be a **dilation**—an enlargement or a reduction—of the other. The symbol ~ means *is similar to*.

Here are two important **properties of similar figures**:

- Corresponding angles are congruent.
- Pairs of corresponding sides are **proportional**—that is, in the same ratio. The ratio of lengths of corresponding sides is called the **scale factor**.

Proportions are reviewed in Chapter 2.

Dilations are reviewed in Chapter 9.

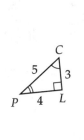 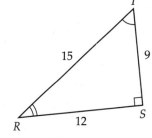

△PLC ~ △RST
Scale factor is 1 : 3

A correct **statement of similarity** for polygons, such as △PLC ~ △RST on the preceding page, names the corresponding vertices in the same order. If △PLC ~ △RST, then:

Corresponding angles	Corresponding sides
∠P ≅ ∠R	$\dfrac{PL}{RS} = \dfrac{4}{12} = \dfrac{1}{3}$
∠L ≅ ∠S	$\dfrac{LC}{ST} = \dfrac{3}{9} = \dfrac{1}{3}$
∠C ≅ ∠T	$\dfrac{PC}{RT} = \dfrac{5}{15} = \dfrac{1}{3}$

The scale factor in this example is $\dfrac{1}{3}$, or 1 : 3.

> The scale factor is sometimes called the ratio of similitude.

As with congruence, an important application of similarity is showing that two triangles are similar, given information about corresponding parts.

Proving Triangles Similar

- **Side-side-side: SSS.** Two triangles are similar if their corresponding sides are in proportion. △ABC ~ △A'B'C' because $\dfrac{a}{a'} = \dfrac{b}{b'} = \dfrac{c}{c'}$.

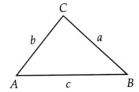
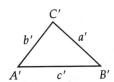

- **Angle-angle: AA.** Triangles are similar if two angles of one triangle are congruent to two angles of the other. Below, △ABC ~ △A'B'C' because each has a 40° angle and an 80° angle. For the sum of the angles to equal 180°, the third angle in each is 60°. Thus all three pairs of angles are congruent.

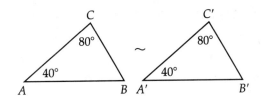

- **Right triangles: acute angle–acute angle.** Two right triangles are similar if an acute angle of one triangle is equal to an acute angle of the other. Below, right triangles ABC and DEF each have 90° and 50° angles, so the third angle for each must be 40°. (This is simply an application of AA similarity for triangles in general.)

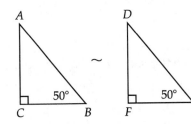

Congruence and Similarity

- **Side-angle-side: SAS.** Two triangles are similar if the ratio of two sides of one is proportional to the ratio of the two corresponding sides of the other, and the included angles are congruent. Below, $\angle B \cong \angle S$ and $\dfrac{AB}{RS} = \dfrac{BC}{SV}$ or $\dfrac{15}{3} = \dfrac{25}{5}$, so $\triangle ABC \sim \triangle RSV$.

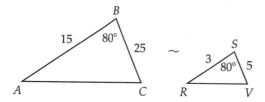

Overlapping Triangles

- A line parallel to one side of a triangle and intersecting the other two sides cuts off a triangle similar to the given triangle. Below, if $\overline{DE} \parallel \overline{AC}$, then $\triangle ABC \sim \triangle DBE$:

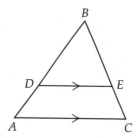

All corresponding angles are congruent, corresponding sides are in proportion, and the segments of the intersected sides are in proportion:

$\angle A \cong \angle BDE$, $\angle B \cong \angle B$, and $\angle C \cong \angle BED$ $\qquad \dfrac{AB}{DB} = \dfrac{BC}{BE} = \dfrac{AC}{DE}$, $\qquad \dfrac{BD}{DA} = \dfrac{BE}{EC}$

(Therefore, if a line divides two sides of a triangle proportionally, that line must be parallel to the third side.)

- If a line segment joins the **midpoints** of two sides of a triangle and is parallel to the third side, that length of the line segment is half the length of the third side. Below, $DE \parallel AC$ and $\triangle DBE \sim \triangle ABC$, so $DE = \dfrac{1}{2}AC$.

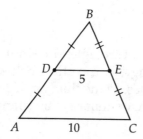

Note: If two triangles are similar, then any pair of corresponding line segments, such as altitudes, medians, and angle bisectors, have the same ratio as any pair of corresponding sides.

234 Chapter 6: Foundations of Geometry

MODEL PROBLEMS

1 △I ~ △II and side $a = 16$

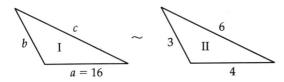

Find the following:
  Scale factor
  Length of side b
  Length of side c

SOLUTIONS

Scale factor $= 16 : 4 = 4 : 1$
Side $b = 4 \cdot 3 = 12$
Side $c = 4 \cdot 6 = 24$

2 George, a surveyor, is using similar triangles to find distance ED across a river:

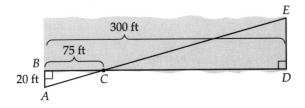

He measures the following: $BD = 300$ feet, $BC = 75$ feet, $AB = 20$ feet. Set up the proportion that George should use, and find ED.

SOLUTION

The proportion is: $\frac{225}{75} = \frac{x}{20}$ where $x = ED$

Cross multiply: $75x = 4{,}500$
Solve: $x = 4{,}500 \div 75 = 60$

Answer: $ED = 60$ feet

3 $DE \parallel AC$. Find x and y.

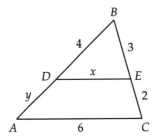

SOLUTION

If $DE \parallel AC$, then $\triangle ABC \sim \triangle DBE$. Set up and solve two proportions.

To find x:
$\frac{3}{5} = \frac{x}{6}$
$5x = 18$
$x = 3\frac{3}{5}$ or 3.6

To find y:
$\frac{4}{y} = \frac{3}{2}$
$3y = 8$
$y = 2\frac{2}{3}$ or $2.\overline{6}$

4 $DE \parallel AC$. Find x.

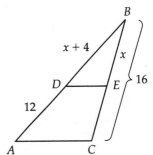

SOLUTION

If $DE \parallel AC$, then $\triangle ABC \sim \triangle DBE$. Set up and solve a proportion.

$\frac{x}{16} = \frac{x+4}{12 + (x+4)}$ or $\frac{x}{16} = \frac{x+4}{(x+16)}$

$16x + 64 = x^2 + 16x$
$x^2 = 64$
$x = \pm 8$

A dimension cannot be negative, so we reject -8.

Answer: $x = 8$

Practice

1. Which rectangles are similar?

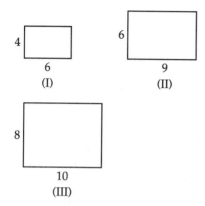

   (1) I, II, and III
   (2) I and II only
   (3) I and III only
   (4) II and III only

2. Which statement is true?

   (1) Corresponding sides of similar figures must be congruent.
   (2) Congruent figures must be similar.
   (3) Any two regular polygons are similar.
   (4) Similar figures must be congruent.

3. Which statement is true?

   (1) All rectangles are similar.
   (2) All rhombuses are similar.
   (3) All isosceles trapezoids are similar.
   (4) All squares are similar.

4. Which statement is *false*?

   (1) All equilateral triangles are similar.
   (2) All regular hexagons are similar.
   (3) All right triangles are similar.
   (4) All circles are similar.

5. If △ABC ~ △DEF, which ratio is *not* equal to 3 : 4?

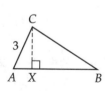 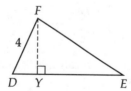

   (1) AB : DE
   (2) m∠B : m∠E
   (3) CX : FY
   (4) BC : EF

6. △MDP ~ △BAC. Find the measure of ∠M, ∠D, ∠B, and ∠C.

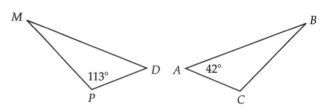

Exercises 7–9: In each case, △I ~ △II. Find the scale factor of △I to △II.

7.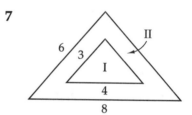

8.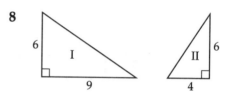

9.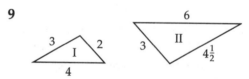

10. △ABC ~ △PMD. Find the lengths of PD and PM.

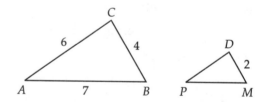

11. ABCD ~ WXYZ. Find WX, XY, and YZ.

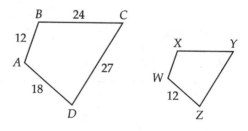

236 Chapter 6: Foundations of Geometry

12 △RMK ~ △BAC. Find RK and BC.

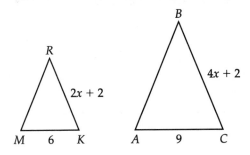

13 △ABC ~ △NQP. Find NQ and NP.

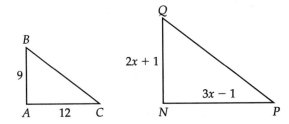

14 The two figures below are similar. Find the length of DS.

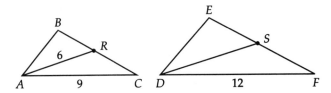

15 △ABC ~ △DEJ. Find these lengths: DJ, BC, EJ.

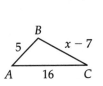 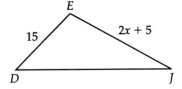

16 In each diagram, the line segment within the triangle is parallel to the third side. Find x in each case:

a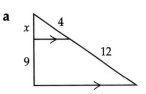

b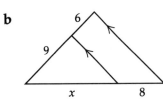

c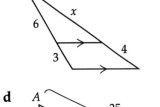

d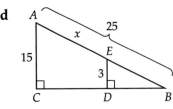

e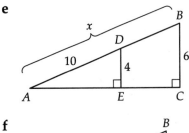

f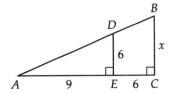

Congruence and Similarity 237

CHAPTER REVIEW

1. If the legs of a right triangle are each $5\sqrt{2}$, what is the length of the hypotenuse?
   - (1) 5
   - (2) 10
   - (3) $10\sqrt{2}$
   - (4) 20

2. A circle is divided into two semicircles by:
   - (1) a tangent
   - (2) a radius
   - (3) a diameter
   - (4) an arc

3. Each surface of a polyhedron is *always*:
   - (1) a base
   - (2) a polygon
   - (3) a regular polygon
   - (4) three-dimensional

4. Slant height ℓ of a circular cone is:
   - (1) a line segment from the vertex to the circumference of the base
   - (2) the distance from the vertex to the center of the base
   - (3) the diameter of the base
   - (4) the lateral surface

5. Find the value of x in each drawing.

   a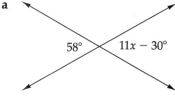

   b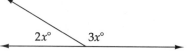

   c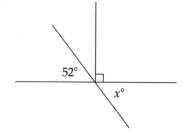

   d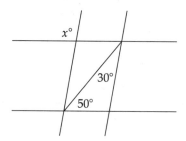

6. a In the figure below, x, y, and z are in the ratio $2 : 3 : 4$. Find the values of x, y, and z.

   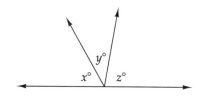

   b In the figure below, $\ell_1 \parallel \ell_2$ and x and y are in the ratio $1 : 2$. Find the values of x and y.

   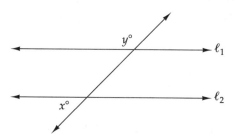

7. a In the figure below, $\ell_1 \parallel \ell_2$. Find the value of y.

   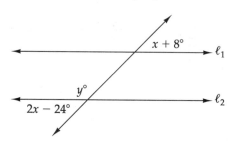

   b In the figure below, $\ell_1 \parallel \ell_2$. Find the value of x.

   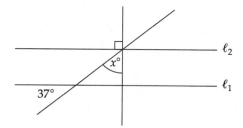

238 Chapter 6: Foundations of Geometry

8 In the figure below, $\ell_1 \perp \ell_2$ and $m = 80$. What is the value of $x - z$?

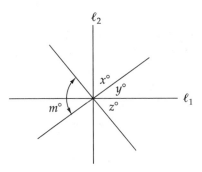

9 Lines ℓ_1, ℓ_2, and ℓ_3 intersect as shown below. What is the value of c in terms of a and b?

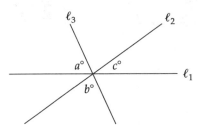

10 Explain why $\triangle ABC \cong \triangle A'B'C'$. Find AC and $A'C'$.

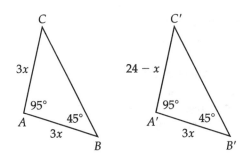

11 $\triangle ABC$ and $\triangle ABD$ below are isosceles triangles with m$\angle CAB = 30$ and m$\angle BDA = 45$. If $AB = AC$ and $AB = BD$, what is m$\angle CAD$?

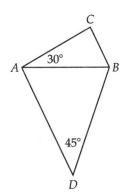

12 Find the sum of the interior angle measures of a convex 12 gon.

13 a Parallelogram *GAME* has a right angle. Explain why *GAME* is a rectangle.

   b Parallelogram *CAKE* has four congruent angles. Explain why *CAKE* is a rectangle.

14 In parallelogram *SONG*, $SO = 2x + 4$, $ON = 3x - 6$, and $NG = 4x - 16$. Explain why *SONG* is a rhombus.

15 If the shorter diagonal, KT, of rhombus *KITE* is 5 and m$\angle KIT = 60$, find the length of a side of the rhombus.

16 If the length of a side of a rhombus is 8 and one of the angles measures 120, find the lengths of each of the diagonals.

17 In rhombus *GOLF*, diagonal GL is congruent to each of the sides. Find the measure of $\angle GOL$.

18 Find x and y in trapezoid *SODA* below.

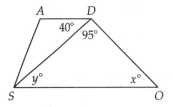

19 Find x and y in trapezoid *TRAP* below.

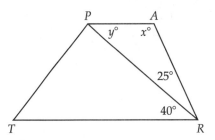

20 Find, w, x, y, and z in isosceles trapezoid *TALK* below.

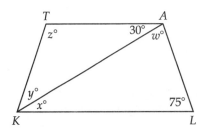

Note: Now turn to page 414 and take the Cumulative Review test for Chapters 1–6. This will help you monitor your progress and keep all your skills sharp.

Chapter 7:
Mathematical Reasoning

Statements, Truth Values, and Negations

Open and Closed Sentences

Logic is the study of deductive reasoning, the process of using mathematical sentences to make decisions. A **mathematical sentence** is always in the form of a simple declarative sentence that states a complete idea.

A **statement** is a sentence that is either true or false. The truth value of a statement is always immediately evident. Questions, commands, or exclamations are sentences that are not statements because they cannot be true or false. If the sentence does not contain variables or replacement pronouns, it is a **closed sentence**. An example of a closed sentence is "Every week has seven days." In this case the statement is true.

Open sentences contain unknown pronouns (*he, she, they, it*) or variables. The truth value of an open sentence cannot be established until that unknown or variable is replaced. Otherwise, open sentences have no truth value. The set of all replacements that will make an open sentence true is called the **solution set** for the open sentence. For example:

Sentence	Unknown	Solution Set
1 $x + 5 = 9$	The variable is x.	The solution set is {4}.
2 He was the sixteenth president of the United States.	The unknown pronoun is *he*.	The solution set is {Abraham Lincoln}.
3 $|x| = 10$	The variable is x.	The solution set is $\{-10, 10\}$.

MODEL PROBLEMS

Label each sentence as a true closed sentence, a false closed sentence, an open sentence, or not a statement.

		SOLUTIONS
1	Sit down and eat your dinner.	Not a statement
2	Why are you wearing a raincoat?	Not a statement
3	All squares are circles.	False closed sentence
4	$6 + 2 = 5$	False closed sentence
5	$4x + 1 = 13$	Open sentence
6	$8 < 4$	False closed sentence
7	A triangle has three sides.	True closed sentence
8	George Washington was the first president of the United States.	True closed sentence
9	$2x - 5 = 3$	Open sentence

Practice

1. "$4 + 1 > 8$" is:
   (1) a true closed sentence
   (2) a false closed sentence
   (3) an open sentence
   (4) not a statement

2. "$x + 3 = 5$" is:
   (1) a true closed sentence
   (2) a false closed sentence
   (3) an open sentence
   (4) not a statement

3. "A hexagon has four sides" is:
   (1) a true closed sentence
   (2) a false closed sentence
   (3) an open sentence
   (4) not a statement

4. "Get out!" is:
   (1) a true closed sentence
   (2) a false closed sentence
   (3) an open sentence
   (4) not a statement

5. "Are you leaving now?" is:
   (1) a true closed sentence
   (2) a false closed sentence
   (3) an open sentence
   (4) not a statement

6. "$x + 3 = 5$ for $x = 2$" is:
   (1) a true closed sentence
   (2) a false closed sentence
   (3) an open sentence
   (4) not a statement

7. Write a sentence that has no truth value.

8. Write a statement about triangles that is true.

9. a Write an open sentence.

   b Rewrite the sentence so that it is a true statement.

   c Rewrite the sentence so that it is a false statement.

Statements, Truth Values, and Negations

Exercises 10–13: Use the replacement set {algebra, geometry, trigonometry, statistics} to rewrite each sentence as a true closed statement.

10 It is the study of analyzing data.

11 It is the study of operations on sets of numbers.

12 It is the study of triangles.

13 It is the study of shapes and sizes.

Negations

A statement and its negation have opposite truth values. In other words, if the statement "A week has seven days" is true, then its **negation**, "A week does not have seven days," is false.

Most explanations of logical relationships assign letters to statements, providing a helpful shorthand for writing them. The most common letters used are p, q, and r. For instance, if p represents "The White House is in Washington, D.C.," then we can say that p is a true statement. The shorthand symbol for negation is $\sim$. If p is a true statement, then $\sim p$ is false. If p is false, then $\sim p$ is true.

Negations can be written in several ways. For example, look at the negation of the statement p: The SAT is given to all fourth-graders. (false statement)

- $\sim p$: It is not true that the SAT is given to all fourth-grade students. (true statement)

- $\sim p$: It is not the case that the SAT is given to all fourth-grade students. (true statement)

- $\sim p$: The SAT is not given to all fourth-graders. (true statement)

A double negation has the same meaning as the original statement. For example "It is not true that I am not hungry" has the same meaning as "I am hungry."

Note that the same symbol $\sim$ is used for both similarity and negation. Look at the context to be sure you use the correct meaning.

 MODEL PROBLEM

Write the negation of the statement q: "Taking the SAT is a requirement for graduation."

SOLUTION

The negation can be stated in several different ways.

$\sim q$: It is not true that the SAT is a requirement for graduation.

$\sim q$: Is not the case that the SAT is a requirement for graduation.

$\sim q$: The SAT is not a requirement for graduation.

242 Chapter 7: Mathematical Reasoning

1. What is the negation of the statement "$4 + 3 > 2$"?
   (1) $4 + 3 < 2$
   (2) $4 + 3 > 1$
   (3) $4 + 3 = 7$
   (4) $4 + 3$ is not greater than 2.

2. What is the negation of the statement "The coat is blue"?
   (1) The coat is green.
   (2) The coat is not blue.
   (3) The coat is sometimes blue.
   (4) The coat is black.

3. If the negation of a statement is true, what is the truth value of the original statement?
   (1) true
   (2) false
   (3) no truth value
   (4) cannot be determined

4. What is the negation of the statement "$3 + 6 = 7$"?
   (1) $3 + 6 \neq 7$
   (2) $3 + 6 = 9$
   (3) $-3 + 6 = -7$
   (4) $3 + 6$ equals 7

Exercises 5–10: Write the negation of each statement and state the truth value of the original statement and the truth value of its negation.

5. Chicago is a city in Indiana.

6. January is a winter month.

7. The area of a rectangle is length times width.

8. Each state has two senators.

9. All real numbers are rational.

10. The sun rises in the west.

Compound Statements

Compound statements can be made by joining simple statements with connectives such as *and, or,* and *if/then.*

Consider the following simple statements.
   It is raining outside.
   I stay at home.

From these two simple statements three basic compound statements can be formed.

- It is raining outside *and* I stay at home.
- It is raining outside *or* I stay at home.
- *If* it is raining outside, *then* I stay at home.

Conjunctions

A **conjunction** is a compound sentence that is formed by connecting two simple sentences using the word *and*. The sentence p and q can be represented in symbols as $p \wedge q$.

   p: Today I go to math class.
   q: I am wearing a blue sweater.
   Conjunction ($p \wedge q$): Today I go to math class *and* I am wearing a blue sweater.

For a conjunction to be true, both parts must be true. If either or both parts are false, the conjunction is false.

Examples

1. p: Texas is larger than New Jersey. (true statement)
   q: Alaska is larger than Colorado. (true statement)
   $p \wedge q$: Texas is larger than New Jersey *and* Alaska is larger than Colorado. (true statement)

2. p: Veterans Day is in August. (false statement)
   q: Mother's Day is in May. (true statement)
   $p \wedge q$: Veterans Day is in August *and* Mother's Day is in May. (false statement)

3. p: Harry Truman is the president of the United States. (false statement)
   q: Napoleon Bonaparte is the president of France. (false statement)
   $p \wedge q$: Harry Truman is the president of the United States *and* Napoleon Bonaparte is the president of France. (false statement)

It is often helpful to summarize the truth values for a conjunction in a truth table.

p	q	$p \wedge q$
T	T	T
F	T	F
T	F	F
F	F	F

Disjunctions

A **disjunction** is a compound sentence that is formed by connecting two simple sentences using the word *or*. The sentence *p or q* can be represented in symbols as $p \vee q$. For the disjunction to be false, both parts must be false. When both parts or either part is true, the disjunction is true.

Using the same three sets of simple sentences discussed for conjunctions, you can see how the truth values differ when *and* is replaced with *or*.

Examples

1. p: Texas is larger than New Jersey. (true statement)
   q: Alaska is larger than Colorado. (true statement)
   $p \vee q$: Texas is larger than New Jersey *or* Alaska is larger than Colorado. (true statement)

2. p: Veterans Day is in August. (false statement)
   q: Mother's Day is in May. (true statement)
   $p \vee q$: Veterans Day is in August *or* Mother's Day is in May. (true statement)

3. p: Harry Truman is the president of the United States. (false statement)
   q: Napoleon Bonaparte is the president of France. (false statement)
   $p \vee q$: Harry Truman is the president of the United States *or* Napoleon Bonaparte is the president of France. (false statement)

The commutative property holds for conjunctions and disjunctions so that $p \wedge q$ has the same truth value as $q \wedge p$ and $p \vee q$ has the same truth value as $q \vee p$.

The truth values for conjunctions and disjunctions can be summarized in a truth table.

p	q	$p \wedge q$	$p \vee q$
T	T	T	T
F	T	F	T
T	F	F	T
F	F	F	F

MODEL PROBLEM

Two is an even number.

Six is an odd number.

Write a conjunction using these two statements.

(1) Two is an even number and six is an odd number.

(2) Two is an even number or six is an odd number.

(3) If two is an even number, then six is an odd number.

(4) If two is not an even number, then six is not an odd number.

Answer: (1)

 Practice

Exercises 1–4: Choose the statement that has the same truth value as the given statement.

1 $4 + 7 = 3$ or $3 + 7 = 10$

  (1) $4 + 7 = 3$ and $3 + 7 = 10$
  (2) $3 + 7 = 10$ and $4 + 7 = 3$
  (3) $3 + 7 = 10$ or $4 + 7 = 3$
  (4) $4 + 7 \ne 3$ and $3 + 7 \ne 10$

2 Paris is the capital of France and is located in South America.

  (1) Paris is not the capital of France and is located in South America.
  (2) Paris is the capital of France and is not located in South America.
  (3) Paris is the capital of France or it is located in South America.
  (4) Paris is the capital of France or it is not located in South America.

3 5 is an odd number or $6 + 4 = 12$.

  (1) 5 is an odd number and $6 + 4 = 12$.
  (2) 5 is not an odd number or $6 + 4 \ne 12$.
  (3) 5 is not an odd number and $6 + 4 = 12$.
  (4) 5 is not an odd number or $6 + 4 = 12$

4 Daylight saving time ends in October and the clock is turned back one hour.

  (1) Daylight saving time does not end in October and the clock is turned back one hour.
  (2) Daylight saving time does not end in October and the clock is not turned back one hour.
  (3) Daylight saving time does not end in October or the clock is not turned back one hour.
  (4) Daylight saving time ends in October or the clock is not turned back one hour.

Exercises 5–9: Find the truth value of the compound sentence. Explain how you arrived at your answer.

5. Twelfth-graders are seniors and hamburger is a soft drink.

6. Twelve is a prime number or 36 is a perfect square number.

7. Nine is an odd number and π is an irrational number.

8. Fifteen is not evenly divisible by 4 or 16 is evenly divisible by 4.

9. July 4th is not a holiday or apples are fruits.

10. Suppose p = Every square has four sides.
    q = Every triangle has three sides.

    Write a disjunction using these two statements and describe its truth value. Explain your reasoning.

11. Let p = The sum of the measures of the angles of every triangle is 180°.
    q = Every triangle has four angles.

    Write a conjunction using these two statements and describe its truth value. Explain your reasoning.

Conditionals

A **conditional** is a compound sentence that is formed by connecting two simple sentences using the words *If . . . , then . . . .* In a standard conditional, the sentence between the words *if* and *then* is called the **premise**, the **hypothesis**, or the **antecedent**. The sentence after the word *then* is called the **conclusion** or the **consequent**. A conditional statement can be represented in symbols as $p \rightarrow q$ and is read "If p then q" or "p implies q."

Examples

1. Hypothesis (p):   A triangle is equilateral.
   Conclusion (q):   All of the angles have the same measure.
   Conditional ($p \rightarrow q$):   *If* a triangle is equilateral, *then* all of the angles have the same measure.

2. Hypothesis (p): $3x = 12$
   Conclusion (q): $x = 4$
   Conditional ($p \rightarrow q$):   *If* $3x = 12$, *then* $x = 4$.

3. Hypothesis (p):   I bring a jacket.
   Conclusion (q):   I will be warm.
   Conditional ($p \rightarrow q$):   *If* I bring a jacket, *then* I will be warm.

Conditional statements are not always written in "*If . . . then . . . ."* form. However, they can be rephrased to "*If . . . then . . . ."* form.

Examples

1. A quadratic equation is easy to solve if it is factored.
   Conditional form: *If* a quadratic equation is factored, *then* it is easy to solve.
   Hypothesis: A quadratic equation is factored.
   Conclusion: It is easy to solve.

2. The sides of a rhombus are congruent.
   Conditional form: *If* the figure is a rhombus, *then* the sides are congruent.
   Hypothesis: The figure is a rhombus.
   Conclusion: The sides are congruent.

3 The measure of an exterior angle of a triangle is found by adding the measures of the two nonadjacent interior angles.

Conditional form:	*If* you add the measures of the two nonadjacent interior angles of a triangle, *then* you find the measure of the exterior angle.
Hypothesis:	You add the measures of the two nonadjacent interior angles of a triangle.
Conclusion:	You find the measure of the exterior angle.

The truth value of a conditional statement is false if and only if the hypothesis is true and the conclusion is false. The following examples show the correct truth values if today is Friday.

Examples

1. p: Today is Friday. (true statement)
   q: Tomorrow is Saturday. (true statement)
   $p \rightarrow q$: If today is Friday, then tomorrow is Saturday. (true statement)

2. p: Today is Friday. (true statement)
   q: Tomorrow is Monday. (false statement)
   $p \rightarrow q$: If today is Friday, then tomorrow is Monday. (false statement)

3. p: Today is Tuesday. (false statement)
   q: Tomorrow is Saturday. (true statement)
   $p \rightarrow q$: If today is Tuesday, then tomorrow is Saturday. (true statement)

4. p: Today is Thursday. (false statement)
   q: Tomorrow is Monday. (false statement)
   $p \rightarrow q$: If today is Thursday, then tomorrow is Monday. (true statement)

The following table summarizes the truth values for conditionals:

p	q	$p \rightarrow q$
T	T	T
T	F	F
F	T	T
F	F	T

> The last two lines of this truth table usually invite some discussion. Since the hypothesis is false, the statement cannot be disproved.

 MODEL PROBLEM

Both parts of the statement "If Jane lives in California, then she lives in Los Angeles" are true. Choose the statement that is *false*.

(1) If Jane doesn't live in California, then she doesn't live in Los Angeles.

(2) If Jane lives in Los Angeles, then Jane lives in California.

(3) If Jane doesn't live in Los Angeles, then she doesn't live in California.

(4) If Jane lives in Los Angeles, then she does not live in California.

Answer: (4) In this case the hypothesis is true and the conclusion is false. Therefore, the conditional is false, as shown in the truth table above.

1. Identify the hypothesis in the conditional statement.

   You will feel better if you take this medicine.

   (1) You will feel better.
   (2) You take this medicine.
   (3) You will not feel better.
   (4) You will not take this medicine.

2. Identify the conclusion in the conditional statement.

   I will stay in this afternoon if the teacher assigns homework.

   (1) I will stay in this afternoon.
   (2) The teacher assigns homework.
   (3) I will not stay in this afternoon.
   (4) The teacher doesn't assign homework.

Exercises 3–5: Identify the truth value to be assigned to the sentence.

3. If $3^2 = 9$, then $3^3 = 27$.

   (1) true
   (2) false
   (3) no truth value
   (4) cannot be determined

4. If $13 + 1 > 14$, then $17 + 1 > 18$.

   (1) true
   (2) false
   (3) no truth value
   (4) cannot be determined

5. "If x is divisible by 6, then it is divisible by 3" is true for every integer x.

   (1) true
   (2) false
   (3) no truth value
   (4) cannot be determined

Exercises 6–10: Identify each statement as a negation, conjunction, disjunction, or conditional.

6. If 6 is an even number, 9 is greater than 6.

7. Chicago is a city and it is in the state of Indiana.

8. If it is January, it must be cold.

9. The figure is a rectangle or its area is length times width.

10. All real numbers are not rational.

Exercises 11–15: Find a value of x that will make the statement *false*. Explain your reasoning.

11. If x is divisible by 6, then it is divisible by 4.

12. If x is a prime number, then it is odd.

13. If the slope of a line is x, the line goes from upper left to lower right.

14. If x is a perfect square, then it is even.

15. If the square of x is rational, then x is rational.

Exercises 16 and 17: Identify the truth value to be assigned to the conditional. Explain your reasoning.

16. If every rectangle has only three sides, then $2 + 3 = 6$.

17. $4 > 10$ if $10 < 15$.

18. Given the statements

    Hypothesis: The sun is shining.
    Conclusion: It is not raining.

    Describe the possible truth values of the hypothesis, the conclusion, and the conjunction formed from these simple statements.

Converses, Inverses, and Contrapositives

Given the conditional statement

   If I am 18 years old, I can vote.

Hypothesis: I am 18 years old.
Conclusion: I can vote.

Three new conditionals can be formed from the given statement.

The **converse** of the given conditional is formed by interchanging its hypothesis and its conclusion. The converse of the original statement is

   If I can vote, then I am 18 years old.

The **inverse** of the given conditional is formed by negating both its hypothesis and its conclusion. The inverse of the original statement is

   If I am not 18 years old, then I cannot vote.

The **contrapositive** of the given conditional is formed by interchanging and negating its hypothesis and conclusion. The contrapositive of the original statement is

   If I cannot vote, then I am not 18 years old.

The truth values of these statements can best be summarized in a truth table where p is the hypothesis and q is the conclusion. The symbol $\sim$ indicates negation and $\rightarrow$ indicates a conditional statement. The truth value of each conditional is determined by the truth value of its components.

p	q	$\sim p$	$\sim q$	Conditional $p \rightarrow q$	Converse $q \rightarrow p$	Inverse $\sim p \rightarrow \sim q$	Contrapositive $\sim q \rightarrow \sim p$
T	T	F	F	T	T	T	T
T	F	F	T	F	T	T	F
F	T	T	F	T	F	F	T
F	F	T	T	T	T	T	T

Two statements are **logically equivalent** when they always have the same truth value. Note that the truth values in the conditional and contrapositive columns are always the same. This means that the statements are logically equivalent. The truth values in the converse and inverse columns are also always the same, so these statements are also logically equivalent.

Knowing the truth value of a conditional statement does not imply anything about the truth value of the converse or the inverse.

Examples

Conditional	If two angles are consecutive angles of a parallelogram, the sum of their measures is 180°.	True
Converse	If the sum of the measures of two angles is 180°, the angles are consecutive angles of a parallelogram.	Not always true
Inverse	If two angles are not consecutive angles of a parallelogram, the sum of their measures is not 180°.	Not always true

MODEL PROBLEM

Form the converse, inverse, and contrapositive of the conditional statement.

If the polygon has only three sides, then the polygon is a triangle.

SOLUTION

Hypothesis: The polygon has only three sides.
Conclusion: The polygon is a triangle.
Converse: If the polygon is a triangle, then the polygon has only three sides.
Inverse: If the polygon does not have only three sides, then the polygon is not a triangle.
Contrapositive: If the polygon is not a triangle, then the polygon does not have only three sides.

In this special case, all the above statements are true.

 Practice

1. Choose the converse of the statement "If the traffic light is red, then you stop your car."
   (1) If you stop your car, then the traffic light is red.
   (2) If you do not stop your car, then the light is not red.
   (3) If the traffic light is not red, then you stop your car.
   (4) If the traffic light is red, then you do not stop your car.

2. Choose the contrapositive of the statement "If two numbers are both even, then their sum is even."
   (1) If two numbers are not both even, then their sum is not even.
   (2) If two numbers are not both even, then their sum is even.
   (3) If the sum of two numbers is not even, then the numbers are not both even.
   (4) If the sum of two numbers is not even, then the numbers are both even.

3. Choose the inverse of the statement "If the merry-go-round was oiled, then it will not squeak."
   (1) If the merry-go-round was not oiled, then it will squeak.
   (2) If the merry-go-round was not oiled, then it will not squeak.
   (3) If the merry-go-round was oiled, then it will squeak.
   (4) If the merry-go-round squeaks, then it was not oiled.

4. Choose the converse of the statement "If it is summer, we are on vacation."
   (1) If it is not summer, we are not on vacation.
   (2) If we are on vacation, it is summer.
   (3) It is summer and we are on vacation.
   (4) If we are not on vacation, it is not summer.

5. Choose the contrapositive of the statement "If a triangle has three right angles, it is equilateral."
   (1) If a triangle is equilateral, it has three right angles.
   (2) If a triangle does not have three right angles, it is equilateral.
   (3) If a triangle does not have three right angles, it is not equilateral.
   (4) If a triangle is not equilateral, it does not have three right angles.

6. Choose the inverse of the statement "If I turned on the air conditioner, then the house is cool."
   (1) If the house isn't cool, then I didn't turn on the air conditioner.
   (2) If I didn't turn on the air conditioner, then the house isn't cool.
   (3) If the house isn't cool, then I didn't turn on the air conditioner.
   (4) If the house is cool, then I didn't turn on the air conditioner.

7 p: x is divisible by 5 q: x is divisible by 10

Which of the following is true for $x = 25$?

(1) conjunction, $p \land q$
(2) disjunction, $p \lor q$
(3) conditional, $p \to q$
(4) contrapositive, $\sim q \to \sim p$

Exercises 8–17: Write the converse, the inverse, and the contrapositive for each conditional statement.

8 If it is September, school is open.

9 If the figure is a square, it has four sides.

10 If I have cookies with my lunch, someone will trade desserts.

11 If I typed my paper, my teacher will be able to read it.

12 If I did not set my alarm, I'll be late to school.

13 If I don't go to practice, I will not be in the starting lineup.

14 If I don't make honor roll, I will be unhappy.

15 If I go shopping, I will not have money for the movies.

16 If I live close to school, I won't take a school bus.

17 If I study French, I don't study Latin.

18 Given the statement "The measures of congruent segments are equal."

a Rewrite the statement as a conditional and state its truth value.

b Write the converse and state its truth value.

Biconditionals

A **biconditional** is a compound statement formed by the conjunction of a conditional statement and its converse. A biconditional is written using *if and only if*.

Example

Conditional ($p \to q$):   If two intersecting segments form a right angle, then they are perpendicular.

Converse ($q \to p$):   If two intersecting segments are perpendicular, then they form a right angle.

Biconditional ($p \leftrightarrow q$):   Two intersecting segments form a right angle *if and only if* they are perpendicular.

When both original statements have the same truth value—that is, when they are both true or both false—the biconditional is true. If the statements have different truth values, the biconditional is false. The truth values for the biconditional can be summarized in this truth table.

p	q	$p \to q$	$q \to p$	$p \leftrightarrow q$
T	T	T	T	T
T	F	F	T	F
F	T	T	F	F
F	F	T	T	T

MODEL PROBLEMS

1 Given the following, write the conditional, converse of the conditional, and the biconditional.

p: Two lines are parallel.

q: Two lines never meet no matter how far they are extended.

SOLUTION

Conditional ($p \to q$): If two lines are parallel, then the two lines never meet no matter how far they are extended.

Converse ($q \to p$): If two lines never meet no matter how far they are extended, then the two lines are parallel.

Biconditional ($p \leftrightarrow q$): Two lines are parallel *if and only if* the two lines never meet no matter how far they are extended.

2 For what integer values of x is the statement "$x > 7$ if and only if $x > 5$" false?

SOLUTION

The biconditional statement means

If $x > 7$, then $x > 5$ *and* if $x > 5$ then $x > 7$.

"If $x > 7$, then $x > 5$" is true for any x since 7 is greater than 5, but

"If $x > 5$, then $x > 7$" is not true for $x = 6$ or 7.

In this case, since "If $x > 7$, then $x > 5$" is true and "If $x > 5$, then $x > 7$" is false, the whole statement is false.

Answer: $x = 6$ or 7

3 What is the truth value of the statement "A triangle has a 90° angle if and only if the triangle is a right triangle"?

SOLUTION

This means "if a triangle has a 90° angle, then the triangle is a right triangle" *and* "if the triangle is a right triangle, the triangle has a 90° angle." Both statements are true, so the biconditional is true.

4 Biconditionals can be used to show that statements are logically equivalent.

Show that the two statements are equivalent.

Conditional ($p \to q$): If a number is even, then the number is divisible by 2.

Disjunction ($\sim p \lor q$): A number is not even or the number is divisible by 2.

SOLUTION

p: A number is even.

q: The number is divisible by 2.

The two statements being tested are $p \to q$ and $\sim p \lor q$.

The biconditional to test is $(p \to q) \leftrightarrow (\sim p \lor q)$.

To be equivalent, two statements must always have the same truth values and the biconditional must always be true.

The truth table below can be used to show the truth values of the two statements and the biconditional.

In each case, the truth values are the same so the biconditional is true. Therefore, the statements are logically equivalent.

p	q	$\sim p$	$p \to q$	$\sim p \lor q$	$(p \to q) \leftrightarrow (\sim p \lor q)$
T	T	F	T	T	T
T	F	F	F	F	T
F	T	T	T	T	T
F	F	T	T	T	T

Practice

1. Identify the truth value of the statement.
   $x + 4 = 28$ if and only if $x = 24$.
   (1) true for any value of x
   (2) false for any value of x
   (3) true for some values of x, false for others
   (4) not a statement

2. Identify the truth value of the statement.
   $x + 3 = 9$ if and only if $x + 5 = 12$.
   (1) true for any value of x
   (2) false for any value of x
   (3) true for some values of x, false for others
   (4) not a statement

3. Identify the truth value of the statement.
   In Euclidean geometry any two lines are skew if and only if the lines never meet.
   (1) true
   (2) false
   (3) truth cannot be determined
   (4) not a statement

4. The statement "I am sleepy if and only if I did not get eight hours of sleep" is logically equivalent to:
   (1) If I am sleepy, then I did not get eight hours of sleep *and* if I did not get eight hours of sleep, then I am sleepy.
   (2) I am sleepy or I did not get eight hours of sleep.
   (3) I am sleepy if I did not get eight hours of sleep.
   (4) I am sleepy and I did not get eight hours of sleep.

5. The statement "If Paul buys chicken, he roasts it" is logically equivalent to:
   (1) Paul buys chicken or he roasts it.
   (2) Paul buys chicken and he does not roast it.
   (3) Paul does not buy chicken or he roasts it.
   (4) Paul does not buy chicken and he does not roast it.

6. A biconditional statement is formed by:
   (1) the disjunction of two conditionals
   (2) the conjunction of a conditional and its contrapositive
   (3) the conjunction of a conditional and its inverse
   (4) the conjunction of a conditional and its converse

7. Write this statement as a conditional statement that is logically equivalent to it:
   Either I don't have money or I will earn it.

8. Write the converse of the statement "If today is Thursday, tomorrow is Friday" and give the truth value of the converse.

9. Write the converse of the statement "If a number is a repeating decimal, it is rational." Use the conditional and the converse to write a biconditional statement. For what numbers is the biconditional false?

10. Write the contrapositive of the statement "If two angles are complementary, the sum of their measures is 45°." Describe a pair of angles that would make both statements false. Describe another pair that would make both statements true.

Biconditionals

CHAPTER REVIEW

1. What is the truth value of the statement?
   $9 + 4 > 10$ or $7 - 3 = 2$
   (1) true
   (2) false
   (3) no truth value
   (4) cannot be determined

2. What is the truth value of the statement?
   If $6 + 7 = 11$, then $9 - 6 > 4$.
   (1) true
   (2) false
   (3) no truth value
   (4) cannot be determined

3. If a statement is false, what is the truth value of the negation of its negation?
   (1) true
   (2) false
   (3) no truth value
   (4) cannot be determined

4. Given the statements
   Harry is a horse.   (true)
   Harry is an animal.   (true)
   Find the truth value of the statement
   Harry is not a horse and Harry is not an animal.
   (1) true
   (2) false
   (3) no truth value
   (4) cannot be determined

5. Given the true statement "If I do not set my alarm, I will be late for school," which statement must also be true?
   (1) If I am late for school, I did not set my alarm.
   (2) If I set my alarm, I will not be late for school.
   (3) If I am not late for school, then I did set my alarm.
   (4) If I am late for school, then I set my alarm.

6. What can be said about the statement "If the figure is a square, it is a rhombus" and its converse "If the figure is a rhombus, it is a square"?
   (1) The statement is always true but its converse cannot be determined.
   (2) The statement is always false but its converse is always true.
   (3) Both the statement and its converse are always false.
   (4) Both the statement and its converse are always true.

Exercises 7 and 8: Identify the hypothesis and the conclusion.

7. The square of the longest side of a triangle is equal to the sum of the squares of the other sides if the triangle has a 90° angle.

8. If $\angle A$ and $\angle B$ are alternate interior angles, they are congruent.

Exercise 9–13: Use the statements below.
Daffodils are animals.   (false)
Daises are flowers.   (true)
Find the truth value of each statement:

9. Daffodils are animals and daisies are flowers.

10. Daffodils are not animals and daisies are flowers.

11. Daffodils are animals or daisies are flowers.

12. Daffodils are animals or daisies are not flowers.

13. If daffodils are not animals then daisies are not flowers.

14. For the conditional statement "If the altitude bisects the base, the triangle is isosceles," write the inverse, the converse, and the contrapositive. Then use the conditional and its converse to write a biconditional statement.

Exercises 15–20: For each statement, state the truth value of the conditional and its converse. If the conditional and the converse are both always true, write the biconditional. If either statement can be false, provide an example that makes it false.

15. If a number is an integer, then it is rational.

16. If a number is a perfect square, then it is not prime.

17. If the roots of a quadratic equation are equal, its parabola is tangent to the x-axis.

18. If a number is rational, then it is real.

19. If $x = \sqrt{32}$, then $x = 4\sqrt{2}$.

20. If a relation passes the vertical line test, then it is a function.

Note: Now turn to page 416 and take the Cumulative Review test for Chapters 1–7. This will help you monitor your progress and keep all your skills sharp.

Chapter 8:
Measurement and Coordinate Geometry

Measurement and Dimensional Analysis

Systems of Measurement and Conversions

Various kinds of mathematical problems involve measures such as **space**, **weight**, **mass**, **angles**, and **time**. Some measures, called **rates**, combine two or more basic measures. For example, speed combines a measure of distance with a measure of time, as in miles per hour.

In many of the situations reviewed in this chapter, spatial measures are important. How we measure **space** depends on the nature of the region being measured. We measure the *length* of a one-dimensional region, the *area* of a two-dimensional region, and the *volume* of a three-dimensional region.

- **Length**: The basic measure of space is **length** (l). When we measure length, we are measuring *one* dimension—a line segment. Therefore, the result is expressed in **linear units**. Linear measures are used for line segments (such as the sides of a triangle), and also for curved lines (such as the circumference of a circle).

- **Area**: For a two-dimensional region (plane), we measure **area** (A), the amount of flat space enclosed within the region. Area may refer to a plane figure, or it may be the **surface area** (SA)—the "outside"—of a solid figure. To calculate area, we multiply length in one dimension by length in the other dimension. Length times length is length *squared*, so the product is in **square units**. For example:

   1 ft $\times$ 1 ft = 1 square foot or 1 ft^2 1 m $\times$ 1 m = 1 square meter or 1 m^2

- **Volume**: The amount of space within a three-dimensional region (solid) is called **volume** (V). To calculate volume, we multiply length in the first dimension by length in the second dimension by length in the third dimension. Length times length times length is length *cubed*, so the product is in **cubic units**. For example:

   1 ft $\times$ 1 ft $\times$ 1 ft = 1 cubic foot or 1 ft^3 1 m $\times$ 1 m $\times$ 1 m = 1 cubic meter or 1 m^3

- **Liquid measure**: Liquid measures, sometimes called **capacity**, are special measures of volume. Examples are the liter and the quart.

For measuring space and weight or mass, there are two main standard systems:
- The **customary system**, sometimes called the **English system**, is used mostly in the United States.
- The **metric system** is used in most of the rest of the world.

Conversion from one measurement to another means finding equivalent values in different units. In converting measures within either the metric or the customary system, we use a **conversion factor**:
- To convert from a smaller unit to a larger unit, *divide* by the conversion factor.
- To convert from a larger unit to a smaller unit, *multiply* by the conversion factor.

In the metric system, which is based on the decimal system, units increase and decrease by factors of 10. Therefore, conversion factors are powers of 10. The basic metric units of measure are:
- Length—**meter** (m)
- Mass—**gram** (g)
- Liquid measure (capacity)—**liter** (L). One liter is 1 cubic decimeter or 1 dm^3.

The names of other units start with prefixes indicating how each unit is related to the basic one.

Measures in the Metric System

Prefix	Meaning	Value Relative to Basic Unit
kilo (k)	thousand	1000 or 10^3
hecto (h)	hundred	100 or 10^2
deka (dk)	ten	10
unit: **meter (m)** **gram (g)** **liter (L)**	one	1
deci (d)	one-tenth	$\frac{1}{10}$ or 0.1 or 10^{-1}
centi (c)	one-hundredth	$\frac{1}{100}$ or 0.01 or 10^{-2}
milli (m)	one-thousandth	$\frac{1}{1000}$ or 0.001 or 10^{-3}

Here are some examples of metric conversion factors.

Length: 100 cm = 1 m, so the conversion factor for centimeters and meters is 100 or 10^2.

Mass: 1000 mg = 1 g, so the conversion factor for milligrams and grams is 1000 or 10^3.

Liquid measure: 1000 L = 1 kl, so the conversion factor for liters and kiloliters is 1000 or 10^3.

Liquid measure to volume: 1 milliliter (ml) = 1 cubic centimeter (cc) and 1 kiloliter (kl) = 1 cubic meter (m^3), so these conversion factors are all 1.

In the customary system, the major units of length are the **inch** (in.), the **foot** (ft), the **yard** (yd), and the **mile** (mi). Liquid measures include the **pint** (pt), **quart** (qt), and **gallon** (gal). Units of weight include the **ounce** (oz) and the **pound** (lb).

Measures in the Customary System

Length	12 inches = 1 foot 36 inches = 1 yard 3 feet = 1 yard 5,280 feet = 1 mile 1,760 yards = 1 mile
Area	1 square foot = 12^2 square inches = 144 square inches 1 square yard = 3^2 square feet = 9 square feet
Volume	1 cubic foot = 12^3 cubic inches = 1,728 cubic inches 1 cubic yard = 3^3 cubic feet = 27 cubic feet
Liquid measure (capacity)	3 teaspoons = 1 tablespoon 16 tablespoons = 1 cup 1 cup = 8 fluid ounces 2 cups = 1 pint 2 pints = 1 quart 4 quarts = 1 gallon
Weight	16 ounces = 1 pound 2,000 pounds = 1 ton

Here are some examples of conversion factors in the customary system:

Length: 12 inches = 1 foot, so the conversion factor for inches and feet is 12

Weight: 16 ounces = 1 pound, so the conversion factor for ounces and pounds is 16

Liquid measure: 2 pints = 1 quart, so the conversion factor for pints and quarts is 2

Some conversions are common to both systems. A **degree of arc** (°), which is $\frac{1}{360}$ of a circle, is subdivided into **minutes** (') and **seconds** (''):

- 1 degree = 60 minutes   1 minute = $\frac{1}{60}$ degree

- 1 minute = 60 seconds   1 second = $\frac{1}{60}$ minute

Units of **time** include the **second** (sec), the **minute** (min), the **hour** (hr), the **day** (d), the **week**, the **month**, and the **year** (yr):

- 60 seconds = 1 minute
- 60 minutes = 1 hour
- 24 hours = 1 day
- 7 days = 1 week
- 12 months = 1 year

Measurement and Dimensional Analysis

MODEL PROBLEMS

1 Convert:

35 feet to inches

SOLUTIONS

We are converting from a larger unit (feet) to a smaller unit (inches), so we multiply by the conversion factor, 12:

$35 \cdot 12 = 420$

Answer: 35 feet = 420 inches

35 meters to centimeters

We are converting from larger (meters) to smaller (centimeters), so we multiply by the conversion factor, 10^2:

$35 \cdot 10^2 = 35 \cdot 100 = 3500$

Answer: 35 m = 3500 cm

Shortcut: Think of 35 m as 35.00 m and just move the decimal point 2 places to the right.

120 minutes to degrees

We are converting from smaller (minutes) to larger (degrees), so we divide by the conversion factor, 60:

$120' \div 60 = 2$

Answer: $120' = 2°$

6 miles to feet

We are converting from larger (miles) to smaller (feet), so we multiply by the conversion factor, 5,280:

$6 \cdot 5,280 = 31,680$

Answer: 6 miles = 31,680 feet

45 tons to pounds

We are converting from larger (tons) to smaller (pounds), so we multiply by the conversion factor, 2,000:

$45 \cdot 2,000 = 90,000$

Answer: 45 tons = 90,000 pounds

138 grams to kilograms

We are converting from smaller (grams) to larger (kilograms), so we divide by the conversion factor, 10^3:

$138 \div 10^3 = 138 \div 1000 = 0.138$

Answer: 138 g = 0.138 kg

Shortcut: Think of 138 g as 138.0 g and just move the decimal point 3 places to the left.

2 Convert 60 miles per hour (mph) to feet per second.

SOLUTION

This problem involves rates. We need to convert from miles to feet and from hours to seconds. Express 60 mph as $\frac{60 \text{ miles}}{1 \text{ hour}}$. Think:

5,280 feet = 1 mile, so $\frac{5,280 \text{ feet}}{1 \text{ mile}}$ is equal to one and will not change our rate if we multiply.

1 hour = 3,600 seconds, so $\frac{1 \text{ hour}}{3,600 \text{ seconds}}$ is also equal to one.

258 Chapter 8: Measurement and Coordinate Geometry

Now multiply and cancel the labels for miles and hours:

$$\frac{60 \text{ miles}}{1 \text{ hour}} \times \frac{5{,}280 \text{ feet}}{1 \text{ mile}} \times \frac{1 \text{ hour}}{3{,}600 \text{ seconds}} = \frac{60 \times 5{,}280 \text{ feet}}{3{,}600 \text{ seconds}} \approx \frac{88 \text{ feet}}{1 \text{ second}}$$

Answer: 88 feet per second

3 The dimensions of a block, in inches, are 3 by 4 by 6. How many such blocks can be stacked in a storage cube that measures 1 cubic foot?

SOLUTION

Since the cube is 1 foot by 1 foot by 1 foot, it is 12 inches by 12 inches by 12 inches. Each dimension of the block is a factor of 12, so we know that the blocks will stack in the storage cube with no wasted space.

volume of storage cube = 12 × 12 × 12 cubic inches = 1,728 cubic inches

volume of each block = 3 × 4 × 6 cubic inches = 72 cubic inches

Then:

$$\frac{1{,}728 \text{ cubic inches}}{72 \text{ cubic inches}} = 24$$

Answer: 24 blocks can be stored in the cube.

Practice

1 A carton of candy contains 12 ounces. How many cartons are needed to make up an order for 3 pounds of candy?

(1) 2
(2) 4
(3) 6
(4) 8

2 If 1 ounce of silver costs $5.50, what will $2\frac{1}{2}$ pounds of silver cost?

(1) $88
(2) $176
(3) $200
(4) $220

3 One bag of potatoes weighs 5 kilograms. Another bag weighs 4300 grams. The total weight of the two bags is:

(1) 54,300 g
(2) 4305 g
(3) 5.43 kg
(4) 9.3 kg

4 A timed math exam takes 80 minutes. Express this in hours.

(1) $2\frac{1}{3}$ hours

(2) $1\frac{2}{3}$ hours

(3) $1\frac{1}{3}$ hours

(4) $\frac{2}{3}$ hour

Measurement and Dimensional Analysis

5. One degree of arc = $\frac{1}{360}$ of a circle. How many seconds (") are there in a circle?
   (1) 3,600
   (2) 21,600
   (3) 1,296,000
   (4) 77,760,000

6. A **microgram** (μg) is one-millionth of a gram. The conversion factor for micrograms and grams is:
   (1) 10^{-4}
   (2) 10^{-5}
   (3) 10^{-6}
   (4) 10^{-7}

7. Freddie can run a mile in 5 minutes. How many feet can he run in 2 minutes?

8. In the former English monetary system, 1 pound was equal to 20 shillings, and 1 shilling was equal to 12 pence. One guinea was equal to 21 shillings. Show how to convert:
   a 15 pounds to shillings
   b 15 pounds to pence
   c 240 pence to pounds
   d 12 guineas to pounds

9. A strip of metal is 0.27 foot long. What is its length to the nearest inch?

10. Marielle drove the TransCanada Highway at 80 kilometers per hour.
    a How many meters per minute did she travel?
    b How many meters per second did she travel?

11. The volume of a sandbox is 45 cubic yards. To fill the sandbox, Sean uses a mechanical scoop that holds 1 cubic foot. How many times must he fill the scoop?

12. One mural can be painted with 2 pints of paint. How many murals can be painted with 3 gallons?

13. A storage tank is a rectangular prism 1 meter by 2 meters by 4 meters. How many liters will it hold?

14. A truck will hold 1 ton of cargo. How many crates can be loaded if each crate weighs 200 pounds?

15. When the Kahns moved, their packing boxes were 6 feet high, 3 feet wide, and 2 feet deep. The interior of their van was 2 yards by 2 yards by 8 yards. How many boxes could they carry in the van?

16. How many blocks with an edge measuring 5 centimeters can be arranged to form a shape 1 cubic meter in volume?

17. On Mother's Day, traffic at the Tappan Zee Bridge was backed up for 15 miles. Assuming a 2-lane highway leading onto the bridge and an average car length of 30 feet, how many cars were involved in the backup?

18. A cube has an edge measuring 6 inches. How many such cubes are needed to have a total volume of 1 cubic foot?

Perimeter and Circumference

The **perimeter** (*P*) of a polygon, the distance around it, is the sum of the measures of its sides. Perimeter is expressed in linear units, such as inches, feet, and centimeters.

To find the perimeter of a polygon that is *not* regular or is not a rectangle, add the measures of all the sides. For example, the following figures are a scalene triangle and a general quadrilateral.

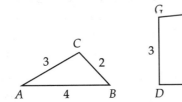

For the triangle: $P = 4 + 3 + 2 = 9$

For the quadrilateral: $P = 3 + 3 + 4 + 5 = 15$

To find the perimeter of a regular polygon or a rectangle, we can of course add as above, but formulas provide a faster method.

- In a **rectangle**, opposite sides are congruent. Two sides have a measure we call the **length** (l), and the other two sides have a measure we call the **width** (w). Thus the **formula for perimeter of a rectangle** is:

  $P = 2l + 2w$

To find the perimeter of a rectangle, substitute the measures for length and width in this formula. The rectangle below is an example:

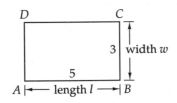

$P = 2l + 2w$
$P = 2(5) + 2(3) = 10 + 6$
$P = 16$

- A **regular polygon** (such as a square) is equilateral, so to find its its perimeter we multiply the measure of one side s by the number n of sides. Thus the **formula for perimeter of a regular n-sided polygon** is:

  $P = ns$

To find the perimeter of a regular polygon, substitute the measure of a side and the number of sides in this formula. The regular hexagon below is an example.

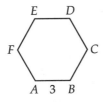

$P = ns$
$P = 6 \cdot 3$
$P = 18$

This equation also works for other equilateral polygons, such as a rhombus.

Measurement and Dimensional Analysis

The **circumference** (C) of a circle, the distance around it, is measured in linear units and is found with a formula. In every circle, the ratio of the circumference C to the diameter d is a constant, called **pi** and represented by the symbol π:

$$\frac{C}{d} = \pi$$

Since the diameter is twice the radius r, this ratio can also be written:

$$\frac{C}{2r} = \pi$$

- Thus the **formula for the circumference of a circle** is:

  $C = \pi d$ or $C = 2\pi r$

Pi is an irrational number, 3.1415926 . . . , which is usually approximated as 3.14 or $\frac{22}{7}$.

When you substitute a value for π, the calculation is approximate, so use the symbol $\approx$.

You can use the π key on a calculator unless a problem specifies one of these values or specifies that the answer should be left in terms of π.

MODEL PROBLEMS

1 A building has the shape of a regular decagon. The length of each outer wall is 7 meters. What is the perimeter of the building?

SOLUTION

Use the formula for perimeter of a regular polygon, $P = ns$, substituting 10 for n and 7 for s:

$P = ns$

$P = 10$ meters • 7

$P = 70$ meters

2 The circumference of a circle is 10 inches. Find its diameter to the nearest tenth of an inch.

SOLUTION

Use the formula $C = \pi d$ and solve for d. In this case, $C = 10$ inches, so:

10 inches $= \pi d$

$d = \frac{10}{\pi}$ inches

The problem does not specify a value for π, so we can proceed in more than one way.

Use $\pi \approx 3.14$:	Use $\pi \approx \frac{22}{7}$:	Use π key on calculator:
$d = \frac{10}{\pi}$	$d = \frac{10}{\pi}$	$d = \frac{10}{\pi}$
$d \approx \frac{10}{3.14} \approx 3.1847133$	$d \approx 10 \div \frac{22}{7} \approx 10 \cdot \frac{7}{22} \approx \frac{70}{22}$	$d \approx 3.183098862\ldots$
This rounds to 3.2.	$d \approx 3.181818\ldots$	This rounds to 3.2.
	This rounds to 3.2.	

Answer: Diameter is about 3.2 inches.

3 Rhombus *ABCD* has diagonals that measure 6 centimeters (cm) and 8 cm. Find the perimeter of the rhombus.

SOLUTION

Sketch the figure:

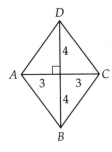

The diagonals of a rhombus are perpendicular to each other and bisect each other, so these diagonals form 4 congruent right triangles with legs that are 3 cm and 4 cm. We can use Pythagorean triples (3, 4, 5) to find that each hypotenuse measures 5 cm.

Then:

$P = ns$

$P = 4(5) = 20$

Answer: Perimeter is 20 cm.

 Practice

1 The perimeter of any polygon can be calculated as:
  (1) $P = $ sum of the sides
  (2) $P = 2l + 2w$
  (3) $P = ns$ where s is a side and n is number of sides
  (4) all of the above

2 Pi (π) is the ratio of:
  (1) the diameter of a circle to the circumference
  (2) the circumference to the diameter
  (3) the radius to the circumference
  (4) the circumference to the radius

3 You want to find the distance around a plane figure, given only one length. In which case is this impossible?
  (1) circle
  (2) square
  (3) trapezoid
  (4) equiangular triangle

4 When we use a value for π, the result of our calculations will always be:
  (1) exact
  (2) approximate
  (3) 3.14 or $\frac{22}{7}$
  (4) a circumference

5 Given the circumference of a circle, you need to find the radius. Use:
  (1) $r = \pi d$
  (2) $r = 2\pi d$
  (3) $r = \frac{C}{2\pi}$
  (4) $r = \frac{2\pi}{C}$

6 A rectangle has a width of 5 mm and a diagonal of 13 mm. Find the perimeter.

7 The base of an isosceles triangle is 8 cm long. The height of the triangle is 3 cm. Find the perimeter.

Measurement and Dimensional Analysis

8. The shorter diagonal of a rhombus measures 10, and the longer diagonal measures 24. Find the perimeter of the rhombus.

9. Each exterior angle of a regular polygon measures 36°. The perimeter is 140 cm. Find the length of a side.

10. A square banner for a school play shows a logo within a circle inscribed in the banner. The radius of the circle is 5 feet. Find the circumference of the circle and the perimeter of the banner.

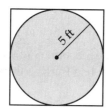

11. A square is inscribed in a circle. A side of the square is $2\sqrt{2}$. Find the circumference of the circle. Leave the answer in terms of π.

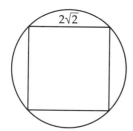

12. The radius of a circle is 6. Find the perimeter of a quarter-circle. Leave the answer in terms of π.

13. The figure below is made up of 2 squares and 2 quarter-circles. Find its perimeter.

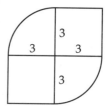

14. The swimming pool shown below has a semicircular shallow end. Find the perimeter of the pool.

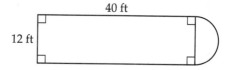

15. Find the perimeter of the shaded region.

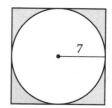

Area

The **area** (*A*) of a plane figure or surface is the number of square units (such as ft² or m²) in the region enclosed by the figure. Formulas used to find area are summarized in the table that follows.

Remember these definitions.

- **Base** (*b*): The **base of a polygon** is the side from which its height is measured.

  The **base of a triangle** is the side opposite any vertex.

  Bases of a trapezoid are a pair of parallel sides.

- **Height** (*h*): The **height of a triangle** is a perpendicular line segment connecting a vertex to the line containing the opposite base.

  The **height of a parallelogram or trapezoid** is a perpendicular line segment connecting its bases.

  Height is also called **altitude**.

- **Length** (*l*) and **width** (*w*): The base and height of a rectangle.

- **Side** (*s*): One of the line segments forming a regular polygon.

264 Chapter 8: Measurement and Coordinate Geometry

Formulas for Area

Figure	Formula	Example
Triangle	$A = \frac{1}{2}bh$	$A = \frac{1}{2} \times 10 \times 4 = 20$
Equilateral triangle	$A = \frac{s^2\sqrt{3}}{4}$	$A = \frac{10^2\sqrt{3}}{4} = \frac{100\sqrt{3}}{4} = 25\sqrt{3}$
Parallelogram	$A = bh$	$A = 10 \times 3 = 30$
Rectangle	$A = lw$	$A = 5 \times 3 = 15$
Rhombus	$A = \frac{1}{2}d_1 \times d_2$	$A = \frac{1}{2} \times 8 \times 6 = \frac{1}{2} \times 48 = 24$
Square	$A = s^2$	$A = 6^2 = 36$

(continues)

Formulas for Area (*continued*)

Figure	Formula	Example
Trapezoid	$A = \frac{1}{2}h(b_1 + b_2)$	$A = \frac{1}{2} \times 4(8 + 20) = 56$
Circle	$A = \pi r^2$	$A = \pi(3)^2 = 9\pi \approx 28.27$

Note: Problems involving area of a circle may specify a value for π $\left(3.14 \text{ or } \frac{22}{7}\right)$ or may say that the answer should be left in terms of π. A problem may also specify a degree of decimal accuracy.

MODEL PROBLEMS

1 Find the area of each figure.

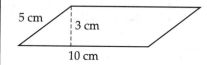

SOLUTIONS

Use the formula for area of a parallelogram $A = bh$. Height h is the dashed line.
$$A = (10)(3) = 30 \text{ cm}^2$$

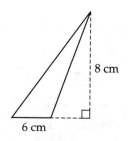

Use the formula for area of a triangle $A = \frac{1}{2}bh$. Height h is the dashed line.
$$A = \frac{1}{2}(6)(8) = 24 \text{ cm}^2$$

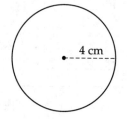

Use the formula for area of a circle $A = \pi r^2$.
$A = \pi(4)^2 = 16\pi \text{ cm}^2$ (exact area)
$A \approx 16(3.14) \approx 50.24 \text{ cm}^2$ (approximate area)

266 Chapter 8: Measurement and Coordinate Geometry

2 Find the area of each region by dividing it as shown. Dimensions are in meters (m).

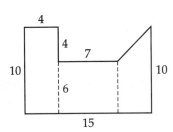

SOLUTIONS

Rectangle: $A = lw = (10 \text{ m})(4 \text{ m}) = 40 \text{ m}^2$

Rectangle: $A = lw = (7 \text{ m})(6 \text{ m}) = 42 \text{ m}^2$

Trapezoid:
$$A = \frac{1}{2}h(b_1 + b_2) = \frac{1}{2}[(4 \text{ m})(10 \text{ m} + 6 \text{ m})] = \frac{1}{2}(4 \text{ m} \cdot 16 \text{ m})$$
$$= \frac{1}{2}(64 \text{ m}^2) = 32 \text{ m}^2$$

Total area = $40 \text{ m}^2 + 42 \text{ m}^2 + 32 \text{ m}^2 = 114 \text{ m}^2$

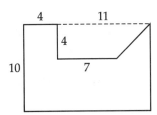

Square: $A = s^2 = 4^2 = (4 \text{ m})(4 \text{ m}) = 16 \text{ m}^2$

Rectangle: $A = lw = (6 \text{ m})(15 \text{ m}) = 90 \text{ m}^2$

Triangle: $A = \frac{1}{2}bh = \frac{1}{2}(4 \text{ m} \cdot 4 \text{ m}) = \frac{1}{2}(16 \text{ m}^2) = 8 \text{ m}^2$

Total area = $16 \text{ m}^2 + 90 \text{ m}^2 + 8 \text{ m}^2 = 114 \text{ m}^2$

The quickest way to find total area is to *subtract* the area of the trapezoid from the area of the rectangle formed by the dashed line.

Rectangle: $A = lw = (15 \text{ m})(10 \text{ m}) = 150 \text{ m}^2$

Trapezoid:
$$A = \frac{1}{2}h(b_1 + b_2) = \frac{1}{2}[(4 \text{ m})(11 \text{ m} + 7 \text{ m})] = \frac{1}{2}(4 \text{ m} \cdot 18 \text{ m})$$
$$= \frac{1}{2}(72 \text{ m}^2) = 36 \text{ m}^2$$

Total area = $150 \text{ m}^2 - 36 \text{ m}^2 = 114 \text{ m}^2$

Practice

1 Which amount is *not* an application of area?

(1) carpeting needed for a living room
(2) extension cord needed to reach from an appliance to an outlet
(3) sod needed for a lawn
(4) laminate needed for a countertop

2 Parallelograms *A*, *B*, and *C* have the same area. Which has the greatest perimeter?

(1) *A*
(2) *B*
(3) *C*
(4) Their perimeters are equal.

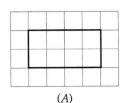
(A)

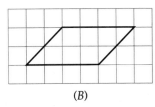
(B)

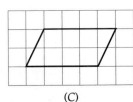
(C)

3 Which statement is *always* true?

   (1) Two rectangles with the same perimeter have the same area.
   (2) Two rectangles with the same area have the same perimeter.
   (3) Two quadrilaterals with the same area have congruent corresponding sides.
   (4) Two quadrilaterals with congruent corresponding sides and angles have the same area.

4 If the diameter of a circle is 10, then $A =$

   (1) 100π
   (2) 50π
   (3) 25π
   (4) 5π

5 Given the area of a circle, you need to find its radius. Use:

   (1) $r = \sqrt{\dfrac{A}{\pi}}$
   (2) $r = \sqrt{\pi A}$
   (3) $r = \dfrac{A}{\pi}$
   (4) $r = \left(\dfrac{A}{\pi}\right)^2$

6 The perimeter of a square field is 12 centimeters. Find its area.

7 One rectangular yard has a length of 36 feet and a width of 24 feet. A second yard with the same area has a length of 32 feet. Find the width of the second yard.

8 One diagonal of a rhombus measures 6 feet, and the area of the rhombus is 48 square feet. Find the measure of the other diagonal.

9 To the nearest tenth of an inch, find the area of the patio below.

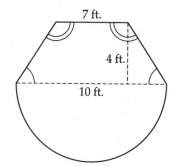

10 The height of a triangle is half as long as its base. The area of the triangle is 36 square inches. Find the base and height.

11 The bases of an isosceles trapezoid are 8 and 12. The bases are 2 units apart. Find the area of the trapezoid.

12 Ms. Green is planting a quarter-circle garden in her square yard. As shown below, one side of the yard is 30 feet long. What part of her yard will *not* be included in the garden? Give your answer to the nearest tenth of a square yard.

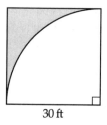

30 ft

13 In a regular polygon, the **apothem** is a perpendicular line segment from the center of the figure to a side. In the regular hexagon below, the apothem is 4 units and each side is $4\sqrt{3}$ units. Find the area of the hexagon in simplest radical form.

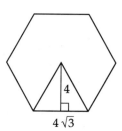

14 The floor plan below is made up of a parallelogram, a right triangle, and a trapezoid. Find the total area.

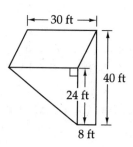

268 Chapter 8: Measurement and Coordinate Geometry

15 A 4-foot-wide sidewalk around an office building must be repaved. The building covers a rectangular space 100 feet by 50 feet. If repaving costs $0.50 per square foot, what is the total cost?

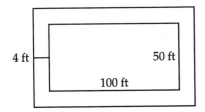

16 On a basketball court, the center circle has a diameter of 4 feet. What is its area to the nearest tenth of a square foot?

17 A circle is inscribed in a square, as shown below. The area of the square is 36 square inches. To the nearest tenth of a square inch, what is the area of the shaded region?

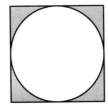

18 The target below consists of two concentric circles. The radius of one circle is 4 inches, and the radius of the other circle is 8 inches. To the nearest tenth of a square inch, what is the area of the shaded region?

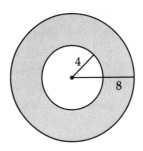

19 Using an overhead projector, a teacher displays a circular chart on a screen. By adjusting the focus, the teacher changes the radius of the circle from 3 feet to 4 feet. Find the percent increase in the area of the circle, to the nearest tenth of a percent.

20 The cost of making a cheese pizza is 4 cents per square inch. If Town Pizzeria sells a 9-inch-diameter pizza for $10, what is the profit made on each pizza?

Surface Area and Volume

For three-dimensional (solid) figures, we can find surface area and volume.

Surface Area The **surface area** (SA), or **total surface area**, of a solid figure is the sum of the areas of all its surfaces, or faces. For figures that have a base or bases and a lateral surface or lateral faces, SA includes the base or bases. The **lateral area** (LA) of these figures does *not* include the base or bases—in other words, for such figures SA is lateral area plus the base or bases.

If you need to review solids again, see Chapter 6.

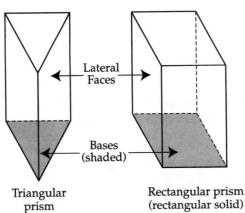

Measurement and Dimensional Analysis

To find the surface area of a polyhedron, we can follow these steps:

- Step 1: Describe the size and shape of each face.
- Step 2: Find the area of each face.
- Step 3. Find the sum of these areas.

For some polyhedrons, we can use a formula for surface area. We also use formulas to find the surface area of cylinders, cones, and spheres. These formulas are reviewed here.

Surface Area of a Right Prism A common example of a right prism is a rectangular prism. To understand how a rectangular prism is formed, consider this diagram:

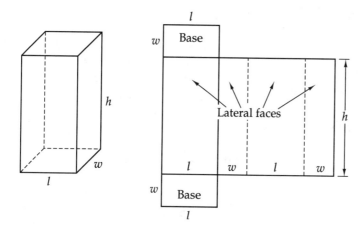

Lateral area LA of a right prism is the sum of the areas of the lateral faces. If the edges are laid out side by side, a rectangle is formed with the base equal to the perimeter of the base of the solid and the height equal to the height of the solid. Therefore, a **formula for lateral area of a right prism** with base perimeter P and height h is:

LA = Ph

Surface area SA of a prism is the sum of lateral area plus 2 times the area of the base. The area of the base, B, is found using the correct formula for the area of the polygon. Therefore, a **formula for surface area of a right prism** with lateral area LA and base area B is:

SA = LA + 2B

An alternative **formula for surface area of a rectangular prism** (rectangular solid) with length l, width w, and height h is:

SA = $2(lw + lh + hw)$

A cube is defined by one measure, the length of an edge e. Area of each face is therefore e^2. Since there are six faces, we have this **formula for surface area of a cube**:

SA = $6e^2$

Surface Area of a Regular Right Pyramid In a regular right pyramid, the base is a regular polygon, so the lateral faces are congruent triangles. To find its surface area, we use **slant height**, which is the altitude of a lateral face:

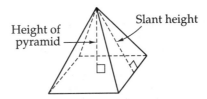

The **formula for lateral area of a regular right pyramid** with base perimeter P and slant height ℓ is:

$$LA = \frac{1}{2}P\ell$$

Therefore, the **formula for surface area of a regular right pyramid** with base area B is:

$$SA = \frac{1}{2}P\ell + B$$

Surface Area of a Right Circular Cylinder As the number of sides of a regular polygon increases, the polygon looks more and more like a circle. The lateral faces of a prism with such a polygon as a base look more and more like a single curved surface. When the base has an infinite number of sides, the result is a right circular cylinder. This cylinder has 2 congruent circular bases and a continuous curved edge that is perpendicular to both bases:

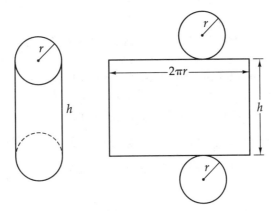

The **formula for lateral area of a right circular cylinder** with radius r and height h is:

$\quad LA = 2\pi rh \quad$ or $\quad LA = Ch$ where C is the circumference of the base

The **formula for surface area of a right circular cylinder** is the sum of the lateral area and twice the base area:

$\quad SA = 2\pi rh + 2\pi r^2 \quad$ or $\quad SA = Ch + 2B$ where B is base area

Surface Area of a Right Circular Cone For the right circular cone, like the pyramid, we use slant height to find lateral area:

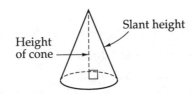

The **formula for lateral area of a right circular cone** with radius r and slant height ℓ is:

$LA = \pi r \ell$

The formula for **surface area of a right circular cone** is lateral area plus base area:

$SA = \pi r \ell + \pi r^2$

Or:

$SA = \dfrac{1}{2}C\ell + B$ where C is base circumference and B is base area

Surface Area of a Sphere   The sphere has no base and thus has only surface area.

The **formula for surface area of a sphere** with radius r is:

$SA = 4\pi r^2$

MODEL PROBLEMS

Find surface area for each figure.   **SOLUTIONS**

This is a rectangular solid, so, in square units:
$SA = 2(lw + lh + hw)$
$SA = 2[(12 \times 6) + (6 \times 4) + (4 \times 12)] = 2(72 + 24 + 48) = 288$

Find perimeter of base. Base is a right triangle, and the Pythagorean triple is 6, 8, 10, so $P = 6 + 8 + 10 = 24$.
Then $LA = 24 \times 12 = 288$ square units.
Base area $B = \dfrac{1}{2} \times 6 \times 8 = 24$ square units
$SA = (2 \times 24) + 288 = 336$ square units

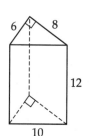

Use the formula for surface area of a cone:
$SA = \pi r \ell + \pi r^2$
$SA = \pi(5)(12) + \pi(5)^2 = 60\pi + 25\pi = 85\pi \approx 266.9$
$SA \approx 267$ square units

Use the formula for surface area of a sphere:
$SA = 4\pi r^2$
$SA = (4)(\pi)(5)^2 = 100\pi \approx 314$ square units

Use the formula for surface area of a cylinder:
$SA = 2\pi rh + 2\pi r^2$
$SA = (2 \cdot \pi \cdot 3 \cdot 10) + (2 \cdot \pi \cdot 3^2) = 60\pi + 18\pi = 78\pi$
$SA \approx 244.92 \approx 245$ square units

Volume The amount of space enclosed by a three-dimensional figure is its **volume** (V). Volume is measured in cubic units, such as in.³ or cm³. We can use the methods reviewed here to find the volume of certain polyhedrons, and the volume of cylinders, cones, and spheres.

Volume of a Right Prism or a Right Cylinder To find the volume of any right prism or right circular cylinder:

- Step 1. Express all the dimensions of the solid using the same linear unit of measure, such as inches, feet, or centimeters.
- Step 2. Find the area of one base. Call this area B.
- Step 3. Find the height h of the solid.
- Step 4. Use this **formula for volume of a right prism or circular cylinder**:
  $V = Bh$

Note also these alternative formulas:

- For the volume of a **rectangular solid** (a rectangular prism), with length l, width w, and height h, we can use the formula:
  $V = lwh$
- For the **volume of a cube** with edge e, we can use this formula:
  $V = e^3$
- For the volume of a **right circular cylinder** with radius r, we can use this formula:
  $V = \pi r^2 h$

Volume of a Right Pyramid or Cone To find the volume of a pyramid or cone, we use its height h, the perpendicular distance from the vertex to the center of the base:

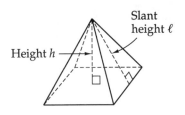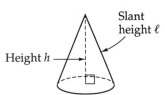

Be careful not to confuse height h of a pyramid or cone with its slant height ℓ.

The **formula for volume of a right pyramid or cone** is one-third of the product of base area B and height h:

$$V = \frac{1}{3}Bh$$

In the case of a **cone** with radius r, this formula can also be written as:

$$V = \frac{1}{3}\pi r^2 h$$

This formula is often found by experiment. Students are given right cones and right cylinders with the same base and height. They find that if they fill a cone and pour the contents into a cylinder, they must do this 3 times to fill the cylinder. Alternatively, they find that the contents of a cylinder will fill 3 cones.

Measurement and Dimensional Analysis **273**

Volume of a Sphere To find the volume of a sphere, we need only one measure, its radius. The formula for **volume of a sphere** with radius r is:

$$V = \frac{4}{3}\pi r^3$$

Remember that problems involving the volume of cylinders, cones, and spheres may specify a value for π or may require an answer to be left in terms of π.

MODEL PROBLEMS

1 A prism has a trapezoidal base with parallel sides measuring 12 inches and 8 inches. The parallel sides are 4 inches apart. The height of the trapezoidal prism is 6 inches. Find its volume.

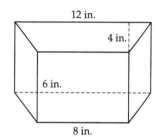

SOLUTION

First find the area B of a base of the prism, using the formula for area of a trapezoid:

$B = \frac{1}{2}h(b_1 + b_2)$ where b_1 and b_2 are bases of the trapezoid

$B = \frac{1}{2} \cdot 4 \text{ in.} \cdot (12 \text{ in.} + 8 \text{ in.}) = 40 \text{ in.}^2$

Now use the formula for volume of a prism:

$V = Bh$

$V = (40 \text{ in.}^2)(6 \text{ in.}) = 240 \text{ in.}^3$

Answer: Volume is 240 in.^{3}.

2 A juice can is a cylinder. The radius of the base is 2.5 centimeters, and the height of the can is 12 centimeters. Find the volume to the nearest cubic centimeter.

SOLUTION

Use the formula for volume of a cylinder, substituting the given values and using the π key on your calculator:

$V = \pi r^2 h$

$V = \pi(2.5 \text{ cm})^2(12 \text{ cm}) \approx 235.6 \text{ cm}^3$

Answer: Volume to the nearest cubic centimeter is 236 cm^3.

 Practice

1 If the surface area of a cube is 96 cm^2, what is the length of a side?

(1) 3 cm
(2) 4 cm
(3) 5 cm
(4) 6 cm

2 The volume of a sphere is $\frac{500\pi}{3}$. What is its radius?

(1) 25π
(2) 25
(3) 5π
(4) 5

3 What is the surface area of this rectangular prism?

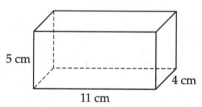

(1) 119 cm^2
(2) 220 cm^2
(3) 238 cm^2
(4) 375 cm^2

4. Each side of the base of a square pyramid is 6 cm. Total surface area is 132 cm². What is the area of each lateral face?

   (1) 21 cm²
   (2) 22 cm²
   (3) 23 cm²
   (4) 24 cm²

5. The height of a cylinder is 16, and the area of a base is 25π square units. In cubic units, the volume is:

   (1) 25π
   (2) 50π
   (3) 200π
   (4) 400π

Exercises 6 and 7: (a) Name the polygon. Find its: (b) lateral area, (c) surface area, and (d) volume.

6.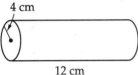

7.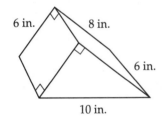

8. A right cylinder is 8 inches high. The circumference of the base is 6π inches. Find:

   a lateral area
   b surface area
   c volume

9. A cereal box is a rectangular prism 30 centimeters high. The sides of the base measure 8 centimeters and 25 centimeters. Find:

   a lateral area
   b surface area
   c volume

10. A swimming pool is 8 feet deep, 25 feet long, and 12 feet wide. The water in it comes up to a level 2 feet below the top edge. Find the volume of water in the pool.

11. The volume of a rectangular prism is 54 cubic inches, and its height is 3 inches. Find the area of the base of the prism.

12. The volume of a cone is 27π cubic centimeters. The radius of the base is 3 centimeters. Find the height.

13. The surface area of an athletic training ball is k cm². The volume of the ball is k cm³. Find the radius of the ball.

14. A sphere with a radius of 10 centimeters is in a cube-shaped box with a 20-centimeter edge. What percent of the box does the ball fill?

15. A **hemisphere** is half of a sphere. Shown below is a hemisphere with a radius of 2 inches, attached to a cone with a height of 9 inches. Find the volume of the figure formed.

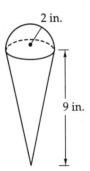

Measures of Similar Figures

Two polygons (plane figures) or two polyhedrons (solids) are **similar** when their corresponding angles are congruent and their corresponding lengths are proportional—that is, in the same ratio. This ratio is called the **scale factor**.

Two circular cylinders, or two circular cones, are similar if their radii and heights are proportional.

Measurement and Dimensional Analysis **275**

When a figure is defined by only one linear measure, all figures of that type are similar. Thus:

- Any two regular polygons with the same number of sides are similar. (The defining linear measure is side *s*.)

- In a **regular polyhedron**, all the faces are congruent. Any two regular polyhedrons with the same number of faces are similar. (The defining linear measure is edge *e*.)

- All circles are similar. All spheres are similar. (The defining linear measure is radius *r*.)

If we know the scale factor of corresponding linear dimensions of similar plane figures, we can find the ratio of their areas by squaring their linear dimensions. For solids, if we know the scale factor of corresponding linear dimensions, we can find the ratios of their volumes by cubing their linear dimensions.

Similar Figures

If the ratio of corresponding linear measures is $\frac{a}{b}$, then:		
The ratio of perimeters or circumferences is $\frac{a}{b}$.	The ratio of areas, lateral areas, or surface areas is $\frac{a^2}{b^2}$.	The ratio of volumes is $\frac{a^3}{b^3}$.

For related reviews, see Chapter 2 (ratios and proportions), Chapter 5 (direct variation), and Chapter 6 (similarity and scale factors).

These relationships in similar figures are an example of **direct variation**:

- *Perimeter* varies directly as the length of a side.
- *Circumference* varies directly as the length of the radius.
- *Area, lateral area, or surface area* varies directly as the *square* of a side or radius.
- *Volume* varies directly as the *cube* of a side or radius.
- The *scale factor* is the **constant of variation**.

As noted in Chapter 6, similar figures are important in problems involving scale drawings.

 MODEL PROBLEMS

1 The ratio of the radii of two similar cylinders is $\frac{1}{3}$. The surface area of the smaller figure is 7π square units. Find the surface area of the larger figure in terms of π.

SOLUTION

Let *s* = surface area of the larger cylinder. Area varies as the square of the radius, so:

$$\frac{\text{surface area of small cylinder}}{\text{surface area of large cylinder}} = \frac{1^2}{3^2} = \frac{1}{9}$$

Therefore:

$$\frac{7\pi}{s} = \frac{1}{9}$$

Cross multiply:

$$s = 9 \cdot 7\pi = 63\pi$$

Answer: Surface area of the larger cylinder is 63π square units

276 Chapter 8: Measurement and Coordinate Geometry

2 The ratio of the lateral areas of two rectangular prisms is $\frac{9}{4}$. The height of the larger prism is 12 centimeters. Find the height of the smaller prism.

SOLUTION

Lateral area increases as the square of height. Therefore, if the ratio of lateral areas is $\frac{9}{4}$, the ratio of heights is:

$$\frac{\text{height of larger prism}}{\text{height of smaller prism}} = \frac{\sqrt{9}}{\sqrt{4}} = \frac{3}{2}$$

Let h = height of smaller prism. Set up a proportion, cross multiply, and solve:

$$\frac{12}{h} = \frac{3}{2}$$
$$3h = 24$$
$$h = 8$$

Answer: Height of the smaller prism is 8 centimeters.

3 The ratio of the edges of two cubes is $\frac{3}{2}$. The volume of the smaller cube is 32 cubic inches. Find the volume of the larger cube.

SOLUTION

Volume increases as the cube of an edge, so the ratio of the volumes is:

$$\frac{\text{volume of larger cube}}{\text{volume of smaller cube}} = \frac{3^3}{2^3} = \frac{27}{8}$$

Let V = volume of larger cube. Set up a proportion, cross multiply, and solve:

$$\frac{V}{32} = \frac{27}{8}$$
$$8V = 27 \times 32 = 864$$
$$V = \frac{864}{8} = 108$$

Answer: Volume of the larger cube is 108 cubic inches.

Practice

1 If each edge of a cube is doubled, the volume is multiplied by:

(1) 2
(2) 4
(3) 6
(4) 8

2 The radius of a cone is doubled and the height remains the same. The volume is multiplied by:

(1) 2
(2) 4
(3) 6
(4) 8

3 Which pair of figures are *not* similar?

(1) cube I with edge = 2 cm
    cube II with edge = 3 cm

(2) sphere I with radius = 3 cm
    sphere II with radius = 4 cm

(3) rectangular prism I:
    6 cm by 8 cm by 12 cm
    rectangular prism II:
    9 cm by 12 cm by 18 cm

(4) cylinder I:
    base area = 10 cm² and height = 3 cm
    cylinder II:
    base area = 40 cm² and height = 12 cm

4 The length and width of a rectangle are both increased by 10 percent. By what percent does the area increase?

(1) 10 percent
(2) 11 percent
(3) 20 percent
(4) 21 percent

5 Which match is correct?

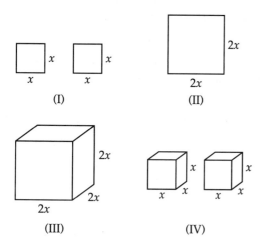

(1) I represents $(2x)^2$.
(2) II represents $2x^3$.
(3) III represents $(2x)^3$.
(4) IV represents $2x^2$.

6 The sides of a triangle measure 8, 10, and 13 inches. Find the perimeter of a similar triangle in which the shortest side measures 24 inches.

7 Frank told Mary that all squares are similar and all circles are similar. Explain why Frank is correct.

8 The height of a rectangle is 4 inches and the base is 3 inches. The height of a similar rectangle is 12 inches. Find the perimeter of the larger rectangle.

9 The perimeters of two similar parallelograms are 9 centimeters (cm) and 18 cm. The area of the smaller parallelogram is 5 cm². Find the area of the larger parallelogram.

10 The ratio of the areas of two similar polygons is 9 to 16. What is the ratio of their perimeters?

11 The ratio of the volumes of two spheres is 1 to 8. What is the ratio of their radii?

12 The ratio of the volumes of two similar right cylinders is 64 to 27. Find the ratio of the areas of the bases.

13 The ratio of the lateral areas of two rectangular prisms is 5 to 2. Find the ratio of the heights of the prisms in simplest radical form.

14 The perimeter of a hexagon is 48 units. The length of the sides of a second hexagon is 4 units. Find the ratio of the area of the larger hexagon to the area of the smaller hexagon.

15 Two cylindrical water tanks have the same dimensions. One is filled to a height of 4 feet, and the other is filled to a height of 8 feet. What is the ratio of the amount of water in the two tanks?

16 A **blueprint** is a scale drawing. The scale on a blueprint of a closet is 1 centimeter = 0.5 meter. Find the actual dimensions of the closet if its dimensions on the blueprint are 6.8 centimeters long by 6.4 centimeters wide.

17 A blueprint has a scale of 1 inch = 12 feet. Find the actual length represented by each length on this blueprint:

a $\frac{3}{4}$ inch

b $1\frac{1}{2}$ inches

c $2\frac{1}{8}$ inches

18 A blueprint has a scale of 1 inch = 12 feet. Find the length on the blueprint needed to represent each length:

a 6 feet
b 24 feet
c 16 feet

19 A scale model of the ship *Queen Elizabeth II* measures 32 inches long. If the scale is 1 inch = 30 feet, find the length of the ship.

20 The wingspan of the Concorde supersonic jet plane is 25.6 meters. If a model of the Concorde is built to a scale of 1 centimeter = 2 meters, find the wingspan of the model.

Error in Measurement

Even when great care is taken, error in measurement sometimes occurs. Although human error can be the cause of an inaccurate measurement, often the error is due to the limitations of the measuring equipment. For example, we can use a ruler to measure something to the nearest $\frac{1}{2}$ inch or—with a steady eye—to the nearest $\frac{1}{16}$ inch. But if we used this ruler to measure a pencil that was exactly $6\frac{3}{70}$ inches long, we would only be able to say that it looked very close to 6 inches. The fraction is too small to measure, so our measurement has an error of $\frac{3}{70}$.

Margin of Error Margin of error is a term that describes the maximum difference expected between a measurement and the true value. A margin of error is sometimes written with a plus-or-minus sign (±). A smaller margin of error implies that the measurement is more exact. For our ruler above, the margin of error might be $\pm\frac{1}{32}$ inch.

MODEL PROBLEMS

1. The population of a town is 4,560 when rounded to the nearest 10 people. What is the range of numbers for the actual population? What is the margin of error?

SOLUTION

If there were 4,555 people, 4,555 would round up to 4,560: 4,560 − 4,555 = 5

If there were 4,564 people, 4,564 would round down to 4,560: 4,564 − 4,560 = 4

The range for the population is from 4,555 to 4,564. The maximum possible error is 5, so the margin of error is ± 5.

2. A survey says that 59% of people prefer sneaker brand A to sneaker brand B, with a margin of error of 10%. What is smallest possible percentage of people who prefer brand A to brand B?

SOLUTION

59% − 10% = 49%

It is possible that only 49% of people actually prefer brand A sneakers.

Percentage Error Knowing a margin of error is often not enough to tell how accurate a measurement is. A margin of error of 10 is large when compared with a measurement of 1 and small when compared with a measurement of 100,000,000. Therefore, the amount of error is often shown as a percentage of the true value.

To Find the Percentage Error in a Measurement

- Record the measured reading from the measuring equipment.
- Record the accepted value of the reading, taken from more reliable or trusted sources.
- Find the difference between the measured reading and the accepted value.
- Divide that difference by the accepted value.
- Change the decimal value to a percent.

MODEL PROBLEM

A 200-pound weight was placed on a bathroom scale. The scale dial displayed 210 pounds. If the scale is consistently off by the same percentage, how much does a man really weigh if his weight displayed on this scale is 168 pounds?

SOLUTION

Write the formula for percentage error.

$$\text{Error} = \left| \frac{\text{measured value} - \text{accepted value}}{\text{accepted value}} \right| \times 100\%$$

Substitute and evaluate.

$$= \left| \frac{210 - 200}{200} \right| \times 100\% = 5\%$$

Every measure on the bathroom scale is 5% too high.

actual weight $\times$ 1.05 = observed weight

Substitute the observed weight.

actual weight $\times$ 1.05 = 168

Solve.

$$\text{actual weight} = \frac{168}{1.05} = 160 \text{ pounds}$$

Answer: 160 pounds

Initial Error in Subsequent Calculations When we substitute a measurement for a variable in an equation, we must be careful with our answer. The small error in our measurement may become a larger error in the answer. It is often useful to substitute the highest and lowest possible true values of the measurement in the equation to see how exact our answer truly is.

 MODEL PROBLEM

Shanayda has a small aquarium that she wants to fill with water. The dimensions of the aquarium, rounded to the nearest centimeter, are 16 centimeters, 10 centimeters, and 120 centimeters. What is the maximum number of liters of water that she can put in and still be sure that the aquarium will not overflow?

(1) 16 liters
(2) 17 liters
(3) 19 liters
(4) 20 liters

SOLUTION

This problem is asking for the smallest possible volume of the aquarium. If Shanayda pours in more than this smallest volume, the tank might overflow.

To find the smallest value, we need to start with the smallest possible true values for the measurements. Since each measurement is to the nearest centimeter, the true values can be at most 0.5 centimeter less than the measurements.

16 cm $\times$ 10 cm $\times$ 120 cm becomes 15.5 cm $\times$ 9.5 cm $\times$ 119.5 cm.

Use a calculator.   15.5 cm $\times$ 9.5 cm $\times$ 119.5 cm

17596.375 cm³ or 17596.375 ml

Convert to liters.   17.596375 L

Shanayda can pour at most 17.596375 liters into the aquarium without an overflow. The answer choice of 17 is the closest she can come.

Answer: (2)

Error in Different Dimensions In geometric calculations, error increases when dimension increases, such as when linear measure is used to find area or volume. Error decreases when dimension decreases, such as when area or volume is used to find length. To understand how this relationship works, study the table below.

Measurement and True Value	Percent Error of Measurement	Calculations	Percent Error of Calculation
Example 1			
Length of side of cube: measured: 9 inches true: 10 inches	$\left\|\dfrac{10-9}{10}\right\| = \dfrac{1}{10}$ $= 10\%$	Volume of cube: $9 \times 9 \times 9 = 729$ $10 \times 10 \times 10 = 1{,}000$	$\left\|\dfrac{1{,}000 - 729}{1{,}000}\right\| = \dfrac{271}{1{,}000}$ $= 0.271 = 27.1\%$ ERROR INCREASED
Example 2			
Volume of cube: measured: 900 cu in. true: 1,000 cu in.	$\left\|\dfrac{1{,}000-900}{1{,}000}\right\| = \dfrac{100}{1{,}000}$ $= 10\%$	Length of a side: $\sqrt[3]{900} = 9.655$ $\sqrt[3]{1{,}000} = 10$	$\left\|\dfrac{10 - 9.655}{10}\right\| = \dfrac{0.345}{10}$ $= 0.0345 = 3.45\%$ ERROR DECREASED

Note: Since we are working with measurements, we can ignore the negative square roots.

 Practice

1. In a physics experiment, the acceleration of gravity was found to be 9.5 meters per second squared. The accepted acceleration of gravity to this degree of accuracy is 9.8 meters per second squared. Find the approximate percentage error.

    (1) 96.9%
    (2) 3.16%
    (3) 3.06%
    (4) 1.03%

2. A senator polled the people in his district to discover their opinion on two versions of a crime bill. The poll indicated that 45% preferred bill A, 35% preferred bill B, and 20% were undecided. The poll had a margin of error of ±5%. Which of the following is a possibility for the true opinions of the people in the senator's district?

    (1) 40% for bill A, 30% for bill B, and 30% undecided
    (2) 40% for bill A, 40% for bill B, and 20% undecided
    (3) 50% for bill A, 40% for bill B, and 10% undecided
    (4) 50% for bill A, 25% for bill B, and 25% undecided

3 Douglas measured his backyard and found it to be 26 feet by 43 feet. He calculated the area and bought enough new sod to cover the yard. When he got home, he found that his measurements were each 2 feet too long. How much excess sod will he have?

(1) 142 square feet
(2) 134 square feet
(3) 86 square feet
(4) 52 square feet

4 Tracy uses the measure of a side of a cube to calculate the volume. She finds the volume to be 216 cubic inches. Later, she realizes that the true volume of the cube is 343 cubic inches. To the nearest tenth of a percent, what was the error of her measure *for the length of the side*?

(1) 14.3%
(2) 16.7%
(3) 37.0%
(4) 58.8%

5 An odometer measures the distance a vehicle travels. The odometer on Eric's car read 65,480 miles before he left Long Island. When he arrived at his cousin's house in Orange County, his odometer read 65,565 miles. The trip distance was actually 90 miles. On a trip from New York City to Albany, his odometer display changed from 66,450 miles to 66,620 miles. Assuming that the odometer is always incorrect by the same percentage, how far did he actually travel?

6 Harry was stopped on the New York State Thruway for driving at an excessive speed. The officer's radar recorder indicated a speed of 80 mph. Harry claimed that although he was driving too fast, his speedometer recorded a speed of only 70 miles per hour. If the radar has a margin of error of ±3 mph, what is the minimum percentage error of Harry's speedometer?

Exercises 7–10: The temperature of Maddy's oven is always off by the same percentage error. She must set the oven temperature to 400° when she uses a recipe that calls for roasting at 375°. Finish the chart for Maddy so that she can easily convert from recipe temperature to displayed oven temperature. Round all temperatures to the *nearest 5 degrees*.

	Desired temperature	Setting on Maddy's Oven
7	325	
8	350	
	375	400
9	400	
10	425	

Exercises 11–12: A man weighed 150 pounds and his wife weighed 125 pounds on his doctor's accurate scale. Wearing identical clothes, the man later weighed himself on a scale in the shopping center. The shopping center scale displayed a weight of 153 pounds.

11 If the shopping center scale was off by a constant *amount*, what would his wife weigh on the shopping center scale?

12 If the shopping center scale was off by a constant *percentage error*, what would his wife weigh on the shopping center scale?

Triangles and Measurement

Proportions in a Right Triangle

If an altitude is drawn to the hypotenuse of a right triangle, the three theorems on the next page are always true.

Right Triangle Proportions

The two interior triangles are similar to each other and to the original triangle.	(figure: right triangle ABC with altitude CD, subdividing into triangles I and II)	$\triangle I \sim \triangle II$ $\triangle I \sim \triangle ABC$ $\triangle II \sim \triangle ABC$
The altitude is the **mean proportional** between the measures of the segments of the hypotenuse.	(figure: right triangle with altitude m, segments x and y)	$\dfrac{x}{m} = \dfrac{m}{y}$ or $m^2 = xy$
The measure of each leg of the original triangle is the **mean proportional** between the hypotenuse and the segment of the hypotenuse that is adjacent to that leg.	(figure: right triangle with legs b and a, altitude, segments x and y, hypotenuse h)	$\dfrac{h}{b} = \dfrac{b}{x}$ or $b^2 = hx$ and $\dfrac{h}{a} = \dfrac{a}{y}$ or $a^2 = hy$

Note: If x is the **mean proportional** between a and b, the proportion is true:

$\dfrac{a}{x} = \dfrac{x}{b}$

MODEL PROBLEMS

1 In right triangle ABC, altitude CD is drawn to the hypotenuse AB. If $AD = 4$ and $DB = 9$, find the lengths of (a) CD, (b) AC, and (c) CB.

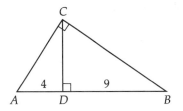

SOLUTIONS

Use a proportion.   Substitute.   Cross multiply.   Find the positive square root of each side.

a $\dfrac{\text{left segment}}{\text{altitude}} = \dfrac{\text{altitude}}{\text{right segment}}$ $\dfrac{4}{CD} = \dfrac{CD}{9}$ $(CD)^2 = 36$ $CD = 6$

b $\dfrac{\text{hypotenuse}}{\text{leg}} = \dfrac{\text{leg}}{\text{adjacent segment}}$ $\dfrac{13}{AC} = \dfrac{AC}{4}$ $(AC)^2 = 4 \times 13$ $AC = 2\sqrt{13}$

c $\dfrac{\text{hypotenuse}}{\text{leg}} = \dfrac{\text{leg}}{\text{adjacent segment}}$ $\dfrac{13}{CB} = \dfrac{CB}{9}$ $(CB)^2 = 9 \times 13$ $CB = 3\sqrt{13}$

Answer: $CD = 6$, $AC = 2\sqrt{13}$, $CB = 3\sqrt{13}$

2 In right triangle *ABC*, altitude *CD* is drawn to the hypotenuse *AB*. If $CD = 6$, $AD = x$, and $DB = x + 5$, find the lengths of *AD* and *DB*.

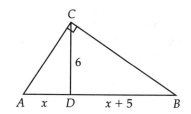

SOLUTION

Set up a proportion.	$\dfrac{AD}{CD} = \dfrac{CD}{BD}$
Substitute known values.	$\dfrac{x}{6} = \dfrac{6}{x+5}$
Cross multiply.	$x(x+5) = 36$
Distribute the *x*.	$x^2 + 5x = 36$
Subtract 36 from each side.	$x^2 + 5x - 36 = 0$
Factor.	$(x+9)(x-4) = 0$
Set each factor equal to zero and solve.	$(x+9) = 0$ or $(x-4) = 0$
A negative answer is not a measurement.	$x \neq -9$ or $x = 4$

Answer: $AD = 4$ and $DB = x + 5 = 9$

Practice

1 Find the mean proportional between 4 and 25.

(1) 10
(2) 14.5
(3) 15
(4) 20

Exercises 2–4: Use the diagram below.

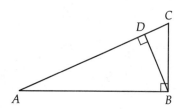

2 If $AD = 7$ and $DC = 3$, find *DB*.

(1) $\sqrt{10}$
(2) 4
(3) $\sqrt{21}$
(4) 5

3 If $CB = 6$ and $AC = 9$, find *DC*.

(1) $3\sqrt{6}$
(2) 7.5
(3) 4
(4) $3\sqrt{2}$

4 If $AD = 7$ and $DC = 4$, find *AB*.

(1) $\sqrt{77}$
(2) $2\sqrt{11}$
(3) $2\sqrt{7}$
(4) $\sqrt{11}$

Exercises 5–8: Use the diagram below.

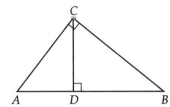

5 If $AD = 5$ and $DB = 12$, what is the value of CD?

6 If $CD = 3\sqrt{5}$ and $DB = 9$, what is the value of AD?

7 If $BC = 3\sqrt{2}$ and $DB = 3$, what is the value of AB?

8 If $AD = 5$ and $DB = 6$, what is the value of AC?

Exercises 9–12: Find the values of x, y, and z. Any radical answer should be left in simplest radical form.

9
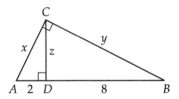

10
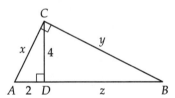

11
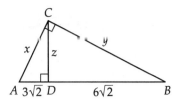

12
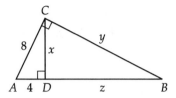

Exercises 13–14: Find the values of x, y, and z to the nearest tenth.

13
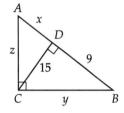

14
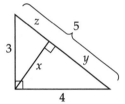

The Pythagorean Theorem

As reviewed in Chapter 6, the Pythagorean theorem states that in a right triangle, the sum of the squares of the legs equals the square of the hypotenuse, or $a^2 + b^2 = c^2$. Conversely, if the lengths of any triangle satisfy the Pythagorean theorem, then the triangle is a right triangle.

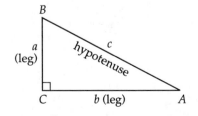

MODEL PROBLEMS

1 If the hypotenuse of a triangle is 4 inches more than one leg and 2 inches more than another, then what is the perimeter of the triangle?

SOLUTION

Let x represent the hypotenuse. Then the legs can be represented by $x - 2$ and $x - 4$.

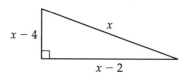

The Pythagorean theorem can help us find x.

Write the Pythagorean theorem.	$a^2 + b^2 = c^2$
Substitute.	$(x - 4)^2 + (x - 2)^2 = x^2$
Square each binomial.	$(x^2 - 8x + 16) + (x^2 - 4x + 4) = x^2$
Combine like terms.	$2x^2 - 12x + 20 = x^2$
Subtract x^2 from each side.	$x^2 - 12x + 20 = 0$
Factor.	$(x - 10)(x - 2) = 0$
Set each factor equal to zero and solve.	$(x - 10) = 0$ or $(x - 2) = 0$
	$x = 10$ or $x = 2$

Check: When $x = 2$, then $x - 4 = -2$. Since a length can never be negative, we must reject this value. When $x = 10$, then $x - 2 = 8$ and $x - 4 = 6$.

Add the sides. $10 + 8 + 6 = 24$

Answer: The perimeter is 24 inches.

2 To the nearest tenth of an inch, find the base of a rectangle whose diagonal is 25 inches and whose height is 16 inches.

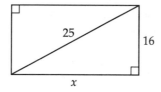

SOLUTION

Since the part of the triangle below the diagonal is a right triangle, we can use the Pythagorean theorem.

Write the Pythagorean theorem.	$a^2 + b^2 = c^2$
Substitute.	$16^2 + x^2 = 25^2$
Evaluate exponents.	$256 + x^2 = 625$
Simplify.	$x^2 = 625 - 256$
	$x^2 = 369$
	$x = \sqrt{369}$ or 19.2

Answer: The base to the nearest tenth is 19.2 inches long.

Practice

1. Which of the following are the sides of a right triangle?
   (1) 1, 2, 3
   (2) 1.5, 2, 2.5
   (3) 3.5, 4.5, 5.5
   (4) 5, 6, 7

2. Two college roommates, Henry and Harry, leave college at the same time. Henry travels south at 30 miles per hour and Harry travels west at 40 miles per hour. How far apart are they at the end of one hour?
   (1) 50 miles
   (2) 60 miles
   (3) 70 miles
   (4) 80 miles

3. The legs of one right triangle have a ratio of 5 : 12. If the hypotenuse is 65 feet in length, what is the length of the shorter leg?
   (1) 63.88 feet
   (2) 60 feet
   (3) 25 feet
   (4) 13 feet

4. A ladder 39 feet long leans against a building and reaches the bottom ledge of a window. If the foot of the ladder is 15 feet from the foot of the building, how high is the window ledge above the ground? (Hint: Buildings usually make right angles with the ground.)
   (1) 24 feet
   (2) 24.2 feet
   (3) 36 feet
   (4) 41.8 feet

Exercises 5–8: a and b represent the legs of a right triangle, and c represents the hypotenuse. Find the missing side. If the answer is not an integer, then find the answer both in simplest radical form and to the nearest tenth.

5. $a = 8, b = 15$

6. $a = 20, c = 25$

7. $a = 5, c = 15$

8. $b = 15, c = 17$

9. In an isosceles triangle, each of the equal sides is 26 centimeters long and the base is 20 centimeters long. Find the area. (Hint: First, draw the altitude to the base.)

10. If a right triangle has sides of lengths $x + 9$, $x + 2$, and $x + 10$, find the value of x.

11. If a rectangle has a diagonal of 20 feet and a side of 12 feet, find the perimeter and area of the rectangle.

12. If the base of an isosceles triangle is 12 inches and the altitude is 8 inches, what is the perimeter of the triangle?

Special Right Triangles

Any family of similar triangles has fixed ratios of sides. Two common triangles have ratios that are useful to memorize. These triangles are referred to by their angle measures: the 45°–45°–90° triangle and the 30°–60°–90° triangle.

The 45°–45°–90° Triangle A triangle with angle measures of 45°, 45°, and 90° is an isosceles right triangle. Since the triangle is isosceles, the ratio comparing the legs is 1 : 1. If we let a triangle have legs of length 1 and use the Pythagorean theorem, we find that the hypotenuse is $\sqrt{1^2 + 1^2}$ or $\sqrt{2}$. Therefore the ratio of sides for a 45°–45°–90° triangle is $1 : 1 : \sqrt{2}$.

To Find the Lengths of the Sides of a 45°–45°–90° Triangle

- If you know the length of a leg, you can find the hypotenuse by multiplying by $\sqrt{2}$.
- If you know the length of the hypotenuse, you can find the leg by dividing by $\sqrt{2}$.
- The two legs have the same length.

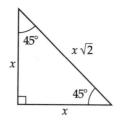

MODEL PROBLEMS

1 To the nearest tenth, find the distance from second base to home plate on a baseball diamond where the bases are 90 feet apart.

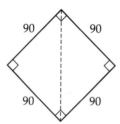

SOLUTION

The line from second base to home plate splits the field into two isosceles right triangles, with legs 90 feet long. The distance is simply $90 \times \sqrt{2}$ or approximately 127.3 feet.

2 If the hypotenuse of a right isosceles triangle is 10, find the exact length of the legs.

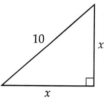

SOLUTION

To find the legs, simply divide 10 by $\sqrt{2}$.

$$\frac{10}{\sqrt{2}} = \frac{10}{\sqrt{2}} \cdot \frac{\sqrt{2}}{\sqrt{2}} = \frac{10\sqrt{2}}{2} = 5\sqrt{2}$$

The 30°–60°–90° Triangle A triangle with angle measures of 30°, 60°, and 90° has a ratio of sides of $1 : \sqrt{3} : 2$. This ratio is easy to verify using an equilateral triangle.

In the equilateral triangle shown, $AB = BC = AC = 10$.

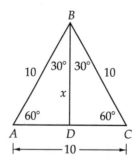

The altitude $\overline{BD}$ bisects the base so that $AD = DC = 5$. Using the Pythagorean theorem for triangle ABD, we can find BD:

$$(AD)^2 + (BD)^2 = (AB)^2$$
$$5^2 + (BD)^2 = 10^2$$
$$25 + (BD)^2 = 100$$
$$(BD)^2 = 75$$
$$BD = \sqrt{75} = \sqrt{25 \cdot 3} = 5\sqrt{3}$$

Therefore the sides are, from shortest to longest, 5, 5√3, 10. These sides are in the ratio 1 : √3 : 2. Any triangle with this ratio of sides is a 30°–60°–90° triangle.

To Find the Lengths of the Sides of a 30°–60°–90° Triangle

- If you know the hypotenuse, divide by 2 to find the length of the leg opposite the 30° angle (the short leg).
- If you know the short leg, multiply by 2 to find the hypotenuse. Multiply by √3 to find the leg opposite the 60° angle (the long leg).
- If you know the long leg, divide by √3 to find the short leg.

MODEL PROBLEMS

1 If the perimeter of equilateral triangle ABC is 27, what is the altitude BD rounded to the nearest tenth?

SOLUTION

Draw the situation.

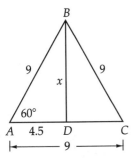

AC is one-third the perimeter.
AD is one-half of AC since $\overline{BD}$ bisects $\overline{AC}$.
BD is the long leg of △ABD and AD is the short leg.

$AC = 27 \div 3 = 9$
$AD = 9 \div 2 = 4.5$
$BD = AD \times \sqrt{3}$
$BD = 4.5\sqrt{3} = 7.79422\ldots$

Answer: $BD \approx 7.8$

2 If the width of a rectangle is 4 inches and the diagonal is 8 inches, what is the length of the rectangle in simplest radical form?

SOLUTION

Draw the situation.

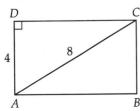

Examine the ratio AC : AD.
This ratio of sides for △ACD corresponds to the 30°–60°–90° relationship.
Find the length of CD.

$8 : 4 = 2 : 1$
AC is the hypotenuse, AD is the short leg, and CD must be the long leg.
$CD = AD \times \sqrt{3}$
$BD = 4\sqrt{3}$

Answer: The length of the rectangle is $4\sqrt{3}$.

Practice

1. If the perimeter of a square is $8\sqrt{2}$, then what is the length of the diagonal?
   - (1) 8
   - (2) $4\sqrt{2}$
   - (3) 4
   - (4) 2

2. If the longest side of an isosceles right triangle is 12 feet long, how long are the other sides?
   - (1) $6\sqrt{2}$ feet and $6\sqrt{2}$ feet
   - (2) 6 feet and $6\sqrt{3}$ feet
   - (3) 6 feet and $6\sqrt{2}$ feet
   - (4) $4\sqrt{3}$ feet and $4\sqrt{3}$ feet

3. If the altitude of an equilateral triangle is $21\sqrt{3}$ inches, what is the perimeter of the triangle?
   - (1) 63 inches
   - (2) 99.4 inches
   - (3) 109.1 inches
   - (4) 126 inches

4. If the legs of a right triangle are $2\sqrt{3}$ and 6, what is the length of the hypotenuse?
   - (1) $4\sqrt{3}$
   - (2) $4\sqrt{6}$
   - (3) $6\sqrt{2}$
   - (4) 12

Exercises 5–10: $\triangle XYZ$ is a 30°–60°–90° triangle. The length of one side is given. Find the lengths of the other two sides in radical form.

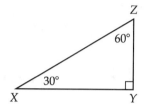

5. $YX = 36$
6. $YZ = 9$
7. $ZX = 6\sqrt{3}$
8. $YZ = 4\sqrt{3}$
9. $YX = 27$
10. $ZX = 10$

Exercises 11–16: $\triangle ABC$ is a 45°–45°–90° triangle. The length of one side is given. Find the lengths of the other two sides in radical form.

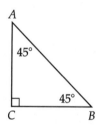

11. $AB = 12$
12. $AC = 3\sqrt{2}$
13. $BC = 4$
14. $AB = 17$
15. $BC = 10\sqrt{2}$
16. $AB = 8\sqrt{3}$

Trigonometric Ratios

The ratio of the lengths of any two sides of a right triangle is called a **trigonometric ratio**. These ratios refer to right triangles only. The three most common ratios are **sine**, **cosine**, and **tangent**. Their abbreviations are *sin*, *cos*, and *tan*.

The following chart of definitions and formulas refers to the right triangle ABC where $\angle C = 90°$.

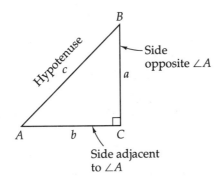

Definition	Formulas	
The **sine** of an acute angle equals the length of the leg opposite the angle divided by the hypotenuse.	$\sin \angle A = \dfrac{\text{leg opposite } \angle A}{\text{hypotenuse}}$	$= \dfrac{a}{c}$
	$\sin \angle B = \dfrac{\text{leg opposite } \angle B}{\text{hypotenuse}}$	$= \dfrac{b}{c}$
The **cosine** of an acute angle equals the length of the leg adjacent the angle divided by the hypotenuse.	$\cos \angle A = \dfrac{\text{leg adjacent to } \angle A}{\text{hypotenuse}}$	$= \dfrac{b}{c}$
	$\cos \angle B = \dfrac{\text{leg adjacent to } \angle B}{\text{hypotenuse}}$	$= \dfrac{a}{c}$
The **tangent** of an acute angle equals the length of the leg opposite the angle divided by the leg adjacent to the angle.	$\tan \angle A = \dfrac{\text{leg opposite } \angle A}{\text{leg adjacent to } \angle A}$	$= \dfrac{a}{b}$
	$\tan \angle B = \dfrac{\text{leg opposite } \angle B}{\text{leg adjacent to } \angle B}$	$= \dfrac{b}{a}$

Remember these with "SOH-CAH-TOA":
$\text{Sin} = \dfrac{\text{Opp}}{\text{Hyp}}$,
$\text{Cos} = \dfrac{\text{Adj}}{\text{Hyp}}$, and
$\text{Tan} = \dfrac{\text{Opp}}{\text{Adj}}$.

There are a number of formulas that show the relationships between the trigonometric functions. Use the table to verify these for yourself.

- $\sin \angle A = \cos \angle B$
- $\sin \angle B = \cos \angle A$
- $\tan \angle A = \dfrac{\sin \angle A}{\cos \angle A}$
- $\tan \angle B = \dfrac{\sin \angle B}{\cos \angle B}$

It is important to know that the value of the trigonometric ratio does not depend on the size of the right triangle. It depends only on the angle. To illustrate, consider three overlapping right triangles that share a common acute angle, $\angle A$. Since all three triangles are similar (by angle-angle similarity), the corresponding sides are proportional.

Triangles and Measurement

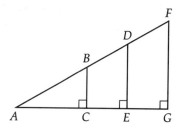

Therefore, $\dfrac{BC}{AB} = \dfrac{DE}{AD} = \dfrac{FG}{AF} = \sin \angle A$. This relationship is true for $\cos \angle A$ and $\tan \angle A$.

Note: When you use a calculator to find trigonometric functions, first be sure that the calculator is in degree mode. The or inverse key will allow you to find angles when you know any trigonometric ratios of the triangle, usually found from the lengths of the sides. See the appendix.

MODEL PROBLEM

A right triangle ABC has sides 5, 12, and 13. Find the sine, cosine, and tangent for $\angle A$ and $\angle B$, rounding your answers to the nearest ten thousandth. Find $\angle A$ and $\angle B$ to the nearest degree.

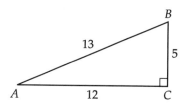

SOLUTION

$\sin \angle A = \dfrac{\text{opp}}{\text{hyp}} = \dfrac{5}{13} = 0.3846$ $\qquad$ $\sin \angle B = \dfrac{\text{opp}}{\text{hyp}} = \dfrac{12}{13} = 0.9231$

$\cos \angle A = \dfrac{\text{adj}}{\text{hyp}} = \dfrac{12}{13} = 0.9231$ $\qquad$ $\cos \angle B = \dfrac{\text{adj}}{\text{hyp}} = \dfrac{5}{13} = 0.3846$

$\tan \angle A = \dfrac{\text{opp}}{\text{adj}} = \dfrac{5}{12} = 0.4167$ $\qquad$ $\tan \angle B = \dfrac{\text{opp}}{\text{adj}} = \dfrac{12}{5} = 2.4000$

Using a calculator to work backward from any of these decimals, we find that $\angle A = 23°$ and $\angle B = 67°$ when rounded to the nearest degree. Since $23° + 67° + 90° = 180°$, our answer checks out.

Practice

1 Which equation is true?

(1) $\sin 30° = \sin 60°$
(2) $\sin 45° = \cos 45°$
(3) $\tan 45° = \cos 45°$
(4) $\sin 90° = \tan 60°$

2 Which statement about trigonometric ratios is true?

(1) The sine of an angle decreases as the angle increases from 0° to 90°.
(2) The cosine of an angle increases as the angle increases from 0° to 90°.
(3) The cosine of an angle decreases as the angle increases from 0° to 90°.
(4) The tangent of an angle decreases as the angle increases from 0° to 90°.

3   Which statement about trigonometric ratios is true?

   (1) The cosine of an angle is always less than or equal to 1.
   (2) The tangent of an angle is always less than or equal to 1.
   (3) The cosine of an angle is always greater than or equal to 1.
   (4) The tangent of an angle is always greater than or equal to 1.

4   If we know that tan $\angle A = 0.75$ and the side adjacent to $\angle A$ is 12 units long, how long is the side opposite $\angle A$?

   (1) 85
   (2) 37
   (3) 16
   (4) 9

Exercises 5–12: Find the trigonometric function or the angle measure as indicated. Use $\triangle RAG$.

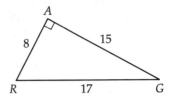

5   sin $\angle R$ to the nearest hundredth
6   cos $\angle R$ to the nearest hundredth
7   tan $\angle R$ to the nearest hundredth
8   $\angle R$ to the nearest degree
9   sin $\angle G$ to the nearest hundredth
10 cos $\angle G$ to the nearest hundredth
11 tan $\angle G$ to the nearest hundredth
12 $\angle G$ to the nearest degree
13 What angle has a tangent of 1?
14 What angle has a sine of $\frac{1}{2}$?
15 What angle has a tangent of $\sqrt{3}$?

Indirect Measure

Very often, it is necessary to find the measurement of a distance or of an angle in a situation where it is impossible to use a ruler or another measuring tool. The height of a tree, the distance across a lake, and the angle of elevation of the sun are examples of measurements that are difficult to find directly. These measurements can be obtained by imagining a right triangle in the situation and applying the principles of trigonometry.

With trigonometry, if you know the length of one side of a right triangle, and either one more side or one angle, then you can find all the lengths and angle measures of the triangle.

To Solve for Parts of a Right Triangle Using Trigonometric Ratios

- Make a sketch of the triangle that contains the given information and label clearly what is given and what needs to be found.
- Decide which ratio connects the given information with the side or angle that needs to be found.
- Substitute the known information in this ratio.
- Solve for the unknown value.

Triangles and Measurement

In many triangle problems, the measured angle is referred to as an *angle of elevation* or *angle of depression*. An **angle of elevation** is the angle formed by a horizontal line and the upward line of sight to an object. An **angle of depression** is the angle formed by a horizontal line and the downward line of sight to an object. The measure of the angle of depression equals the measure of the angle of elevation as they are alternate interior angles.

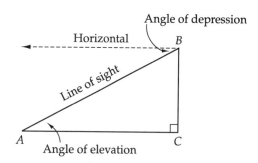

Finding a Length When you need to find the length of the side of a right triangle, choose the trigonometric ratio that compares the side you seek with a side that is given. Once two sides are known, you can find the third side with the Pythagorean theorem instead of trigonometric ratios.

 MODEL PROBLEMS

1. An engineer wishes to estimate the height of a building. He knows that when he stands 225 feet from the base of the building, the angle of elevation to the roof is 18°. To the nearest foot, what is the approximate height of the building?

SOLUTION

Draw and label a right triangle clearly. From the 18° angle, the height of the building on the opposite side is x, and the distance of 225 feet is the adjacent side.

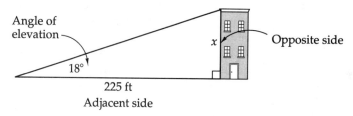

The tangent ratio involves the opposite and adjacent sides. $\tan 18° = \dfrac{x}{225}$

Solve for x. $x = 225(\tan 18°)$

Use a calculator to find the tangent value. $x = 225(0.3249\ldots)$

$x = 73.1069\ldots$

Answer: The building is about 73 feet tall.

2 From the top of a lighthouse 160 feet above sea level, the angle of depression to a boat at sea is 25°. To the nearest foot, what is the horizontal distance from the boat to the base of the lighthouse?

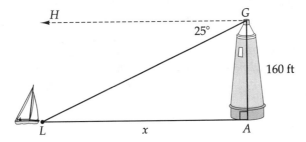

SOLUTION

The angle of depression, $\angle HGL$, is not inside the right triangle LAG.

Find $\angle LGA$, the angle complementary to $\angle HGL$.

$\angle LGA = 90° - \angle HGL$

$\angle LGA = 90° - 25° = 65°$

From $\angle LGA$, x is opposite and 160 is adjacent. We will use the tangent ratio.

$\tan 65° = \dfrac{x}{160}$

Solve for x.

$x = 160(\tan 65°)$

$x = 160(2.1445\ldots)$

$x = 343.1211\ldots$

Answer: The boat is about 343 feet from the base of the lighthouse.

Note: An alternate solution: Use the angle of elevation from the boat, which equals the angle of depression from the lighthouse, and proceed as in Model Problem 1.

Finding an Angle When you need to find an angle in a right triangle, you should use a trigonometric ratio that compares two known lengths. The decimal answer is converted to a degree measure using the inverse key. If one of the acute angles is already known, you can subtract it from 90° to find the other.

MODEL PROBLEMS

1 In $\triangle ABC$, $AC = 10$ and $BC = 6$. Find $\angle A$ to the nearest degree.

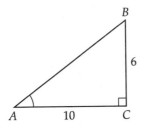

SOLUTION

$\overline{AC}$ is adjacent to $\angle A$ and $\overline{BC}$ is opposite $\angle A$. Therefore, we choose to use tangent of $\angle A$.

$\tan \angle A = \dfrac{\text{opp}}{\text{adj}} = \dfrac{6}{10} = 0.6$

Using a calculator to find the inverse tangent of 0.6, we find that $\angle A$ measures about 30.96°

Answer: $\angle A \approx 31°$

2 If a 20-foot ladder reaches 18 feet up a wall, what angle does the ladder make with the ground, to the nearest degree?

SOLUTION

Sketch the basic details of the situation.

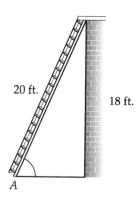

Since we are working with the hypotenuse and a side opposite our angle, we will use the sine ratio:

$$\sin \angle A = \frac{18}{20} = 0.9$$

Using a calculator to find the inverse sine of 0.9, we find that $\angle A$ measures about 64.16°.

Answer: The ladder makes an angle of about 64° with the ground.

3 A pilot takes off at point A and ascends at a fixed angle with the level runway. If he flies a distance of 2,500 yards but covers only a ground distance of 2,200 yards, then what is his angle of ascent?

SOLUTION

Sketch the details of the situation.

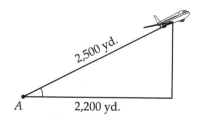

Since we are working with the hypotenuse and a side adjacent to our angle, we will use the cosine ratio:

$$\cos \angle A = \frac{2200}{2500} = 0.88$$

Using a calculator to find the inverse cosine of 0.88, we find that $\angle A$ measures about 28.36°.

Answer: The plane is ascending at an angle of about 28°.

 Practice

1 In $\triangle ABC$, $AB = 5$ and $AC = 4$. What is the measure of $\angle A$ to the nearest degree?

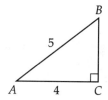

(1) 53°
(2) 51°
(3) 39°
(4) 37°

2 Wire braces are needed for the 80-foot ridgepole of a circus tent. Each brace is supposed to make a 60° angle with the ground. Find, to the nearest foot, how long each brace should be.

(1) 46 feet
(2) 69 feet
(3) 92 feet
(4) 139 feet

3 At a point 40 feet from the foot of an oak tree, the angle of elevation to the top of the tree is 42°. Find the height of the tree to the nearest foot.

   (1) 54 feet
   (2) 44 feet
   (3) 36 feet
   (4) 30 feet

4 In a Rochester parking garage, the levels are 23 feet apart. Each up-ramp to the next level is 144 feet long. Find the measure of the angle of elevation of the incline for each ramp to the nearest tenth of a degree.

   (1) 9.2°
   (2) 9.1°
   (3) 8.1°
   (4) 3.9°

5 Since the distance across Smokey Swamp cannot be measured directly, Margaret (at point M) uses her knowledge of trigonometry. Using the accompanying diagram and the indicated measurements, find, to the nearest meter, the distance across the swamp.

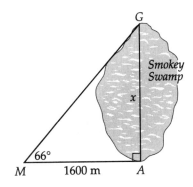

6 If the diagonal of a rectangular sundeck makes an angle of 67° with one side of the deck and the length of the diagonal is 24 feet, what are the dimensions, to the nearest foot, of the sundeck?

7 A ladder leaning against a house makes an angle of 70° with the ground. If the ladder is 25 feet long, find, to the nearest foot.

   a how far the foot of the ladder is from the base of the house
   b how high above the ground the ladder touches the house

8 At a certain time, the angle at the sun between Earth and Jupiter is 90°. The distance from Earth to the sun is 1 astronomical unit and the distance between Jupiter and the sun is 5 astronomical units. At this time, what is the angle separating the sun and Jupiter, when seen from Earth? Round your answer to the nearest degree.

9 An observer lying at the edge of a 200-foot cliff finds that the angle of depression out to a distant farmhouse is 40°. To the nearest foot, how far is the farmhouse from the foot of the cliff?

10 Ken Pollitt's distance from Jericho's water tower is 1,000 feet. If Ken is 6 feet tall and his eyes sight the top of the tower at an angle of 30°, what is the height of the tower?

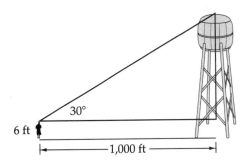

11 A kite is flying at the end of a taut 175-foot string. Patricia, who is 5 feet tall, holds the string, which reaches up into the air at an angle of 75° with the horizontal. Find, to the nearest foot, how high *above the ground* the kite is.

12 From the top of a Maine lighthouse 100 feet above sea level, the angles of depression of two lobster boats in line with the foot of the lighthouse are seen to be 18° and 32°. Find, to the nearest foot, the distance between the boats.

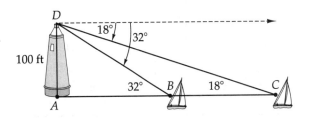

Coordinate Geometry

The Distance Formula

See Chapter 5 to review plotting points in the coordinate plane.

The distance between any two points (or the length of a line segment) in the co-ordinate plane can be used to show that two segments are congruent or to find the perimeter or the area of a geometric figure.

To Find the Distance Between Two Points in the Coordinate Plane

- Locate the ordered pair for each point.
- Draw a vertical line through one point and a horizontal line through the other to form a right triangle.
- Find the distance between the endpoints of each leg of the triangle.
- Use the Pythagorean theorem to find the length of the hypotenuse, which is the distance between the two points.

Pythagorean theorem: $\text{leg}^2 + \text{leg}^2 = \text{hypotenuse}^2$.

 MODEL PROBLEM

Use the accompanying figure to find the distance between $A(-2, 5)$ and $B(4, -3)$

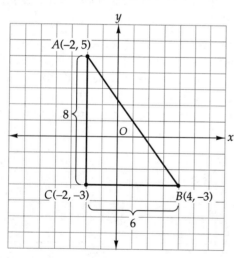

SOLUTION

Draw $\overline{AC}$ and $\overline{CB}$ to form a right triangle.

In right triangle ABC, $(AB)^2 = (AC)^2 + (CB)^2$

$$(AB)^2 = 8^2 + 6^2$$
$$(AB)^2 = 64 + 36 = 100$$
$$AB = \sqrt{100}$$
$$AB = 10$$

In the solution to the previous Model Problem, we *counted squares* to find the lengths of each leg. Below, we repeat the solution using *absolute values* to find the lengths.

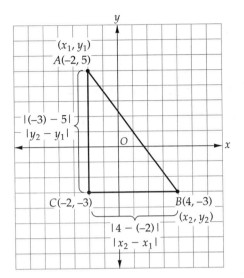

To Use Absolute Value to Find the Distance Between Two Points

- Graph the two points, A and B, on a coordinate grid.
- Draw $\overline{AC}$ and $\overline{CB}$ to form right triangle ABC.
- The length of leg CB is equal to the *absolute value* of the difference between the x-values of the endpoints: $CB = |x_2 - x_1| = |4 - (-2)| = 6$
- The length of leg AC is equal to the *absolute value* of the difference between the y-values of the endpoints: $AC = |y_2 - y_1| = |(-3) - 5| = 8$
- Thus, the distance d between $A(-2, 5)$ and $B(4, -3)$ is the length of segment AB:
$AB = \sqrt{6^2 + 8^2} = \sqrt{36 + 64} = \sqrt{100} = 10$

Having used absolute value and the Pythagorean theorem to find the distance, we can now write a general statement for the distance formula. This formula can be used to find the distance d between any two points (x_1, y_1) and (x_2, y_2) in the coordinate plane.

$$\text{Distance formula} \quad d = \sqrt{(x_2 - x_1)^2 + (y_2 - y_1)^2}$$

 MODEL PROBLEMS

1 Find the length of the line segment that joins $R(-3, 6)$ and $J(0, 2)$.

SOLUTION

Use the distance formula.

$d = \sqrt{(x_2 - x_1)^2 + (y_2 - y_1)^2}$

$d = \sqrt{[0 - (-3)]^2 + (2 - 6)^2} = \sqrt{3^2 + (-4)^2} = \sqrt{9 + 16} = \sqrt{25} = 5$

Answer: $RJ = 5$

2 Find (a) the area and (b) the circumference of a circle if the center is at C(3, −3) and the circle passes through point A(7, 1).

SOLUTION

The radius of the circle is AC. Use the distance formula to find its length.

$d = \sqrt{(x_2 - x_1)^2 + (y_2 - y_1)^2}$

Thus, $d_{AC} = \sqrt{(7-3)^2 + [1-(-3)]^2} = \sqrt{4^2 + 4^2} = \sqrt{16+16} = \sqrt{16 \cdot 2} = 4\sqrt{2}$

a Now substitute $4\sqrt{2}$ for the *radius* in the formula for the area of a circle.

$A = \pi r^2$
$= \pi(4\sqrt{2})^2$
$= \pi(16 \cdot 2) = 32\pi$

b Substitute $4\sqrt{2}$ for the *radius* in the formula for the circumference of a circle.

$C = 2\pi r$
$= 2\pi(4\sqrt{2})$
$= 8\sqrt{2}\pi$

Answer: (a) Area = 32π (b) Circumference = $8\sqrt{2}\pi$

3 Use the distance formula to show that △ABC, with vertices A(−4, 3), B(6, 1), and C(2, −3), is a right triangle.

SOLUTION

For the triangle to be a right triangle, the Pythagorean theorem must hold. Use the distance formula to find the length of each side of the triangle and substitute the values in the Pythagorean theorem.

$d = \sqrt{(x_2 - x_1)^2 + (y_2 - y_1)^2}$

$d_{AB} = \sqrt{[6-(-4)]^2 + (1-3)^2}$

$d_{AB} = \sqrt{10^2 + (-2)^2}$

$d_{AB} = \sqrt{10^2 + 4^2} = \sqrt{104}$

$d_{BC} = \sqrt{(2-6)^2 + [(-3)-1]^2}$

$d_{BC} = \sqrt{(-4)^2 + (-4)^2}$

$d_{BC} = \sqrt{16+16} = \sqrt{32}$

$d_{AC} = \sqrt{[2-(-4)]^2 + [(-3)-3]^2}$

$d_{AC} = \sqrt{6^2 + (-6)^2}$

$d_{AC} = \sqrt{36+36} = \sqrt{72}$

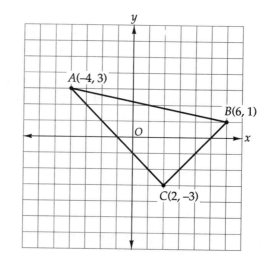

The longest side, the hypotenuse, has to be $\sqrt{104}$. Substitute the values for the legs and the hypotenuse in the Pythagorean theorem.

$a^2 + b^2 = c^2$
$(\sqrt{32})^2 + (\sqrt{72})^2 = (\sqrt{104})^2$
$32 + 72 = 104$
$104 = 104$

Answer: Since the equation for the Pythagorean theorem is true, triangle ABC is a right triangle.

4 Show that quadrilateral NAME, with vertices $N(2, 4)$, $A(-5, 2)$, $M(-2, -1)$, and $E(5, 1)$, is a parallelogram and state the reasons for your conclusion.

SOLUTION

While there are other methods of showing that NAME is a parallelogram (such as slope, or slope and distance), just the distance formula can be used.

$d_{NA} = \sqrt{(-5-2)^2 + (2-4)^2} = \sqrt{(-7)^2 + (-2)^2} = \sqrt{49+4} = \sqrt{53}$

$d_{AM} = \sqrt{[-2-(-5)]^2 + (-1-2)^2} = \sqrt{3^2 + (-3)^2} = \sqrt{9+9} = \sqrt{18}$

$d_{ME} = \sqrt{[5-(-2)]^2 + [1-(-1)]^2} = \sqrt{7^2 + 2^2} = \sqrt{49+4} = \sqrt{53}$

$d_{EN} = \sqrt{(2-5)^2 + (4-1)^2} = \sqrt{(-3)^2 + (3)^2} = \sqrt{9+9} = \sqrt{18}$

Thus, $NA = ME = \sqrt{53}$ and $AM = EN = \sqrt{18}$

Answer: Since both pairs of opposite sides are congruent, quadrilateral NAME is a parallelogram.

Practice

Exercises 1–6: Find the distance between each pair of points. (Answers may be left in simplest radical form.)

1 (2, 5) and (6, 8)
  (1) $\sqrt{7}$
  (2) 5
  (3) $\sqrt{105}$
  (4) $\sqrt{233}$

2 (−6, 4) and (0, −4)
  (1) $2\sqrt{7}$
  (2) 6
  (3) 10
  (4) $4\sqrt{10}$

3 (−6, 4) and (4, 4)
  (1) $\sqrt{6}$
  (2) $\sqrt{8}$
  (3) $\sqrt{68}$
  (4) 10

4 (−6, −2) and (−6, 4)
  (1) $2\sqrt{5}$
  (2) 6
  (3) $2\sqrt{29}$
  (4) $2\sqrt{35}$

5 (−3, −5) and (−2, −6)
  (1) $\sqrt{2}$
  (2) $\sqrt{5}$
  (3) $\sqrt{18}$
  (4) $\sqrt{130}$

6 (2, 20) and (3, 5)
  (1) 4
  (2) $\sqrt{226}$
  (3) $\sqrt{298}$
  (4) $5\sqrt{26}$

7 Find the coordinates of the points A and B in each figure.

a
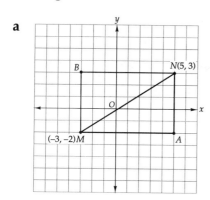

b
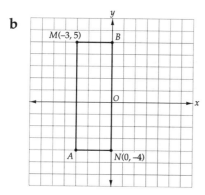

c
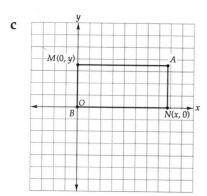

Exercises 8–11: Identify the triangle with the given vertices as isosceles or scalene. Explain your reasoning.

8 $P(2, 3)$, $A(5, 7)$, $T(1, 4)$

9 $A(3, 4)$, $B(5, -3,)$ $C(-2, 2)$

10 $M(-6, 2)$, $A(5, -1)$, $D(4, 4)$

11 $W(7, -1)$, $I(2, -2)$, $T(4, 9)$

12 Show that the triangle with vertices $(0, 0)$, $(2, 0)$, and $(1, \sqrt{3})$ is equilateral. Explain your reasoning.

13 What is the length of the diameter of a circle if the center is at point $(1, 2)$ and passes through point $(4, -2)$?

14 The vertices of quadrilateral $ABCD$ are $A(3, 2)$, $B(3, -1)$, $C(7, -1)$, and $D(7, 2)$. Show that the diagonals are congruent. Explain your reasoning.

15 Use the distance formula to show that triangle ABC, with vertices $A(-1, -2)$, $B(3, 2)$, and $C(1, 4)$, is a right triangle. Explain your reasoning.

The Midpoint Formula

The midpoint of a line segment can be used to show that a line segment is bisected by another line, to find the center of a circle, to aid in finding congruent segments, and to find the endpoint of a median.

> The **midpoint formula** states that for any two points $A(x_1, y_1)$ and $B(x_2, y_2)$ the midpoint of $\overline{AB}$ has coordinates $\left(\dfrac{x_1 + x_2}{2}, \dfrac{y_1 + y_2}{2}\right)$.

Since the midpoint of line segment $\overline{AB}$ lies halfway between the endpoints A and B, the midpoint is the average of the x-values and the average of the y-values.

MODEL PROBLEMS

1. Find the coordinates of M, the midpoint of the segment joining $A(1, 3)$ and $B(9, 5)$.

SOLUTION

Substitute the coordinates of the points in the midpoint formula.

$$M = \left(\frac{x_1 + x_2}{2}, \frac{y_1 + y_2}{2}\right)$$

$$M = \left(\frac{1 + 9}{2}, \frac{3 + 5}{2}\right) = (5, 4)$$

Answer: M (midpoint) $= (5, 4)$

2. If $A(-5, 4)$ and $B(x, y)$ are the endpoints of segment AB and $M(-2, 1)$ is the midpoint of $\overline{AB}$, then what are the coordinates of B?

SOLUTION

Substitute the known values in the midpoint formula.

$$M = \left(\frac{x_1 + x_2}{2}, \frac{y_1 + y_2}{2}\right)$$

$$M(-2, 1) = \left(\frac{-5 + x}{2}, \frac{4 + y}{2}\right), \text{ so that}$$

the x-coordinate of $M = -2 = \frac{-5 + x}{2}$

and

the y-coordinate of $M = 1 = \frac{4 + y}{2}$.

Solve for x: $-4 = -5 + x$ Solve for y: $2 = 4 + y$
$\phantom{\text{Solve for }x: } x = 1$ $\phantom{\text{Solve for }y: 2 = } y = -2$

Answer: The coordinates of B are $(1, -2)$.

ALTERNATIVE SOLUTION

To find the missing endpoint of a line segment if the midpoint and one endpoint are known, use the LOOP method.

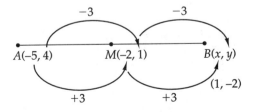

The distance from -5 to -2 is the same as the distance from -2 to x, so add 3 to -2.

Thus, $x = -2 + 3 = 1$.

The distance from 4 to 1 is the same as the distance from 1 to y, so add -3 to 1.

Thus, $y = 1 + (-3) = -2$.

The coordinates of $B(x, y) = (1, -2)$.

3. $\overline{QS}$ is the diameter of a circle whose center is $R(-1, 5)$. If point Q has coordinates $(-3, 2)$, what are the coordinates of point S?

SOLUTION

Draw a diagram and use the LOOP method.

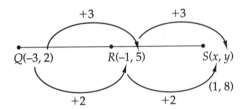

The coordinates of point S are $(1, 8)$.

ALTERNATIVE SOLUTION

The center $R(-1, 5)$ is the midpoint of $\overline{QS}$. Substitute in the midpoint formula.

$$M = \left(\frac{x_1 + x_2}{2}, \frac{y_1 + y_2}{2}\right)$$

$$R(-1, 5) = \left(\frac{-3 + x}{2}, \frac{2 + y}{2}\right), \text{ so that}$$

the x-coordinate of R, $-1 = \frac{-3 + x}{2}$ and the y-coordinate of R, $5 = \frac{2 + y}{2}$.

Solve for x: $-2 = -3 + x$ Solve for y: $10 = 2 + y$
$\phantom{\text{Solve for }x: -2 = } x = 1$ $\phantom{\text{Solve for }y: 10 = } y = 8$

Answer: The coordinates of point S are $(1, 8)$.

Coordinate Geometry

4 If $\triangle ABC$ has vertices $A(7, -3)$, $B(-1, 5)$, and $C(4, 8)$, what is the length, in simplest radical form, of the median from C to $\overline{AB}$?

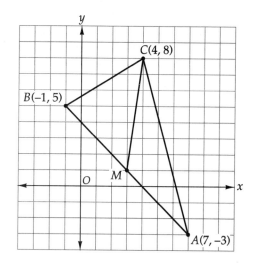

SOLUTION

- Use the midpoint formula to find the coordinates of the midpoint M of $\overline{AB}$.

$$M = \left(\frac{x_1 + x_2}{2}, \frac{y_1 + y_2}{2}\right)$$

$$M = \left(\frac{7 + (-1)}{2}, \frac{-3 + 5}{2}\right) = (3, 1)$$

- Then use the distance formula to find the length of $\overline{CM}$. The coordinates of the points are $C(4, 8)$ and $M(3, 1)$.

$$d = \sqrt{(x_2 - x_1)^2 + (y_2 - y_1)^2}$$

$$d_{CM} = \sqrt{(3 - 4)^2 + (1 - 8)^2} = \sqrt{(-1)^2 + (-7)^2}$$

$$= \sqrt{1 + 49} = \sqrt{50} = 5\sqrt{2}$$

Answer: Median $\overline{CM} = 5\sqrt{2}$

 Practice

Exercises 1–3: Find the midpoint of the line segment that connects each pair of points.

1 $(-2, 7)$ and $(5, -9)$

  (1) $(-3.5, 8)$
  (2) $(1.5, -1)$
  (3) $(2.5, 2)$
  (4) $(3.5, 6)$

2 $(-1, 0.5)$ and $(-4, 3.5)$

  (1) $(-0.25, -0.25)$
  (2) $(0.25, 1.75)$
  (3) $(1.25, -1.75)$
  (4) $(-2.5, 2)$

3 $(-2.7, -5.3)$ and $(-1.3, 5.5)$

  (1) $(-4, 2.1)$
  (2) $(-2, 0.1)$
  (3) $(0.7, 5.4)$
  (4) $(1.4, -3.3)$

Exercises 4–6: Find the coordinates of endpoint B when midpoint M of $\overline{AB}$ and endpoint A are given.

4 $A(4, -2)$ and $M(-2, -1)$

  (1) $(-8, 0)$
  (2) $(-6, 1)$
  (3) $(0, -4)$
  (4) $(1, -1.5)$

5 $A(-3, 2)$ and $M(7, 11)$

  (1) $(2, 6.5)$
  (2) $(10, 9)$
  (3) $(11, 22)$
  (4) $(17, 20)$

6 $A(-5, 1)$ and $M(4, -6)$

  (1) $\left(-\frac{1}{2}, -\frac{5}{2}\right)$
  (2) $(3, 13)$
  (3) $(9, -7)$
  (4) $(13, -13)$

7. Find the midpoint of the line segment that connects points $(3a, 7r)$ and $(7a, 3r)$.

8. Find endpoint B of the segment AB where M, the midpoint of $\overline{AB}$, has coordinates $(2x, 3y)$ and endpoint A has coordinates (x, y).

9. $\overline{AB}$ is the diameter of a circle whose center is $C(4, 7)$. If point A has coordinates $(-2, 3)$, what are the coordinates of point B?

10. The vertices of rhombus $GLAD$ are $G(-1, 1)$, $L(1, -3)$, $A(3, 1)$, and $D(1, 5)$. Draw the graph of the rhombus, find the midpoint of each side, and connect the midpoints in order. What kind of quadrilateral is formed?

11. If $\triangle SAD$ has vertices $S(-4, -3)$, $A(4, -1)$, and $D(-2, 3)$, find (a) the coordinates of M (the midpoint of $\overline{SA}$) and (b) the length of median $\overline{DM}$.

12. The vertices of triangle BAD are $B(1, 3)$, $A(7, 5)$, and $D(9, -3)$. If E is the midpoint of $\overline{BA}$ and F is the midpoint of $\overline{AD}$, show that (a) $\overline{EF}$ is parallel to $\overline{BD}$ and (b) $EF = \frac{1}{2}BD$.

13. If the coordinates of the vertices of $\triangle DAN$ are $D(-3, 0)$, $A(-5, -4)$, and $N(-7, 0)$, find (a) the slope of side $\overline{DA}$, (b) the slope of the altitude from N to side $\overline{DA}$, (c) the midpoint of $\overline{DA}$, and (d) the equation of the line drawn from N to the midpoint of $\overline{DA}$.

Exercises 14–16: Find the length of the median from point A to $\overline{BC}$ for each triangle ABC.

14. $A(2, 6)$, $B(-2, -4)$, $C(6, -2)$

15. $A(-2, 5)$, $B(-4, 1)$, $C(2, -4)$

16. $A(5, 4)$, $B(-1, -1)$, $C(4, -2)$

17. In $\triangle PQR$, M is the midpoint of $\overline{PQ}$. Show that $PM = MQ = RM$ when the coordinates of the vertices are $P(6, 0)$, $Q(0, -8)$, and $R(0, 0)$.

18. The coordinates of quadrilateral $ABCD$ are $A(5, 3)$, $B(3, 5)$, $C(-3, 5)$, and $D(-1, 3)$. Using coordinate geometry, show that the diagonals of quadrilateral $ABCD$ bisect each other.

Finding Areas in Coordinate Geometry

The areas of polygons in coordinate geometry can be found with the usual area formulas. It is especially easy to find areas when one or more sides of the figure are parallel to either of the axes. However, when the polygon does not have sides parallel to either of the axes, the problem of finding the area is more difficult and the method is more complex.

To Find the Area of a Polygon in the Coordinate Plane

- Plot the points and draw the figure.
- Enclose the figure in a rectangle that has the vertical and horizontal lines of the coordinate graph paper as sides and touches as many of the vertices of the figure as possible.
- Find the area of the rectangle.
- Find the area of each of the simple right triangles or quadrilaterals that are outside the given figure.
- Add the areas found in the previous step and subtract that total from the area of the rectangle. The answer is the area of the original polygon.

MODEL PROBLEMS

1 Find the area of triangle *RST* whose vertices are *R*(−5, 4), *S*(2, 1), and *T*(6, 5).

SOLUTION

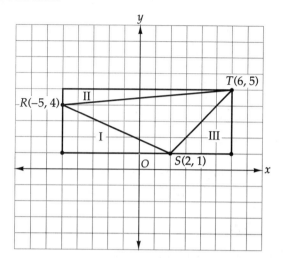

To find the area of the triangle, draw the rectangle as shown in the accompanying diagram. Then find the difference between the area of the rectangle and the triangles created outside triangle *RST*.

The dimensions of the rectangle are 11 units by 4 units, so the area is 44 square units. Label the three triangles within the rectangle outside triangle *RST* I, II, and III and find their areas.

Area of $\triangle$I = $\dfrac{7 \times 3}{2}$ = 10.5 square units

Area of $\triangle$II = $\dfrac{11 \times 1}{2}$ = 5.5 square units

Area of $\triangle$III = $\dfrac{4 \times 4}{2}$ = 8 square units

The sum of the areas of triangles I, II, and III is 10.5 + 5.5 + 8 or 24 square units.

Subtract this total from the area of the rectangle.

Answer: The area of $\triangle RST$ = 44 − 24 = 20 square units.

2 The vertices of quadrilateral *PLUM* are *P*(−4, −1), *L*(2, 3), *U*(5, 1), and *M*(4, −2). Find the area of quadrilateral *PLUM*.

SOLUTION

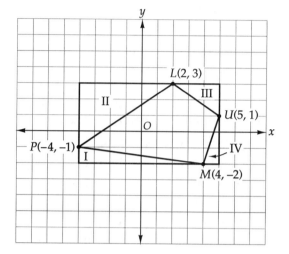

Plot the points and draw quadrilateral *PLUM*. Enclose the quadrilateral with a rectangle and label the triangles formed outside the quadrilateral I, II, III, and IV.

The dimensions of the rectangle are 9 units by 5 units. The area of the rectangle is 45 square units.

The areas of the triangles are:

Area $\triangle$I = $\dfrac{8 \times 1}{2}$ = 4

Area $\triangle$II = $\dfrac{6 \times 4}{2}$ = 12

Area $\triangle$III = $\dfrac{3 \times 2}{2}$ = 3

Area $\triangle$IV = $\dfrac{3 \times 1}{2}$ = 1.5

The sum of the areas of triangles I, II, III and IV is 20.5 square units.

Answer: The area of quadrilateral *PLUM* = 45 − 20.5 = 24.5 square units.

Practice

1. What is the distance between the points whose coordinates are $(3, -2)$ and $(-4, 5)$?
   (1) $\sqrt{10}$
   (2) $5\sqrt{2}$
   (3) $7\sqrt{2}$
   (4) 14

2. What is the length of the diameter of a circle if the endpoints have the coordinates $(-1, -1)$ and $(2, 3)$?
   (1) 5
   (2) 7
   (3) $\sqrt{5}$
   (4) $\sqrt{7}$

3. What is the length of RT if the coordinates of the points are $R(-3, 2)$ and $T(4, 1)$?
   (1) $2\sqrt{2}$
   (2) $\sqrt{10}$
   (3) $4\sqrt{3}$
   (4) $5\sqrt{2}$

4. The endpoints of the diameter of a circle are $L(-3, 7)$ and $M(4, 0)$. What is the midpoint of the diameter?
   (1) $\left(-3\frac{1}{2}, 3\frac{1}{2}\right)$
   (2) $\left(-1\frac{1}{2}, 1\frac{1}{2}\right)$
   (3) $\left(\frac{1}{2}, 3\frac{1}{2}\right)$
   (4) $(2, -2)$

5. Find the area of triangle ABC with vertices $A(-6, -4)$, $B(4, 2)$, and $C(0, 4)$.

6. If the vertices of triangle WIN are $W(2, 1)$, $I(4, 7)$, and $N(8, 3)$, what is the area of $\triangle WIN$?

7. The vertices of triangle NAQ have the coordinates $N(2, 3)$, $A(6, 0)$, and $Q(12, 8)$. Find the area of $\triangle NAQ$.

8. The coordinates of the vertices of $\triangle ABC$ are $A(1, 7)$, $B(7, 6)$, and $C(3, 11)$. Find (a) the area of $\triangle ABC$, (b) the length of side $\overline{AC}$, and (c) the length of the median from vertex B to side $\overline{AC}$.

9. The coordinates of the vertices of $\triangle PAL$ are $P(1, 2)$, $A(7, 5)$, and $L(8, -4)$. Find (a) the area of $\triangle PAL$, (b) the length of side $\overline{PA}$ (to the nearest tenth), and (c) the length of the altitude from vertex L to side $\overline{PA}$.

10. Find the area of quadrilateral $CASH$ whose vertices are $C(-2, 3)$, $A(3, 7)$, $S(8, 6)$, and $H(12, 4)$.

11. What is the area of quadrilateral $RISK$ if the coordinates of the vertices are $R(2, 2)$, $I(4, 6)$, $S(10, 12)$, and $K(8, 4)$?

12. Find the area of trapezoid $ABCD$ if the vertices are $A(1, 5)$, $B(7, 3)$, $C(2, -4)$, and $D(-7, -1)$.

13. Find the area of pentagon $SPICE$ whose vertices are $S(4, 0)$, $P(5, 5)$, $I(0, 9)$, $C(2, 4)$, and $E(0, -3)$.

14. If the coordinates of the vertices of polygon $PEACH$ are $P(1, 1)$, $E(10, 4)$, $A(7, 8)$, $C(2, 9)$, and $H(-3, 3)$, what is the area of pentagon $PEACH$?

Coordinate Geometry

Coordinate Geometry Proofs

Coordinate geometry proofs should be arranged in an orderly step-by-step format with reasoned conclusions clearly stated at the end of the algebraic or arithmetic procedures. Of course, many of the problems can be proven by more than one method.

General Methods of Proof and Formulas

To Prove	Formula to Use
that line segments are congruent, show that the lengths are equal.	Distance formula $d = \sqrt{(x_2 - x_1)^2 + (y_2 - y_1)^2}$
that line segments bisect each other, show that the midpoints are the same.	Midpoint formula $M = \left(\dfrac{x_1 + x_2}{2}, \dfrac{y_1 + y_2}{2}\right)$
that lines are parallel, show that the slopes are equal.	Slope formula $m = \dfrac{y_2 - y_1}{x_2 - x_1}$ $m_1 = m_2$
that lines are perpendicular, show that the slopes are negative reciprocals.	Product of slopes $= -1$ $m_1 \cdot m_2 = -1$ or $m_1 = -\dfrac{1}{m_2}$

The chart below gives methods that can be used to prove that a quadrilateral belongs to a specific category.

To Prove a Figure is a	Methods
Parallelogram	Show *one* of the following: • The diagonals bisect each other. • Both pairs of opposite sides are parallel. • Both pairs of opposite sides are congruent. • One pair of opposite sides is congruent and parallel.
Rectangle	Show that the figure is a parallelogram using any one of the four methods above AND *one* of the following: • The figure has one right angle. • The diagonals are congruent.
Rhombus	Show that the figure is a parallelogram using any one of the four methods above AND *one* of the following: • The diagonals are perpendicular. • Two adjacent sides are congruent.

Square	Show that the figure is a rectangle
	   AND
	two adjacent sides are congruent.
	   OR
	Show that the figure is a rhombus
	   AND
	one angle is a right angle.

(continues)

To Prove a Figure is a	Methods
Trapezoid	Show that the quadrilateral has only one pair of opposite sides parallel.
Isosceles trapezoid	Show that the figure is a trapezoid AND *one* of the following: • The diagonals are congruent. • The legs are congruent.
Right trapezoid	Show that the figure is a trapezoid AND one of the legs is perpendicular to a base of the trapezoid.

MODEL PROBLEMS

1 In right triangle *PAC*, the vertices are $P(-2, -1)$, $A(4, 7)$, and $C(-4, 3)$. Prove that the median to the hypotenuse $\overline{CM}$ is equal to one-half of the hypotenuse.

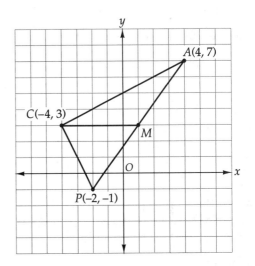

SOLUTION

Plan: Step 1: Find the midpoint M of hypotenuse $\overline{AP}$. Step 2: Find the length of the median $\overline{CM}$. Step 3: Find the length of hypotenuse $\overline{AP}$ and take half of it. Then compare the lengths.

Step 1: Use the midpoint formula to find the midpoint of hypotenuse $\overline{AP}$.

$$M = \left(\frac{x_1 + x_2}{2}, \frac{y_1 + y_2}{2}\right)$$

$$M = \left(\frac{4 + (-2)}{2}, \frac{7 + (-1)}{2}\right) = (1, 3)$$

Step 2: Use the distance formula to find the length of the median.

$$d = \sqrt{(x_2 - x_1)^2 + (y_2 - y_1)^2}$$

$$CM = \sqrt{(1 - (-4))^2 + (3 - 3)^2} = \sqrt{25 + 0} = 5$$

Step 3: Use the distance formula to find the length of the hypotenuse.

$$d = \sqrt{(x_2 - x_1)^2 + (y_2 - y_1)^2}$$

$$AP = \sqrt{(-2 - 4)^2 + (-1 - 7)^2} = \sqrt{36 + 64} = \sqrt{100} = 10$$

Thus, half the hypotenuse = 5.

Answer: Median $CM = 5 = \frac{1}{2} \times$ (hypotenuse)

Coordinate Geometry

2 The vertices of triangle *ABC* are *A*(−3, 1), *B*(4, 2), and *C*(−2, −1). (a) Prove that △*ABC* is a right triangle. (b) Using the information found in (a), find the area of triangle *ABC*.

SOLUTION

Plan: (a) Find the length of each side of △*ABC* and test the measures in the Pythagorean theorem, and (b) apply the formula for the area of any triangle.

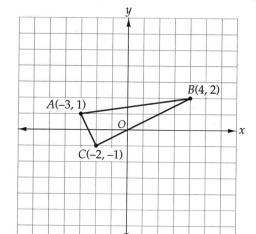

a Use the distance formula to find the lengths of the sides of △*ABC*.

$$d = \sqrt{(x_2 - x_1)^2 + (y_2 - y_1)^2}$$

$$d_{AB} = \sqrt{(4 - (-3))^2 + (2 - 1)^2} = \sqrt{49 + 1} = \sqrt{50}$$

$$d_{BC} = \sqrt{[(-2) - 4]^2 + [(-1) - 2]^2} = \sqrt{36 + 9} = \sqrt{45}$$

$$d_{CA} = \sqrt{[(-3) - (-2)]^2 + [1 - (-1)]^2} = \sqrt{1 + 4} = \sqrt{5}$$

Is it true that $(AB)^2 = (BC)^2 + (CA)^2$? By substitution:

$$(\sqrt{50})^2 = (\sqrt{45})^2 + (\sqrt{5})^2$$

$$50 = 45 + 5 = 50$$

Answer: Since the Pythagorean theorem is true, △*ABC* is a right triangle.

b Area of △*ABC* = $\frac{1}{2}\sqrt{45} \times \sqrt{5} = \frac{1}{2}\sqrt{225} = \frac{1}{2}(15) = 7.5$

Answer: The area of △*ABC* = 7.5 sq. units

3 If the coordinates of the vertices of quadrilateral *CARD* are *C*(−4, 1), *A*(−2, 3), *R*(1, 0), and *D*(−1, −2), prove that *CARD* is a rectangle.

SOLUTION

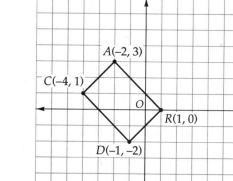

Plan: Step 1: Find the midpoints of the diagonals to show that they bisect each other. Step 2: Find the length of the diagonals to show that they are congruent.

Step 1: Use the midpoint formula to find the midpoint *M* of diagonal $\overline{CR}$ and the midpoint *N* of diagonal $\overline{AD}$.

$$M = \left(\frac{x_1 + x_2}{2}, \frac{y_1 + y_2}{2}\right)$$

$$M = \left(\frac{-4 + 1}{2}, \frac{1 + 0}{2}\right) = \left(\frac{-3}{2}, \frac{1}{2}\right)$$

$$N = \left(\frac{-2 + (-1)}{2}, \frac{3 + (-2)}{2}\right) = \left(\frac{-3}{2}, \frac{1}{2}\right)$$

Since midpoints *M* and *N* are the same, the diagonals bisect each other, and quadrilateral *CARD* is a parallelogram.

Step 2: Use the distance formula to find the lengths of the diagonals.

$$d = \sqrt{(x_2 - x_1)^2 + (y_2 - y_1)^2}$$

$$d_{CR} = \sqrt{(1 - (-4))^2 + (0 - 1)^2} = \sqrt{25 + 1} = \sqrt{26}$$

$$d_{AD} = \sqrt{[(-1) - (-2)]^2 + [(-2) - 3]^2} = \sqrt{1 + 25} = \sqrt{26}$$

Answer: Since the diagonals $\overline{CR}$ are $\overline{AD}$ congruent, parallelogram *CARD* is a rectangle.

4 If the vertices of quadrilateral *PEAR* are $P(-3, 0)$, $E(0, 4)$, $A(5, -6)$, and $R(-1, -4)$, show, using coordinate geometry, that quadrilateral *PEAR* is *not* an isosceles trapezoid.

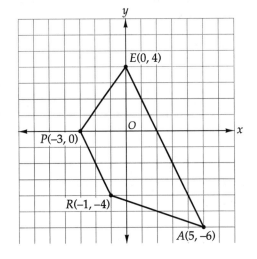

SOLUTION

Plan: Step 1: Find the slopes of all the sides of the quadrilateral and show that it is a trapezoid. Step 2: If quadrilateral *PEAR* is a trapezoid, then find the lengths of the two non-parallel sides and compare them.

Step 1:

$$\text{Slope}_{PE} = \frac{0 - 4}{-3 - 0} = \frac{-4}{-3} = \frac{4}{3}$$

$$\text{Slope}_{EA} = \frac{4 - (-6)}{0 - 5} = \frac{10}{-5} = -2$$

$$\text{Slope}_{AR} = \frac{-6 - (-4)}{5 - (-1)} = \frac{-2}{6} = \frac{-1}{3}$$

$$\text{Slope}_{RP} = \frac{-4 - 0}{-1 - (-3)} = \frac{-4}{2} = -2$$

Since only the slopes of $\overline{EA}$ and $\overline{RP}$ are the same, only one pair of sides (or bases) are parallel. Thus, quadrilateral *PEAR* is a trapezoid.

Step 2: Use the distance formula to find the lengths of the nonparallel sides.

$$d = \sqrt{(x_2 - x_1)^2 + (y_2 - y_1)^2}$$

$$d_{PE} = \sqrt{(0 - (-3))^2 + (4 - 0)^2} = \sqrt{9 + 16} = \sqrt{25} = 5$$

$$d_{RA} = \sqrt{[5 - (-1)]^2 + [(-6) - (-4)]^2} = \sqrt{36 + 4} = \sqrt{40}$$

Since the non-parallel sides (or legs) $\overline{PE}$ and $\overline{RA}$ do not have the same lengths, the trapezoid is not an isosceles trapezoid.

1. The vertices of a rectangle are $(-1, -2)$, $(5, -2)$, $(5, 2)$, and $(-1, 2)$. If the midpoints of the four sides are joined in order, what kind of quadrilateral is formed?

   (1) rectangle
   (2) rhombus
   (3) square
   (4) trapezoid

2. The slope of line AB is $\frac{3}{5}$ and the slope of line CD is $\frac{10}{k}$. If $AB \perp CD$, what is the value of k?

   (1) $-16\frac{2}{3}$
   (2) -6
   (3) 6
   (4) $16\frac{2}{3}$

3. To prove that a quadrilateral is a rhombus you must show that:

   (1) one pair of opposite sides is congruent and parallel
   (2) the diagonals are perpendicular and congruent
   (3) both pairs of opposite sides are parallel and the figure has one right angle
   (4) the diagonals are perpendicular and both pairs of opposite sides are congruent

4. Quadrilateral $TOPS$, with vertices $T(-1, 0)$, $O(3, 3)$, $P(5, 0)$, and $S(3, -3)$, is a:

   (1) rectangle
   (2) rhombus
   (3) square
   (4) trapezoid

5. The vertices of triangle WIN are $W(2, 1)$, $I(4, 7)$, and $N(8, 3)$. Using coordinate geometry, show that $\triangle WIN$ is an isosceles triangle and state the reasons for your conclusion.

6. Triangle NAQ has coordinates $N(2,3)$, $A(6, 0)$, and $Q(12, 8)$. Show, by means of coordinate geometry, that $\triangle NAQ$ is a right triangle, and state the reasons for your conclusion.

7. If the vertices of right triangle ABC are $A(-1, 2)$, $B(3, 8)$, and $C(5, -2)$, use coordinate geometry to prove that point M, the midpoint of the hypotenuse, is equidistant from all three vertices of the triangle.

8. The coordinates of the vertices of triangle KEN are $K(2, 1)$, $E(9, 4)$, and $N(5, 8)$. Show that the median from K to $\overline{EN}$ is perpendicular to $\overline{EN}$.

9. The vertices of $\triangle BIG$ are $B(-4, 3)$, $I(1, 8)$, and $G(6, 3)$. Show that $\triangle BIG$ is an isosceles right triangle. Explain your reasoning.

10. If the vertices of quadrilateral $PQRS$ are $P(1, 1)$, $Q(2, 4)$, $R(5, 6)$, and $S(4, 3)$, then use slopes to show that $PQRS$ is a parallelogram.

11. If the vertices of quadrilateral $SAND$ are $S(-2, -2)$, $A(-1, 2)$, $N(5, 3)$, and $D(4, -1)$, use the midpoint formula to show that $SAND$ is a parallelogram.

12. If the vertices of quadrilateral $LEAP$ are $L(-3, 1)$, $E(2, 6)$, $A(9, 5)$, and $P(4, 0)$, use the distance formula to show that $LEAP$ is a parallelogram.

13. The vertices of quadrilateral $PQRS$ are $P(-2, 1)$, $Q(2, 5)$, $R(6, -1)$, and $S(4, -7)$. If points A, B, C, and D are the midpoints of sides $\overline{PQ}$, $\overline{QR}$, $\overline{RS}$, and $\overline{SP}$, show that $ABCD$ is a parallelogram. Explain your reasoning.

14. The coordinates of the vertices of quadrilateral $BETH$ are $B(-4, -1)$, $E(5, -2)$, $T(2, 3)$, and $H(-7, 4)$. Using coordinate geometry:

    a find the length of the diagonals.
    b show that the diagonals *do* or *do not* bisect each other.
    c giving a reason, show whether $BETH$ *is* or *is not* a parallelogram.
    d giving a reason, show whether $BETH$ *is* or *is not* a rectangle.

15. The coordinates of the vertices of quadrilateral $NICK$ are $N(-3, -1)$, $I(3, 1)$, $C(7, 5)$, and $K(1, 3)$. Use slopes only to show:

    a that $NICK$ is a parallelogram
    b that $NICK$ is or is not a rhombus

16 The coordinates of the vertices of parallelogram KATE are K(−1, 0), A(3, 3), T(6, −1), and E(2, −4). Use coordinate geometry to show that KATE is a square. Give reasons for your conclusion.

17 The vertices of isosceles trapezoid PLAY are P(0, 0), L(10, 0), A(6, 8), and Y(4, 8). If points A, B, C, and D are the midpoints of sides $\overline{PL}$, $\overline{LA}$, $\overline{AY}$, and $\overline{YP}$ respectively, prove that quadrilateral ABCD is a rhombus. Give clear reasons for your conclusion.

18 The coordinates of the vertices of quadrilateral PAUL are P(0, 0), A(6a, 3b), U(3a, 4b), and L(a, 3b).

  a Show that $\overline{PA} \parallel \overline{UL}$, and give reasons for your conclusion.

  b Show that $\overline{PL}$ is not parallel to $\overline{AU}$. Explain your reasoning.

  c What kind of figure is quadrilateral PAUL? Explain your reasoning.

19 The coordinates of the vertices of △BEN are B(−4, 1), E(2, 13), and N(10, 9). Use coordinate geometry to:

  a Find the length of $\overline{BE}$ in simplest radical form.

  b Find the slope of $\overline{BE}$.

  c Find the coordinates of point M, the midpoint of $\overline{EN}$.

  d Write the equation of the line that is parallel to $\overline{BE}$ and passes through point M.

CHAPTER REVIEW

1 If a side of a square is 5, its diagonal is:
  (1) $5\sqrt{2}$
  (2) $5\sqrt{3}$
  (3) 10
  (4) $10\sqrt{2}$

2 A pocket park is a rectangle 33 feet by 44 feet. If you walk diagonally across the park instead of along two adjacent sides, how many feet do you save?
  (1) 11
  (2) 22
  (3) 55
  (4) 77

3 In a right triangle, the length of the altitude to the hypotenuse is 6. The altitude divides the hypotenuse into two segments with lengths of 4 and x. What is the value of x?
  (1) 6
  (2) 8
  (3) 9
  (4) 10

4 The center of a circle is at (−3, 7). The circle passes through (2, −1). What is the length of the radius?
  (1) $\sqrt{61}$
  (2) 11
  (3) $\sqrt{89}$
  (4) 13

5 The midpoint of line segment AB is (−1, 5). If the coordinates of A are (−3, 2), what are the coordinates of B?
  (1) (1, 10)
  (2) (1, 8)
  (3) (0, 7)
  (4) (−5, 8)

6 The vertices of a rectangle are (−2, 3), (4, 3), (4, 7), and (−2, 7). If the midpoints of the four sides are joined in order, what kind of quadrilateral is formed?
  (1) rectangle
  (2) rhombus
  (3) square
  (4) trapezoid

7. A diagonal of a rectangle is 5 centimeters. The height is 1 centimeter more than the width. Find the perimeter.

8. In square units, the difference between the surface area of a right cylinder and its lateral area is 32π. Find the radius of the base.

9. A cup shaped like a right circular cone has a height of 8 inches. The radius of the base is 2 inches. This cup is used to fill a right circular cylinder that has the same height and radius with water. After the cone is emptied once, what volume of water is needed to finish filling the cylinder? Give your answer to the nearest cubic inch.

10. The ratio of the surface area of two similar polyhedrons is 4 : 1. The volume of the smaller polyhedron is 10 cubic inches. Find the volume of the larger polyhedron.

11. The dimensions of a trash dumpster are 3 feet by 3 feet by 6 feet. How many dumpsters are necessary to carry 150 cubic yards of trash?

12. An odometer measures how far a bicycle has traveled by multiplying the circumference of a wheel times the number of revolutions the wheel makes. Suppose that your bike has a wheel circumference of 200 centimeters, but the odometer is set for 210 centimeters. If you ride for 20 kilometers, what will the odometer show?

13. The length of the hypotenuse of a 30°–60°–90° triangle is 40. In simplest radical form, what are the lengths of the legs?

14. A 20-foot flagpole casts a 25-foot shadow at the same time that a telephone pole casts a 32-foot shadow. Find the height of the telephone pole to the nearest tenth of a foot.

15. At a point 43 feet from the base of a statue, the angle of elevation to the top is 52°. To the nearest tenth of a foot, how tall is the statue?

16. A 24-foot ladder, leaning against a house, reaches the bottom of a window that is 20 feet above the ground. To the nearest degree, find the angle the ladder makes with the house.

17. The coordinates of three vertices of rectangle DRAW are $D(-1, 5)$, $R(3, 5)$, and $A(3, 1)$. What are the coordinates of W?

18. The coordinates of the vertices of $\triangle ABC$ are $A(-3, -2)$, $B(2, 2)$, and $C(3, -7)$. If M is the midpoint of $\overline{AB}$ and R is the midpoint of $\overline{BC}$, show that $MR = \frac{1}{2}AC$.

19. The coordinates of $\triangle ABC$ are $A(1, 3)$, $B(8, 4)$, and $C(4, 7)$. Using slopes, show that $\triangle ABC$ is a right triangle.

20. The coordinates of rhombus DAVE are $D(2, 1)$, $A(6, -2)$, $V(10, 1)$, and $E(6, 4)$. Find the following:

   a length of a side of DAVE

   b coordinates of the point of intersection of the diagonals

   c area of DAVE

   d length of the altitude from E to side DA. (Hint: Use the formula for the area of a parallelogram.)

Now turn to page 419 and take the Cumulative Review test for Chapters 1–8. This will help you monitor your progress and keep all your skills sharp.

Chapter 9:
Transformation, Locus, and Construction

This chapter will deal with three applications of geometry on a plane: transformation, locus, and construction.

A **transformation** occurs when the size or position of an original figure is changed. Transformations often involve changes in location, size, and orientation. An algebraic rule, often called a **mapping**, defines a transformation by assigning ordered pairs of the coordinate plane to new locations. If the points of a given figure are labeled with letters, such as A, B, and C, then the corresponding **transformation image** has points that are labeled with the same letters and a prime sign, such as A', B', and C'. The original points, A, B, and C are called the **preimage** points.

A **locus** is the set of points that satisfy a given set of conditions. A locus can be described with an equation or a geometric construction. A graph of the points in a solution set of a linear equation is an example of a locus.

A **geometric construction** is a process much more constrained than simply drawing an object. When a figure is drawn, only two instruments are allowed: the compass and the straightedge. A **compass** is an adjustable V-shaped device used for drawing arcs and measuring distances. A **straightedge** has a straight edge without any ruler-type markings and is used only for drawing lines.

Duplicating an Image

Translations

A **translation** is a transformation in which a figure *slides* a certain distance. A translation shifts each point of the figure in the plane by the same distance and in the same direction. In the accompanying drawing, $\triangle ABC$ is shifted 1.125 inches to $\triangle A'B'C'$.

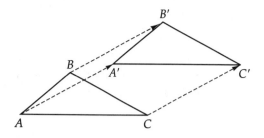

When an image is translated, it changes only in location. All distance and angle measurements of the image are the same as the measurements of the preimage. The image also does not tilt or turn.

Translation Mapping Rule A translation of the x-value a horizontal distance a and the y-value a vertical distance b is written as $(x, y) \xrightarrow{T_{a,b}} (x + a, y + b)$ or $T_{a,b}(x, y) = (x + a, y + b)$.

MODEL PROBLEMS

1 What is the image of point $P(-3, 2)$ under the transformation $T_{-2, 6}$?

SOLUTION

$T_{-2, 6}$ means add -2 to the x-value (-3) and add 6 to the y-value (2). Thus, the image of point P is $(-5, 8)$.

Answer: $(-5, 8)$

2 A transformation maps $P(x, y)$ onto $P'(x - 4, y + 2)$. Under the same transformation, what are the coordinates of Q', the image of $Q(2, -3)$?

SOLUTION

Since the given transformation is a translation $T_{-4, 2}$, the point $Q(2, -3)$ is mapped onto Q' by adding -4 to the x-value (2) and 2 to the y-value (-3). Hence, the coordinates of Q' are $(-2, -1)$.

Answer: $(-2, -1)$

3 If the coordinates of the vertices of $\triangle ABC$ are $A(-4, -1)$, $B(-1, 5)$, and $C(2, 1)$, then what is the image of $\triangle ABC$ under the translation $T_{4, 3}$?

SOLUTION

Translation $T_{4, 3}$ means we must add 4 to each x-value and 3 to each y-value. Thus, $A(-4, -1)$ is mapped onto $A'(0, 2)$, $B(-1, 5)$ onto $B'(3, 8)$, and $C(2, 1)$ onto $C'(6, 4)$.

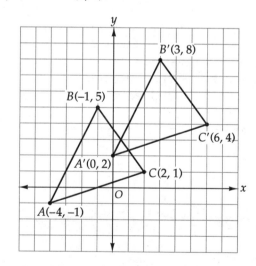

Practice

1. The image of $H(-1, 3)$ under the translation $T_{2,-1}$ is:
   (1) $(1, 2)$
   (2) $(-3, 4)$
   (3) $(-2, -3)$
   (4) $(-2, 5)$

2. The transformation $T_{-2,3}$ maps the point $(5, -5)$ onto the point whose coordinates are:
   (1) $(-10, -15)$
   (2) $(3, -2)$
   (3) $(8, -7)$
   (4) $(3, -8)$

3. If translation T maps point $P(-3, 1)$ onto point $P'(5, 5)$, which is the translation T?
   (1) $T_{2,4}$
   (2) $T_{2,6}$
   (3) $T_{8,6}$
   (4) $T_{8,4}$

4. If the transformation $T_{x,y}$ maps point $M(1, -3)$ onto point $M'(-5, 8)$, what is the value of x?
   (1) 4
   (2) 1
   (3) -5
   (4) -6

Exercises 5 and 6: Express the translation in the form $T_{a,b}$.

5. $(x, y) \rightarrow (x, y + 2)$

6. $(x, y) \rightarrow (x - 3, y)$

Exercises 7 and 8: Find the image of the point $(5, -2)$ under the translation.

7. $T_{-1,1}$

8. $T_{-2,-5}$

Exercises 9–12: Find the image or preimage, as specified, of the given point under translation T. Translation T is defined by $(x, y) \rightarrow (x + 3, y - 1)$.

9. Find the image of $(-1, 5)$.

10. Find the image of $(1, -2)$.

11. Find the preimage of $(-1, -3)$.

12. Find the preimage of $(1, 1)$.

Exercises 13 and 14: Find the values of a and b if the translation $T_{a,b}$ maps the first point onto the second.

13. $(2, 3) \rightarrow (5, 5)$

14. $(5, 0) \rightarrow (3, 2)$

15. If a translation maps $(3, 1)$ onto $(-4, 2)$, what is the image of $(4, -1)$ under the same translation?

16. If a translation maps point $(7, -2)$ to point $(0, -3)$, what is the image of point $(0, -2)$ under the same translation?

17. A transformation maps $P(x, y)$ onto $P'(x + 3, y - 5)$. What are the coordinates of Q, whose image under the same transformation is $Q'(7, 2)$?

18. If the coordinates of the vertices of $\triangle KEN$ are $K(8, 5)$, $E(10, -3)$, and $N(-2, 2)$, then what are the coordinates of the vertices for the image of $\triangle KEN$ under the translation defined by $(x, y) \rightarrow (x - 6, y + 4)$?

19. The coordinates of the vertices of $\triangle DEW$ are $D(-3, 5)$, $E(4, 6)$, and $W(0, 2)$. Graph $\triangle DEW$ and complete the following:
    a. Graph $\triangle D'E'W'$, the image of $\triangle DEW$ under the translation $T_{2,-3}$.
    b. Graph $\triangle D''E''W''$, the image of $\triangle D'E'W'$ under the translation $T_{1,4}$.
    c. Name a single transformation that would map $\triangle DEW$ onto $\triangle D''E''W''$.

20. The coordinates of the vertices of quadrilateral $WASH$ are $W(1, -2)$, $A(0, 1)$, $S(3, 4)$, and $H(5, 1)$. Graph quadrilateral $WASH$ and complete the following:
    a. Graph quadrilateral $W'A'S'H'$, the image of quadrilateral $WASH$ under the translation $T_{-3,2}$.
    b. Graph $W''A''S''H''$ the image of quadrilateral $W'A'S'H'$ under the translation $T_{7,-1}$.
    c. Name a single transformation that would map quadrilateral $WASH$ onto quadrilateral $W''A''S''H''$.

Constructing Duplicate or Congruent Figures

To Construct $\overline{CD}$ Congruent to $\overline{AB}$

- Using the straightedge, draw any segment longer than $\overline{AB}$ and mark point C near one end of the segment.
- On $\overline{AB}$, place the point of the compass on A and the pencil point on B.
- Without changing the compass setting, place the point of the compass at C and draw an arc intersecting your line segment.
- Label the point of intersection D.
- $\overline{CD} \cong \overline{AB}$

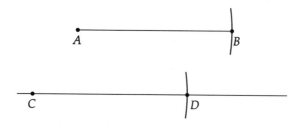

To Construct $\angle FED$ Congruent to $\angle ABC$

- Draw point E and draw a line segment through it with a straightedge.
- Place the point of the compass on B and draw an arc intersecting $\overline{BA}$ and $\overline{BC}$. Label the points of intersection G and H.
- Without changing the compass setting, place the point of the compass at E and draw an arc longer than $\overparen{GH}$ through your line segment. Label the point of intersection D.
- Place the point of the compass on H and the pencil point on G.
- Without changing the compass setting, place the point of the compass at D and make a second arc that intersects the first. Label the intersection F.
- Draw $\overrightarrow{EF}$ with the straightedge.
- $\angle FED \cong \angle ABC$

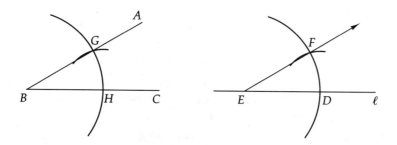

Note: If we were to draw $\overline{DF}$ and $\overline{GH}$, then $\triangle GBH$ would be congruent to $\triangle FED$.

318 Chapter 9: Transformation, Locus, and Construction

Practice

Exercises 1–3: Use only a compass and the figures below.

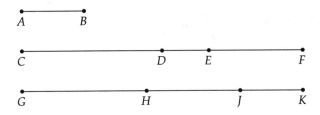

1. Which two line segments are equal?
   - (1) $\overline{CD}$ and $\overline{GJ}$
   - (2) $\overline{CD}$ and $\overline{HK}$
   - (3) $\overline{EF}$ and $\overline{GH}$
   - (4) $\overline{EF}$ and $\overline{HJ}$

2. Which line segment is exactly twice as long as $\overline{AB}$?
   - (1) $\overline{CE}$
   - (2) $\overline{DF}$
   - (3) $\overline{GH}$
   - (4) $\overline{HK}$

3. Which line segment is exactly three times as long as $\overline{AB}$?
   - (1) $\overline{CE}$
   - (2) $\overline{DF}$
   - (3) $\overline{GJ}$
   - (4) $\overline{HK}$

4. Use your compass to determine which two triangles are congruent.

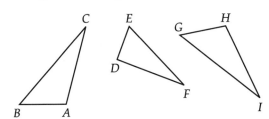

   - (1) △ABC and △DEF
   - (2) △ABC and △GHI
   - (3) △DEF and △GHI
   - (4) △BAC and △GHI

5. Draw a line segment and label it AB. Construct $\overline{AC}$ that is three times the length of $\overline{AB}$.

6. Draw an angle and label it RST. Construct $\overline{SV}$ so that ∠RSV has twice the measure of ∠RST.

7. Draw two line segments. Construct a third line segment that has a length equal to the sum of the two initial segments.

8. Draw two line segments. Construct a third line segment that has a length equal to the difference of the two initial segments.

9. Draw two angles. Construct a third angle that has a measure equal to the difference of the two initial angles.

10. Create parallelogram FADE by constructing △FAD congruent to △DEF.

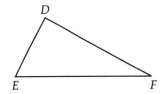

11. Give a geometric reason why the procedure of copying a line works. (Hint: Think of the compass as two sides of a triangle.)

12. Give a geometric reason why the procedure of copying an angle works.

13. How would you use a compass to prove that a figure is a rhombus?

14. How would you use a compass and a straightedge to construct an isosceles triangle?

Duplicating an Image

Reflections and Symmetry

Line Reflection and Symmetry

A **line reflection** is a transformation in which a figure is reflected over a given line as if in a mirror. Each point of the reflection image is the same distance from the line of reflection as the corresponding point in the original figure. The line of reflection is also perpendicular to each line joining a point to its image. In Figure (a), line l is the bisector of $\overline{AA'}$, $\overline{BB'}$, and $\overline{CC'}$. In Figure (b), line m is the bisector of $\overline{AA'}$, $\overline{CC'}$, and $\overline{TT'}$.

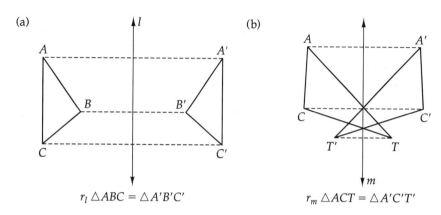

Note: The line of reflection is a **perpendicular bisector** of each line segment joining a point to its image. See the section on perpendicular bisectors later in this chapter for more about these lines.

In a line reflection, the image "flips over" the line but does not tilt. Note that the clockwise order of A, B, and C in Figure (a) is changed in the line reflection to A', C', and B'. Thus, the orientations of $\triangle ABC$ and $\triangle A'B'C'$ are opposite. Despite this difference, all distance and angle measurements of the image are the same as the measurements of the preimage.

Line Reflection Mapping Rules

Reflections in the x-axis	$r_{x\text{-axis}}(x, y) = (x, -y)$
Reflections in the y-axis	$r_{y\text{-axis}}(x, y) = (-x, y)$
Reflections in the $y = x$ line	$r_{y=x}(x, y) = (y, x)$
Reflections in the $y = -x$ line	$r_{y=-x}(x, y) = (-y, -x)$

These rules are applied in the following model problems.

MODEL PROBLEMS

1 Find a reflection in the x-axis of figure ABCD with coordinates A(−4, 3), B(−3, 5), C(4, 6), and D(0, 2).

SOLUTION

Using our mapping rule, the image points are A′(−4, −3), B′(−3, −5), C′(4, −6), and D′(0, −2). Observe that the image of each point is the same perpendicular distance from the x-axis as the original points A, B, C, and D and the x-axis or y = 0 line is the perpendicular bisector of $\overline{AA'}$, $\overline{BB'}$, $\overline{CC'}$, and $\overline{DD'}$.

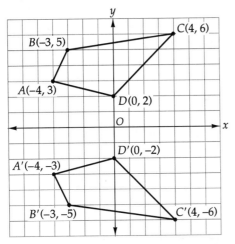

2 Find the reflection in the y-axis of △PQR with coordinates P(−8, −2), Q(−4, 4), and R(−3, −1).

SOLUTION

Using our mapping rule, the image points are P′(8, −2), Q′(4, 4), and R′(3, −1). Again, observe that the image of each point is the same perpendicular distance from the y-axis as the original P, Q, and R points.

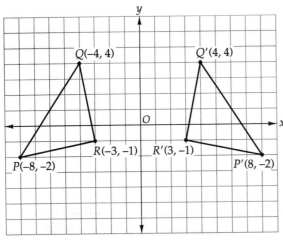

3 Find the reflection across the y = x line of △ABC with coordinates A(2, 1), B(9, 6), and C(5, −4).

SOLUTION

Using our mapping rule, the image points are A′(1, 2), B′(6, 9), and C′(−4, 5).

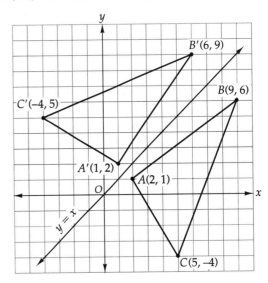

4 Find the reflection in the y = −x line of △PQR with coordinates P(2, 2), Q(5, 3), and R(3, −1).

SOLUTION

Using our mapping rule, the image points are P′(−2, −2), Q′(−3, −5), and R′(1, −3).

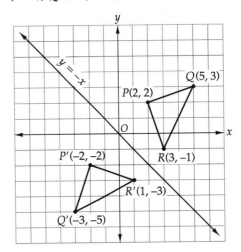

Reflections and Symmetry **321**

For reflections across any vertical or horizontal line, the easiest way to find the image is to use the **counting method**. Count the distance from each point to the line of reflection and then use that distance to find the image point on the other side of the line. In horizontal and vertical reflections, either the x or the y will change, not both.

 MODEL PROBLEMS

1 Find the reflection of the figure whose coordinates are $A(-1, 3)$, $B(4, 6)$, $C(7, 2)$ across the line represented by $x = 2$.

SOLUTION

The line $x = 2$ is the perpendicular bisector of $\overline{AA'}$, $\overline{BB'}$, and $\overline{CC'}$. Count to find the distance between each point A, B, and C and the line $x = 2$. Then use that distance to find points A', B', and C'. Since the y-values remain the same, the image points are $A'(5, 3)$, $B'(0, 6)$, and $C'(-3, 2)$.

2 Find the reflection of the figure whose coordinates are $A(-3, -7)$, $B(3, -4)$, and $C(9, -6)$ across the line represented by $y = -3$.

SOLUTION

The line $y = -3$ is the perpendicular bisector of $\overline{AA'}$, $\overline{BB'}$, and $\overline{CC'}$. Count to find the distance between each point A, B, and C and the line $y = -3$. Then use that distance to find points A', B', and C'. Since the x-values remain the same, the image points are $A'(-3, 1)$, $B'(3, -2)$, and $C'(9, 0)$.

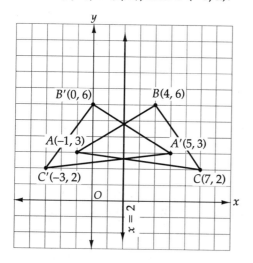
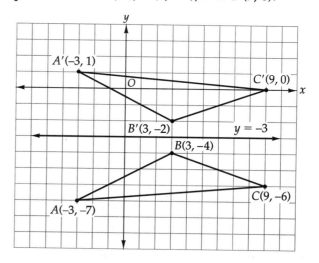

If a figure has **line symmetry**, a line can be drawn through the figure such that both sides of the figure are "mirror images" of each other. Such a line is also called the **axis of symmetry**. There can be more than one line of symmetry in a figure, or there may be none.

One line of symmetry

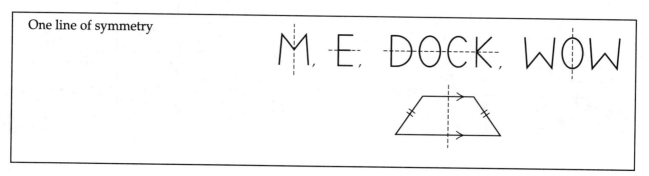

(continues)

322 Chapter 9: Transformation, Locus, and Construction

Two lines of symmetry	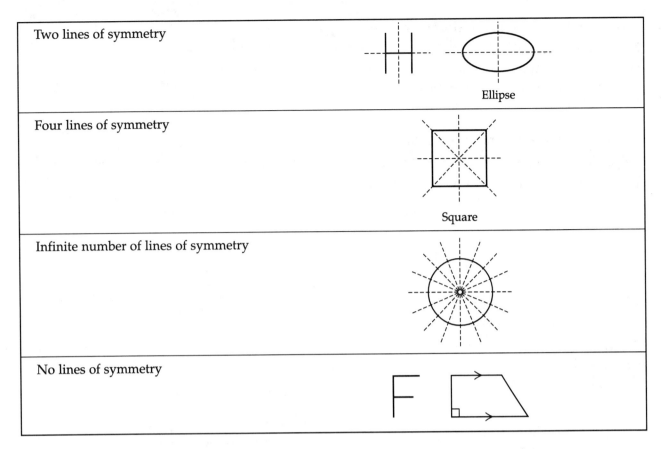
Four lines of symmetry	
Infinite number of lines of symmetry	
No lines of symmetry	

There are many real objects that exhibit symmetry, including butterflies, people's faces, and some works of art and architecture.

MODEL PROBLEM

Draw all the lines of symmetry for this figure.

SOLUTION

There are only two lines of symmetry.

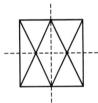

Practice

1. Which letter has vertical line symmetry?
   (1) S
   (2) N
   (3) A
   (4) B

2. Which letter has vertical but not horizontal line symmetry?
   (1) X
   (2) O
   (3) V
   (4) E

Reflections and Symmetry 323

3. What kind of symmetry does the name OTTO have?

(1) only vertical line symmetry
(2) only horizontal line symmetry
(3) both vertical and horizontal line symmetry
(4) neither horizontal nor vertical line symmetry

4. What is the total number of lines of symmetry in a rectangle that is not a square?

(1) 1
(2) 2
(3) 3
(4) 4

5. When point $(-2, 7)$ is reflected in the line $x = 1$, the image is:

(1) $(7, 2)$
(2) $(-2, -5)$
(3) $(4, 7)$
(4) $(0, 7)$

Exercises 6–9: Determine whether each figure is symmetric over the x-axis, the y-axis, both, or neither.

6.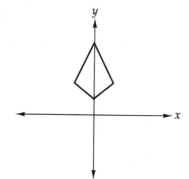

7.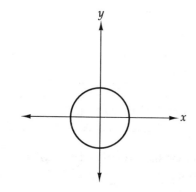

8.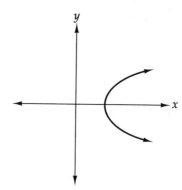

9.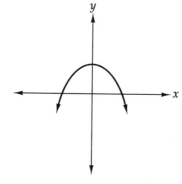

Exercises 10–12: Plot the point and find its image when reflected in the line $y = -2$.

10. $(1, 4)$

11. $(-3, 3)$

12. $(4, -5)$

Exercises 13–15: State the coordinates of the point $(-2, 5)$ under the reflection.

13. in the line $y = x$

14. in the line $y = 4$

15. in the line $x = -2$

Exercises 16–19: The coordinates of the vertices of a polygon are given. Graph and label the polygon and its reflection in the given line.

16. $A(2, 1), B(3, 4), C(-4, 5), r_{x\text{-axis}}$

17. $G(0, 0), H(4, 3), I(1, 4), J(-5, 2), r_{y\text{-axis}}$

18. $A(3, 3), B(8, 5), C(5, 1), r_{y=x}$

19. $Q(-4, -2), Z(-4, 3), D(-2, -1), r_{y=-x}$

20. Using the domain $-3 \leq x \leq 3$, graph and label the parabola $y = x^2$. On the same set of axes, sketch the graph of the image of the parabola under the reflection in the line $y = x$. Label the image $x = y^2$.

324 Chapter 9: Transformation, Locus, and Construction

Point Reflection and Symmetry

If a figure is reflected in or through point P, then P is the *midpoint* of the line segment joining each point to its corresponding image. In the accompanying diagram, △ABC is reflected through point P to △A'B'C'. In the diagram, AP = PA', BP = PB', and CP = PC'. The notation is $r_p(\triangle ABC) = \triangle A'B'C'$.

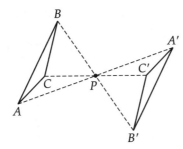

One way to visualize a point reflection is as two line reflections. The first reflection is over a horizontal line through point P. The second is over a vertical line through P. The image after both line reflections is the same as after a single point reflection. Since the only property of a shape changed by a line reflection was the clockwise order of the points, the two line reflections cancel each other out, and the image has all the properties of the original figure.

Point Reflection Mapping Rule A reflection through the origin is written $r_{origin}(x, y) = (-x, -y)$ or $r_O(x, y) = (-x, -y)$, where point O is the origin.

In coordinate geometry, the usual point of reflection is the origin. In the accompanying figure, △BUG with vertices B(2, 1), U(3, 5), and G(6, 3) is reflected through the origin onto the image △B'U'G' with coordinates B'(−2, −1), U'(−3, −5), and G'(−6, −3).

Point reflection can also be defined as a rotation of 180° or a dilation of −1. These other transformations will be discussed later in this chapter.

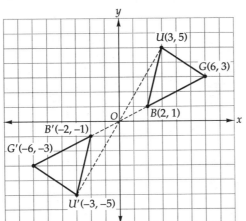

A figure is said to have **point symmetry** if the figure coincides with itself when reflected through a point or when rotated in either direction 180° about a point. The point that is the center of the reflection or rotation is called the **point of symmetry**.

Shapes without point symmetry	F, G, △
Shapes with point symmetry	Z, N, S, ▱

Reflections and Symmetry

Practice

1. What letter has both point symmetry and line symmetry?
   - (1) A
   - (2) H
   - (3) E
   - (4) S

2. *All* isosceles trapezoids have
   - (1) only line symmetry
   - (2) only point symmetry
   - (3) both point and line symmetry
   - (4) neither point nor line symmetry

3. What kind of symmetry does a rhombus have?
   - (1) only point symmetry
   - (2) only line symmetry
   - (3) both point and line symmetry
   - (4) neither point nor line symmetry

4. What is the image of $(k, 2k)$ after a reflection through the origin?
   - (1) $(2k, k)$
   - (2) $(k, -2k)$
   - (3) $(-k, -2k)$
   - (4) $(-2k, -k)$

Exercises 5–8: Identify whether the figure has point symmetry, line symmetry, or both.

5.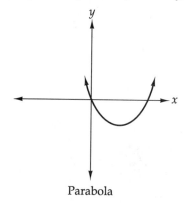
Parabola

6.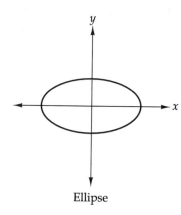
Ellipse

7.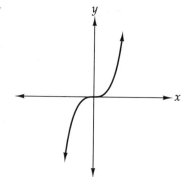
A cubic equation

8.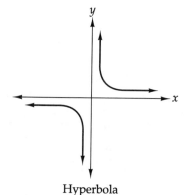
Hyperbola

Exercises 9–13: Find the image of each point under a point reflection in the origin.

9. $(1, 3)$

10. $(3, -1)$

11. $(-9, 0)$

12. $(-2, -6)$

13. $(-4, -4)$

For Exercises 14–17, find the image of (−4, 2) under a point reflection in the given point.

14 (2, 2)

15 (−1, −1)

16 (−1, 5)

17 (3, −4)

18 What is the difference between point reflection and point symmetry? Illustrate your answer.

Constructing a Perpendicular Line

To construct a line perpendicular to $\overleftrightarrow{AB}$ through a given point P,

- With the point of the compass at P, draw an arc that intersects $\overleftrightarrow{AB}$ twice. Label the points of intersection C and D.
- Using points C and D as centers, and using a slightly larger opening on your compass, draw arcs that intersect at point E.
- Use the straightedge to draw a line through E and P.
- $\overleftrightarrow{EP} \perp \overleftrightarrow{AB}$

This construction works whether P lies on $\overleftrightarrow{AB}$ or not.

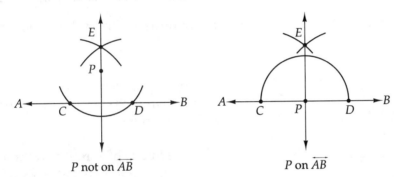

P not on $\overleftrightarrow{AB}$ P on $\overleftrightarrow{AB}$

Since the line joining a point and its image over a line reflection is perpendicular to the line of reflection, a very similar construction to the one above can be used to find the image of line reflection.

To find the image of a point, P, reflected over a line, $\overleftrightarrow{AB}$,

- With the point of the compass at P, draw an arc that intersects $\overleftrightarrow{AB}$ twice. Label the points of intersection C and D.
- Without changing the compass setting and using points C and D as centers, draw arcs on the other side of $\overleftrightarrow{AB}$ that intersect. Label the point of intersection P'.
- $r_{\overleftrightarrow{AB}} P = P'$

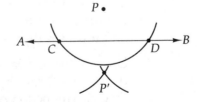

Note: The line connecting P and P' is perpendicular to $\overleftrightarrow{AB}$. Use a compass to measure the shortest distance from P to $\overleftrightarrow{AB}$. Is the distance from P' to $\overleftrightarrow{AB}$ the same?

To Find the Image of a Point P Reflected Over a Point A

- Draw a ray from P through point A.
- Place the point of the compass on A and the pencil point on P.
- Without changing the compass setting or moving the compass point from A, draw an arc intersecting $\overrightarrow{PA}$ opposite point P. Label this point P'.
- $r_A P = P'$

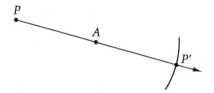

Practice

Exercises 1–3: Refer to the figure below.

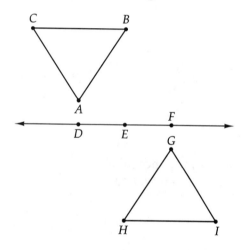

1 The image of B over line $\overleftrightarrow{DF}$ is

(1) E
(2) G
(3) H
(4) I

2 △ABC is the image of △GHI under which reflection?

(1) through point E
(2) through point F
(3) over line $\overleftrightarrow{EF}$
(4) over a line not shown

3 The figure as a whole exhibits which forms of symmetry?

(1) line symmetry
(2) point symmetry
(3) line symmetry and point symmetry
(4) no symmetry

4 If C is the image of D reflected through point B and $\overleftrightarrow{AB}$ is perpendicular to $\overleftrightarrow{CD}$, which statement is true?

(1) A is the image of B under a point reflection through C.
(2) B is the image of A under a line reflection on $\overleftrightarrow{CD}$.
(3) C is the image of D under a line reflection on $\overleftrightarrow{AB}$.
(4) D is the image of C under a point reflection through A.

5 Draw a line. Construct a perpendicular line and create a right triangle.

6 Draw a triangle and construct an altitude. (Hint: An altitude is perpendicular to a side and it passes through the opposite vertex.)

7 Draw a line and label it AB. Construct a perpendicular through point B and label it line l. Find the reflection of A through point B and also through line l. How do these two images of A compare?

8 Draw a line segment and a point. Find the reflection of the line segment through the point.

9 Draw a triangle and a line that cuts through the triangle. Construct the reflection of the triangle over the line.

10 Draw any quadrilateral and label one of the vertices *A*. Construct the point reflection of each vertex through *A* and draw the image of the reflected quadrilateral.

11 Use geometric properties to justify the procedure for constructing a line reflection of a point.

12 Use geometric properties to justify the procedure for constructing a perpendicular line.

Dilations

Dilations in the Coordinate Plane

A **dilation** is a transformation in which the size of a figure is changed and the figure is moved. In the physical world, enlarging and reducing photographs is a typical size transformation. All the angle measures of the image are the same as the measures of the corresponding angles of the original.

Dilation Mapping Rule In a dilation of constant k, where the center of dilation is the origin, the x-values and y-values of each point in the figure are multiplied by the constant to generate the coordinates of its image. The dilation is written $D_k(x, y) = (kx, ky)$. For example:

$$D_3(x, y) = (3x, 3y) \quad \text{and} \quad D_{\frac{1}{2}}(x, y) = \left(\frac{1}{2}x, \frac{1}{2}y\right)$$

In the accompanying figure, the image of $\triangle ABC$ under the dilation D_2 results in the larger triangle $A'B'C'$. The image of $\triangle ABC$ under the dilation D_{-1} results in the triangle $A''B''C''$, which is also a point reflection.

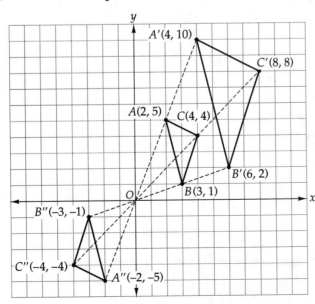

Dilations 329

Note: If $k = 1$, then the image is congruent and is identical to the original.
If $k = -1$ then the image is congruent and is the same as a point reflection.
If $|k| > 1$, then the image is similar and is larger.
If $|k| < 1$, then the image is similar and is smaller.

MODEL PROBLEMS

1 What are the coordinates of the image of point $(2, -4)$ under the dilation D_{-2}?

SOLUTION

Multiply each coordinate by -2. Thus, $(-2 \times 2, -2 \times -4) = (-4, 8)$

Answer: $(-4, 8)$

2 Find the constant or scale factor, k, for the dilation that maps $(-4, 6)$ onto $(-10, 15)$.

SOLUTION

Solve either equation.

$$k(-4) = -10 \quad \text{or} \quad k(6) = 15$$

$$k = \frac{-10}{-4} \qquad k = \frac{15}{6}$$

$$k = 2.5 \qquad k = 2.5$$

Answer: $k = 2.5$

3 $\triangle EAT$ has vertices $E(1, 1)$, $A(1, 4)$, and $T(5, 1)$. (a) What is the image of $\triangle EAT$ under the dilation D_3? (b) What is the ratio of the area of $\triangle EAT$ to the area of $\triangle E'A'T'$?

SOLUTION

a Using the mapping rule $D_3(x, y) = (3x, 3y)$, the coordinates of the image are found by multiplying each x-value and y-value by 3: $D_3 E(1, 1) = E'(3, 3)$, $D_3 A(1, 4) = A'(3, 12)$, and $D_3 T(5, 1) = T'(15, 3)$.

Answer: Image $\triangle E'A'T'$ is $E'(3, 3)$, $A'(3, 12)$, and $T'(15, 3)$

b The ratio of the area of $\triangle EAT$ to the area of $\triangle E'A'T'$ is the ratio of any corresponding segments squared or the ratio of corresponding dilations squared.

$$\frac{\text{Area } \triangle EAT}{\text{Area } \triangle E'A'T'} = \frac{1^2}{3^2} = \frac{1}{9}$$

Answer: $\frac{1}{9}$

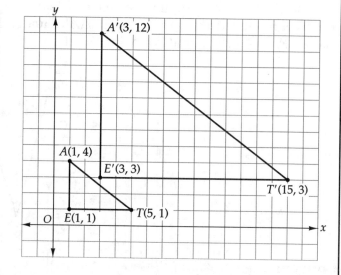

Practice

1. What are the coordinates of the image of point (2, −5) under the dilation D_{-2}?
   (1) (−10, 4)
   (2) (4, −10)
   (3) (−4, 10)
   (4) (10, −4)

2. If the dilation $D_k(2, -4) = (-1, 2)$, the scale factor k is equal to:
   (1) −2
   (2) $-\frac{1}{2}$
   (3) $\frac{1}{2}$
   (4) 2

3. Which mapping represents a dilation?
   (1) $(x, y) \to (y, x)$
   (2) $(x, y) \to (-y, -x)$
   (3) $(x, y) \to (x + 3, y + 3)$
   (4) $(x, y) \to (2x, 2y)$

4. If the area of △ANT is 5 square inches, the area of its image, △A'N'T', under the dilation D_{-2} would be:
   (1) $\frac{5}{2}$ square inches
   (2) 3 square inches
   (3) 10 square inches
   (4) 20 square inches

Exercises 5–7: Find the image of (6, −9) under the dilation.

5. D_{-1}

6. $D_{-\frac{1}{2}}$

7. $D_{\frac{2}{3}}$

Exercises 8–11: △ABC has vertices whose coordinates are A(0, 4), B(−1, 1), and C(−3, 2). Graph △ABC and find its image under each dilation.

8. D_2

9. $D_{-\frac{3}{4}}$

10. D_{-3}

11. D_{-2}

Exercises 12–14: Find the coordinates (x, y).

12. $D_2\left(\frac{1}{2}, \frac{\sqrt{3}}{2}\right) \to (x, y)$

13. $D_4(x, y) \to (2\sqrt{2}, 2\sqrt{2})$

14. $D_3(a, b) \to (x, y)$

15. Transformation D_k maps (−3, 6) to (−1, 2). What is the value of k? What is the image of (−6, −12) under the same transformation?

16. Transformation D_k maps (4, −12) to (−2, 6). What is the value of k? What is the image of (2, 4) under the same transformation?

17. Complete this transformation. If $D_k(2, -3) \to (6, -9)$, then $D_k(-1, 4) \to (?)$.

18. Rectangle GNAT has the following vertices: G(−2, 2), N(8, 2), A(8, −2), and T(−2, −2). What is the image of rectangle GNAT under the dilation $D_{\frac{1}{2}}$?

Constructing Similar Figures

Constructing a similar figure that is larger than its original, such as D_2 or D_{-3}, is easy to do with the construction skills we have covered so far. We would simply create a line segment two or three times as long as one of the sides of the original figure, then copy the angles and draw the other lines.

Dilations 331

To construct a similar figure that is smaller than the original, like $D_{\frac{1}{2}}$ or $D_{-\frac{1}{3}}$, we must first be able to divide a line into fractional parts.

To Divide $\overline{AB}$ into 3 Congruent Parts

- Draw $\overrightarrow{AW}$, longer than $\overline{AB}$, creating $\angle BAW$.
- On $\overrightarrow{AW}$, mark off 3 convenient lengths, AC, CD, and DE, by keeping the compass radius the same.
- Draw $\overline{BE}$.
- Copy $\angle AEB$ and construct congruent angles ADF and ACG.
- $\overline{AG} \cong \overline{GF} \cong \overline{FB}$

The number 3 is arbitrary. Any counting number can be used here.

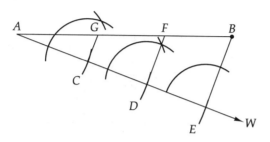

MODEL PROBLEM

Construct a triangle similar to $\triangle JKL$ where each corresponding side is $\frac{1}{3}$ as long as a side of $\triangle JKL$.

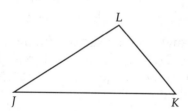

SOLUTION

We must begin by trisecting any one of the sides of $\triangle JKL$. Trisect segment JK, using $\overline{JL}$ as the reference line.

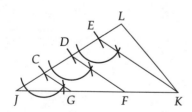

$\overline{JG}$ is $\frac{1}{3}$ as long as $\overline{JK}$. Copy angle JKL onto point G.

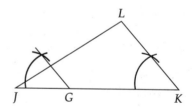

$\triangle JGM$ is similar to $\triangle JKL$.

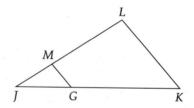

332   Chapter 9: Transformation, Locus, and Construction

Practice

1 You are given $\overline{AB}$ and an isosceles triangle. Which method will allow you to construct a similar isosceles triangle with $\overline{AB}$ as a base?

(1) Construct a perpendicular line through $\overline{AB}$ and copy the sides from the original triangle to connect the endpoints of $\overline{AB}$ to the perpendicular line.

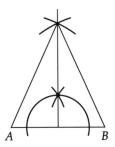

(2) Copy the base angle onto both sides of $\overline{AB}$ and extend the sides until they meet.

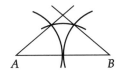

(3) Copy a base angle and a side of the original triangle, extend the second side of the angle to a length equal to $\overline{AB}$, and connect these two sides.

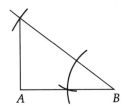

(4) Copy the vertex angle, measure out a length equal to $\overline{AB}$ along each side of the angle, and connect these two points.

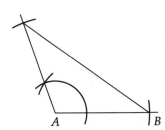

2 Use a straightedge and compass to determine which segment is $\frac{1}{3}$ as long as $\overline{CD}$.

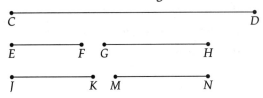

(1) $\overline{EF}$
(2) $\overline{GH}$
(3) $\overline{JK}$
(4) $\overline{MN}$

3 Use a compass to determine which two triangles are similar by comparing their angles.

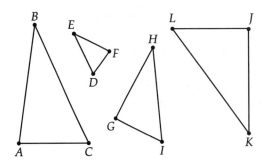

(1) △ABC and △DEF
(2) △ABC and △GHI
(3) △DEF and △GHI
(4) △GHI and △JKL

4 Use a straightedge and compass to determine which figure is an image of A after a dilation of $1\frac{1}{3}$.

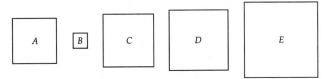

(1) B
(2) C
(3) D
(4) E

Dilations 333

Exercises 5–7: Use the line segment provided.

_____

5. Copy the line segment and divide it into thirds.

6. Recopy the line segment and divide it into fourths.

7. Construct a line segment that is 1.25 as long as the original.

Exercises 8–10: Copy the triangle provided.

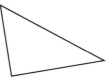

8. Construct an image of the triangle after a dilation of $\frac{1}{3}$.

9. Construct a triangle that has 4 times the area of the original.

10. Construct an image of the triangle after a dilation of $-1\frac{1}{2}$.

Rotations

A **rotation** is a transformation in which a figure is turned around a point called the **point of rotation**. The image of a rotated figure has all the same angle measures and lengths as the original and differs only in position. Rotations that are *counterclockwise* are rotations of positive degree measure. Rotations that are *clockwise* have a negative degree measure. The two rotated right triangles illustrate counterclockwise positive 90° rotation and clockwise −90° rotation.

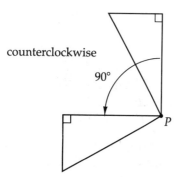
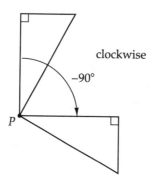

A figure has rotational symmetry if the figure is its own image under a rotation of more than 0° but less than 360°. Rotational symmetry of exactly 180° is the same as point symmetry or a dilation of −1.

Using letters of the alphabet with point symmetry (Z, X, S, O, N, I, H), it is easy to see—by turning the page upside down—that they have rotational symmetry. However, the converse is *not* true. For example, equilateral triangle *ABC* has rotational symmetry of 120°, but it does not have point symmetry.

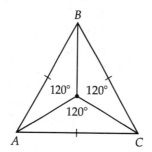

334 Chapter 9: Transformation, Locus, and Construction

All rotations are assumed to be counterclockwise about the origin unless stated otherwise.

Rotation Mapping Rules

Rotation of 90°	$R_{90°}(x, y) = (-y, x)$	
Rotation of 180°	$R_{180°}(x, y) = (-x, -y)$	
Rotation of 270°	$R_{270°}(x, y) = (y, -x)$	
Rotation of 360°	$R_{360°}(x, y) = (x, y)$	

MODEL PROBLEM

What is the image of $(-2, 3)$ under a rotation of 90°? 180°? 360°?

SOLUTION

To find each image, simply substitute -2 for x and 3 for y in the equations above.

$R_{90°}(x, y) = (-y, x)$ $\qquad R_{90°}(-2, 3) = (-3, -2)$

$R_{180°}(x, y) = (-x, -y)$ $\qquad R_{180°}(-2, 3) = (2, -3)$

$R_{360°}(x, y) = (x, y)$ $\qquad R_{360°}(-2, 3) = (-2, 3)$

Practice

1. Which of the following figures does *not* have rotational symmetry?
   (1) square
   (2) trapezoid
   (3) equilateral triangle
   (4) regular pentagon

2. Which figure has 60° rotational symmetry?
   (1) regular hexagon
   (2) regular pentagon
   (3) square
   (4) equilateral triangle

3. Which geometric figure has 72° rotational symmetry?
   (1) regular hexagon
   (2) regular pentagon
   (3) square
   (4) rhombus

4. Which polygon has rotational symmetry of 90°?
   (1) equilateral triangle
   (2) rectangle
   (3) regular pentagon
   (4) regular hexagon

5. Which figure has 120° rotational symmetry?
   (1) equilateral triangle
   (2) square
   (3) regular pentagon
   (4) regular hexagon

6. If the letter F is rotated 180°, which is the resulting figure?

   (1)

   (2)

   (3)

   (4)

Exercises 7–10: In the accompanying figure, point P is the center of the square. Find the image of each of the indicated letters under the given rotation.

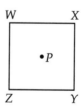

7. $R_{P,90}(W)$
8. $R_{P,180}(Z)$
9. $R_{P,270}(X)$
10. $R_{P,-90}(Y)$

336 Chapter 9: Transformation, Locus, and Construction

Exercises 11–13: Name the coordinates of the image of each point under a counterclockwise rotation of 90°.

11. (5, 1)
12. (−3, 3)
13. (8, −2)

Exercises 14–16: Name the coordinates of the image of each point under a clockwise rotation of 90°.

14. (−3, 2)
15. (−2, −2)
16. (4, 4)

Exercises 17–19: Name the coordinates of the image of each point under a counterclockwise rotation of 270°.

17. (−3, 6)
18. (−6, −2)
19. (5, 5)

20. If the coordinates of $\triangle WHY$ are $W(3, 2)$, $H(8, 2)$, and $Y(5, 10)$, what are the coordinates of $\triangle W'H'Y'$, the image of $\triangle WHY$ after a half-turn or $R_{180°}$?

Properties Under Transformations

Every geometric shape has intrinsic properties, some of which are changed by the transformations in this chapter, and some of which are **preserved**, or not changed. There are six properties to be considered:

- **Angle measures** are preserved when each angle and its image are equal in measure. Example: If $\angle ABC$ is a right angle, then $\angle A'B'C'$ is also a right angle.

- **Betweenness** is preserved when the transformation of a segment and its midpoint results in a segment image that includes the corresponding midpoint image. Example: If $\overline{AD}$ has midpoint M, then $\overline{A'D'}$ has midpoint M'.

- **Collinearity** is preserved when three or more points lie on a straight line and their transformed images also lie on a straight line. Example: If A, G, and E are collinear, then their images A', G', and E' are also collinear.

- **Distance** is preserved when each segment and its transformed image are equal in length. Example: If $QR = 7$, then $Q'R'$ must equal 7.

- **Parallelism** is preserved when images of parallel lines are also parallel. Example: If $\overline{AB} \parallel \overline{DC}$, then the transformed image $\overline{A'B'} \parallel \overline{D'C'}$.

- **Orientation** is preserved when the clockwise or counterclockwise reading of points in a given figure is the same as the image of that figure. Example: If a quadrilateral has clockwise vertices of A, B, C, and D, then the clockwise order of the image's vertices must also read A', B', C', and D'.

Properties Preserved Under Transformations

	Translation	Line Reflection	Point Reflection	Dilation	Rotation
Angle Measure	✓	✓	✓	✓	✓
Betweenness	✓	✓	✓	✓	✓
Collinearity	✓	✓	✓	✓	✓
Distance	✓	✓	✓	X	✓
Parallelism	✓	✓	✓	✓	✓
Orientation	✓	X	✓	✓	✓

A transformation that preserves distance is called an **isometry**. An image under an isometry is congruent to the original figure. Reflections, translations, and rotations also produce figures that are congruent to each other. These transformations are isometries.

A **direct isometry** is an isometry that preserves orientation. Therefore, in a direct isometry distance and orientation must be preserved. Translations, point reflections, and rotations are direct isometries.

An **opposite isometry** is an isometry that changes the orientation from clockwise to counterclockwise or vice versa. A line reflection is an opposite isometry.

 Practice

1. Under which transformation is the area of a triangle *not* equal to the area of its image?

    (1) rotation
    (2) dilation
    (3) line reflection
    (4) translation

2. Which transformation is *not* an example of an isometry?

    (1) line reflection
    (2) translation
    (3) rotation
    (4) dilation

3. $\triangle ABC$ with coordinates $A(4, 0)$, $B(8, 1)$, and $C(8, 4)$ is mapped onto $\triangle A'B'C'$ with coordinates $A'(4, -8)$, $B'(8, -7)$, and $C'(8, -4)$. Which of the following transformations is responsible for this mapping?

    (1) reflection
    (2) translation
    (3) rotation
    (4) dilation

4. Which transformation does *not* preserve orientation?

    (1) $T_{3,-5}$
    (2) $r_{y=x}$
    (3) D_4
    (4) $R_{90°}$

5. A transformation that maps $(2, 3)$ onto $(-2, -3)$ is equivalent to

    (1) rotation $R_{90°}$
    (2) rotation $R_{-90°}$
    (3) dilation D_{-1}
    (4) translation $T_{-2,-3}$

Exercises 6–9: Describe the transformation that maps figure A to figure B.

6.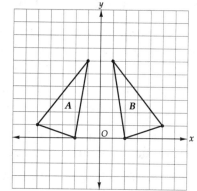

7.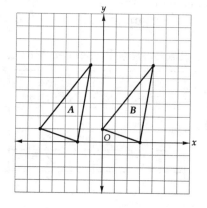

8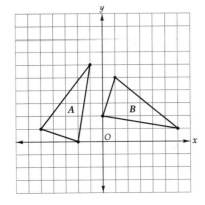

9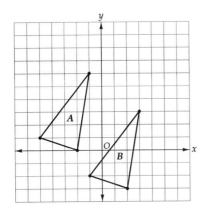

Exercises 10–12: The coordinates of the vertices of △KMZ are K(2, −2), M(5, −2), and Z(3, −4). Graph and label △KMZ and its image under the transformation. State whether the transformation is an isometry and whether it preserves orientation.

10 A: $(x, y) \to (-x, y)$

11 B: $(x, y) \to (x - 3, y + 3)$

12 C: $(x, y) \to (2x, 2y)$

13 If the coordinates of the vertices of △CAR are C(−1, −2), A(0, −4), and R(3, −1), then graph and label △CAR. Be sure to state the coordinates of the vertices for each new image.

   a Graph and label △C'A'R', the image of △CAR after the translation $T_{4,-3}$.

   b Graph and label △C"A"R", the image of △C'A'R' after a reflection in the origin.

   c Graph and label △C'''A'''R''', the image of △C"A"R" after a reflection in the line $y = -x$.

   d Which transformation *does not* preserve orientation: (a), (b), or (c)?

14 The coordinates of the vertices of △KID are K(1, 6), I(2, 9), and D(7, 10). Graph and label △KID.

   a Graph and state the coordinates of △K'I'D', the image of △KID after a reflection over the line $y = x$.

   b Graph and state the coordinates of △K"I"D", the image of △K'I'D' after a reflection over the y-axis.

   c Graph and state the coordinates of △K'''I'''D''', the image of △K"I"D" after the transformation $(x, y) \to (x + 5, y - 3)$.

15 Graph and label clearly △JAR, whose vertices have the coordinates J(2, 0), A(6, −2), and R(4, −4).

   a Graph and state the coordinates of △J'A'R', the image of △JAR after the transformation $(x, y) \to \left(\frac{1}{2}x, \frac{1}{2}y\right)$.

   b Graph and state the coordinates of △J"A"R", the image of △JAR after a counterclockwise rotation of 90° about the origin.

   c Graph and state the coordinates of △J'''A'''R''', the image of △JAR after a reflection in the origin.

Properties Under Transformations   **339**

Locus

A **locus** is the set of points that satisfies a given set of conditions. (The plural of locus is **loci**.) In this chapter, assume that all points and loci lie in the same plane. There are five basic loci: points at a set distance from a fixed point, points equidistant from two points, points at a set distance from a line, points equidistant from parallel lines, and points equidistant from intersecting lines.

A **compound locus** is a locus that involves two or more conditions. To find points that satisfy a compound locus, construct the locus of points for each condition on the same diagram. Make certain to label each locus. Mark the points where the loci intersect. These points satisfy both sets of conditions.

1. Distance From a Point: The Circle

Given a fixed point O and distance d, find all points that are a distance d from point O.

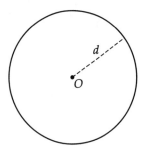

Locus: a circle with center at O and radius d.

To Find a Circle in the Coordinate Plane

- A circle can be defined as the set of all points (x, y) in the plane that are a fixed distance from a given point (h, k) called the **center**. The fixed distance is the **radius** of the circle. Algebraically, the general equation for a circle with center at (h, k) and radius r is derived from the distance formula. The equation of a circle is $(x - h)^2 + (y - k)^2 = r^2$.

To Construct a Circle

- Place the point of the compass at the fixed point, O.
- Stretch the compass out until the pencil tip is at a distance d from O.
- Draw the circle by rotating the compass around point O.

Can you see the similarity between the formula for the circle and the distance formula?

MODEL PROBLEMS

1 Graph the locus of points 3 units from point $A(1, 2)$ and find an equation for the graph.

SOLUTION

The locus is a circle with its center at $(1, 2)$ and a radius of 3. Every point on the circle is 3 units from the center, so $(1 + 3, 2)$ or $(4, 2)$ is a point on the circle.

Construct the circle by putting the point of the compass at $(1, 2)$ and starting the pencil at $(4, 2)$, which is a distance of 3 from the center.

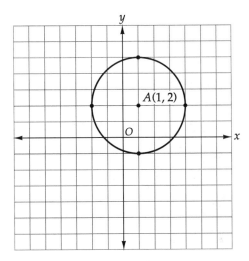

Write the general equation of a circle.	$(x - h)^2 + (y - k)^2 = r^2$
Substitute 1 for h, 2 for k, and 3 for r.	$(x - 1)^2 + (y - 2)^2 = 3^2$
	$(x - 1)^2 + (y - 2)^2 = 9$

2 Write the equation of a circle with its center at the origin and passing through the point $(12, 5)$.

SOLUTION

Write the general equation of a circle.	$(x - h)^2 + (y - k)^2 = r^2$
Since the center is at the origin, (h, k) is $(0, 0)$.	$(x - 0)^2 + (y - 0)^2 = r^2$
	$x^2 + y^2 = r^2$
Substitute $(12, 5)$ for (x, y) in the equation.	$12^2 + 5^2 = r^2$
Solve for r.	$144 + 25 = r^2$
	$169 = r^2$
	$r = 13$
Substitute h, k, and r in the original equation.	$x^2 + y^2 = 13^2$
	$x^2 + y^2 = 169$

If the center (h, k) of the circle is at the origin, the equation collapses to $x^2 + y^2 = r^2$. The x-intercepts are $(-r, 0)$ and $(r, 0)$ and the y-intercepts are $(0, -r)$ and $(0, r)$.

Answer: $x^2 + y^2 = 169$

Locus

3 Graph the equation: $(x - 3)^2 + (y + 2)^2 = 4$

SOLUTION

Write the equation. $\qquad (x - 3)^2 + (y + 2)^2 = 4$

Put the equation in the form $(x - h)^2 + (y - k)^2 = r^2$. $\qquad (x - 3)^2 + [y - (-2)]^2 = 2^2$

Identify the values of h, k, and r. $\qquad h = 3, k = -2, r = 2$

Graph the equation by placing the point of the compass on $(3, -2)$ and starting the pencil at a point 2 units away such as $(5, -2)$.

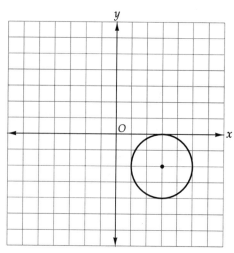

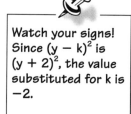

Watch your signs! Since $(y - k)^2$ is $(y + 2)^2$, the value substituted for k is -2.

4 Describe the locus of points inside a circle of radius 3 centimeters and 2 centimeters from the edge of the circle.

SOLUTION

Note: This is a compound locus problem. The word *and* separates the two conditions. Watch for this word in locus questions.

The first locus is the set of all points within a circle of radius 3 centimeters. This region is shaded in the figure below. The second locus is the set of points 2 centimeters from the edge of the circle. Points 2 centimeters from the edge form circles either 2 centimeters closer to or 2 centimeters farther from the center of the circle. Since these circles all have the same center, they are **concentric circles**, one with a radius of 1 centimeter and one with a radius of 5 centimeters from the center.

The two loci intersect only along the graph of the inside concentric circle, of radius 1.

Answer: The locus is a concentric circle of radius 1.

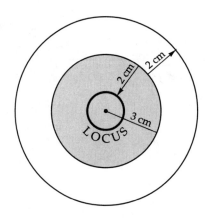

342 Chapter 9: Transformation, Locus, and Construction

5 Two dogs, Archie and Butch, are leashed to stakes that are 4 yards apart. Their leashes allow them to go out to a distance just under 3 yards from their stakes. Janus, the cat, teases the dogs by sitting just outside of their reach. Where can Janus sit to tease both dogs at the same time?

SOLUTION

Reword the question so that it includes the word *and*.

Two points *A* and *B* are 4 yards apart. Find the number of points 3 yards from *A* and 3 yards from *B*.

Construct each locus and label the intersection.

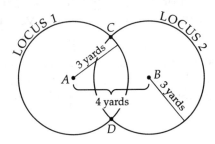

In the next section we will show that these locus points lie on the perpendicular bisector of $\overline{AB}$.

Answer: Janus should sit at one of the two points, *C* or *D*, as shown.

Practice

1 Which equation represents the locus of points 5 units from point (1, 3)?

(1) $(x + 1)^2 + (y + 3)^2 = 25$
(2) $(x + 1)^2 + (y + 3)^2 = 5$
(3) $(x - 1)^2 + (y - 3)^2 = 25$
(4) $(x - 1)^2 + (y - 3)^2 = 5$

2 Identify the center and radius of the circle $(x + 1)^2 + (y - 3)^2 = 16$.

(1) center (1, −3), radius 16
(2) center (−1, 3), radius 16
(3) center (1, −3), radius 4
(4) center (−1, 3), radius 4

3 The locus of points half as far from the center as the points of a given circle is:

(1) a line
(2) a circle
(3) a point
(4) the empty set

4 Describe the locus of points equidistant from two concentric circles whose radii are 8 inches and 14 inches.

(1) one concentric circle of radius 6 inches
(2) two concentric circles of radius 8 inches and 14 inches
(3) one concentric circle of radius 11 inches
(4) one concentric circle of radius 22 inches

5 Points *A* and *B* are 10 inches apart. The locus of points 8 inches from *A* and 4 inches from *B* is:

(1) a point
(2) a line
(3) a pair of points
(4) the empty set

6 Points *R* and *G* are 10 inches apart. The locus of points 13 inches from *R* and 2 inches from *G* is:

(1) a point
(2) 2 points
(3) a circle
(4) the empty set

Locus **343**

Exercises 7–9: State the center and the radius of the circle.

7   $x^2 + y^2 = 36$

8   $x^2 + (y-1)^2 = 9$

9   $(x+1)^2 + y^2 = 1$

Exercises 10–12: Write the equation of a circle with the given center and radius. Graph the circle.

10 Center $(-1, -1)$, radius 3

11 Center $(0, 2)$, radius $2\frac{1}{2}$

12 Center $(-1, 3)$, radius 5.5

13 Write the equation of the circle centered at the origin that passes through $(3, 4)$.

14 Write an equation of the locus of points whose distance from the origin is 4.

Exercises 15–17: Graph the locus of points 2 units from the origin and 4 units from the given point.

15 $(0, -3)$

16 $(0, 6)$

17 $(7, 0)$

18 A mosquito repellent, sprayed upward from an aerosol canister on the ground, covers all points less than or equal to 50 feet from the canister. Find the area on the ground covered by the spray to the nearest ten square feet.

19 Two circular riding paths have centers 1,000 feet apart. Both paths have a radius of 600 feet. In how many places do the paths intersect?

2. Equidistant From Two Points: The Perpendicular Bisector

Given two points, A and B, find all points equidistant from A and B.

$A \bullet \qquad \bullet B$

Locus: The perpendicular bisector, $\overleftrightarrow{CD}$, of $\overline{AB}$.

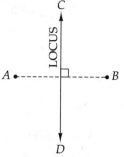

To Find a Perpendicular Bisector of $\overline{AB}$ in the Coordinate Plane

- Find the midpoint of $\overline{AB}$ using the midpoint formula.
- Find the slope of $\overline{AB}$.
- Calculate the negative reciprocal of this slope. This is the slope of the perpendicular line.
- Use the midpoint and slope to write the equation for the perpendicular bisector.

Note: To find a line perpendicular to $\overline{AB}$ that passes through some other point, simply use the coordinates of that point rather than the midpoint of $\overline{AB}$.

 MODEL PROBLEM

Find the points equidistant from $A(-2, 5)$ and $B(4, 5)$.

SOLUTION

Write the midpoint formula.
$$\left(\frac{x_A + x_B}{2}, \frac{y_A + y_B}{2}\right)$$

Substitute the coordinates of A and B.
$$\left(\frac{-2 + 4}{2}, \frac{5 + 5}{2}\right)$$

Simplify.
$$\left(\frac{2}{2}, \frac{10}{2}\right)$$
$$(1, 5)$$

Write the slope formula.
$$m = \frac{y_A - y_B}{x_A - x_B}$$

Substitute the coordinates of A and B.
$$m = \frac{5 - 5}{-2 - 4}$$

Simplify.
$$m = \frac{0}{-6}$$
$$m = 0$$

Since the slope is 0, $\overline{AB}$ is parallel to the x-axis. The perpendicular bisector will be parallel to the y-axis.

The line parallel to the y-axis, passing through the point $(1, 5)$, is the line $x = 1$.

Answer: The set of all points on the line $x = 1$.

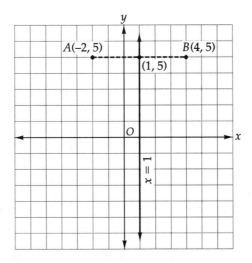

Locus **345**

To Construct the Perpendicular Bisector of $\overline{AB}$

- Open the compass so that the radius is more than half the length of $\overline{AB}$.
- Using point A as a center, draw one long arc through $\overline{AB}$.
- Using point B as a center, draw a second arc with the same radius through $\overline{AB}$ so that it intersects the first arc.
- Label one intersection of the two arcs C. Label the other intersection of the two arcs D.
- Use the straightedge to draw a line through C and D. This is $\overleftrightarrow{CD}$, the perpendicular bisector of $\overline{AB}$.

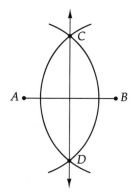

 MODEL PROBLEM

A rural area is receiving funds for a new medical center. Ideally, the center should be equidistant from two town centers. How can the planners start their search for a possible location?

SOLUTION

If points A and B above represent the two town centers, then all locations on the perpendicular bisector $\overleftrightarrow{CD}$ are equidistant from the town centers. The planners should look for locations along this bisector.

 Practice

1. The locus of points equidistant from any two points is:

   (1) one point
   (2) one line
   (3) two points
   (4) two lines

2. Two points L and D are 8 units apart. How many points are equidistant from L and D and also 3 units from L?

   (1) 0
   (2) 1
   (3) 2
   (4) 4

3. The locus of points equidistant from the four vertices of a rectangle is:

   (1) the empty set
   (2) a point
   (3) a line
   (4) a pair of points

4. Which is an equation of the locus of points equidistant from points $(2, 3)$ and $(-6, 3)$?

   (1) $x = -2$
   (2) $y = 3$
   (3) $y - 3 = x + 2$
   (4) $y - 3 = x - 2$

5. Construct an equilateral (or equiangular) triangle. (Hint: Do not change your compass setting.)

6. Write an equation for the locus of points equidistant from $A(-1, -3)$ and $B(-1, 5)$.

7. Find the equation of the line of reflection that makes points $(4, 1)$ and $(-2, 1)$ images of each other. Remember that the line of reflection is the perpendicular bisector of the segment joining a point and its image.

8. Draw a triangle and construct a median. (Hint: A median passes through both the midpoint of a side and the opposite vertex. The midpoint of a side is on its perpendicular bisector.)

9. Draw a triangle and construct a circle *circumscribed* about the triangle. (Hint: The perpendicular bisectors of each of the sides of any given triangle meet at a point—called the **point of concurrency**—which is equidistant from each of the vertices.)

10. A heliport must be built equidistant from three university hospitals. How can planners determine the location of the heliport?

11. Marlene is directing renovation of the town swimming center. As shown in the diagram, there are two pools, with the diving board sides of the pools 70 feet apart. Marlene wants to set up outdoor showers at an equal distance from each diving board. Indicate where the showers could be placed. Explain your reasoning.

12. Earthquake recording stations can record the distance from the station to the *epicenter* of an earthquake. Kevin said that reports from two centers would provide enough information to find an exact location. Is Kevin correct? Make a diagram to support your response.

3. Distance From a Line: Two Parallel Lines

Given $\overleftrightarrow{AB}$ and a distance d, find all points that are a distance d from the line.	Locus: $\overleftrightarrow{CD}$ and $\overleftrightarrow{EF}$, each parallel to $\overleftrightarrow{AB}$, and at a perpendicular distance d from $\overleftrightarrow{AB}$.

To Find a Line Parallel to a Given Line in the Coordinate Plane

- Find a point on the parallel line. (This is usually given.)
- Find the slope of the given line.
- Use the coordinates and slope to write an equation for the line.

Locus 347

MODEL PROBLEMS

1 Find the locus of points 2 units from the line $x = 3$.

SOLUTION

The line $x = 3$ and the lines parallel to it are vertical. The line $x = 1$ is 2 units to the left and the line $x = 5$ is 2 units to the right.

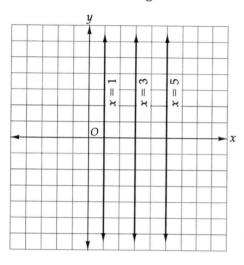

2 Find the locus of points 2 units from line l and 3 units from a point on line l.

SOLUTION

The locus of points 2 units from line l is two lines parallel to line l, each 2 units away. Construct the locus meeting the first condition.

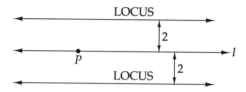

On the same diagram, construct a circle with a radius of 3 units and center at point P on line l.

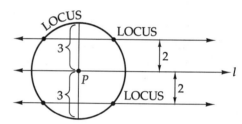

The circle locus intersects each of the 2 line loci twice.

4. Equidistant From Two Parallel Lines: One Parallel Line

Given $\overleftrightarrow{AB} \parallel \overleftrightarrow{CD}$, find all points equidistant from these lines.	Locus: $\overleftrightarrow{MN}$, parallel to the two lines and midway between them.
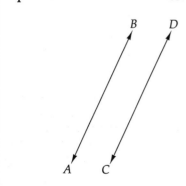	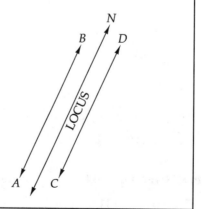

348   Chapter 9: Transformation, Locus, and Construction

To Find a Third Parallel Line Between Two Lines in the Coordinate Plane

- Pick one point on each given line and find the midpoint of a line drawn between them. This point will be on the locus.
- Find the slope of the given lines.
- Using the midpoint and the slope, write an equation for the third line.

 MODEL PROBLEM

Find the points equidistant from the lines $y = 2x + 1$ and $y = 2x - 3$.

SOLUTION

The locus is a line with a slope equal to 2 because it must be parallel to the other two lines. Since it must lie midway between the other lines, its y-intercept must be -1, midway between 1 and -3. Thus the required locus is the line $y = 2x - 1$.

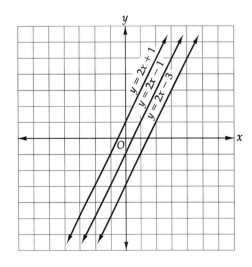

To Construct a Line Through Point P Parallel to $\overleftrightarrow{AB}$

- Draw any line through P intersecting $\overleftrightarrow{AB}$. Label the point of intersection C. Label a point R on the line on the far side of P.
- Copy angle PCB to point P and label the new angle RPS.
- Draw a line through P and S.
- $\overleftrightarrow{PS} \parallel \overleftrightarrow{AB}$

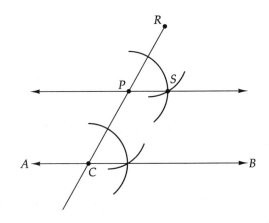

See the section on congruent figures earlier in this chapter for a reminder on how to copy an angle.

Locus

Practice

1. Which statement describes the locus of points 2 units from the x-axis?
   (1) $x = 2$
   (2) $y = -2$
   (3) $x = 2$ or $x = -2$
   (4) $y = 2$ or $y = -2$

2. What is the total number of points that are both 2 units from the x-axis and 3 units from the origin?
   (1) 0
   (2) 1
   (3) 2
   (4) 4

3. What is the total number of points that are both 3 units from the x-axis and 3 units from the origin?
   (1) 0
   (2) 1
   (3) 2
   (4) 4

4. Parallel lines x and y are cut by transversal t. What is the locus of points equidistant from x and y and at a given distance d from transversal t?
   (1) a point
   (2) a line
   (3) a pair of points
   (4) the empty set

5. Find an equation for the locus of points 4 units from the line $y = 6$.

6. Write an equation for the locus of points equidistant from the lines $y = 6$ and $y = 2$.

7. Find an equation for the locus of points equidistant from the lines $y = -x + 5$ and $y = -x + 1$.

8. How many points are 5 centimeters from a line and 8 centimeters from a point P on the line?

9. How many points are 5 centimeters from a line and 5 centimeters from a point P on the line?

10. How many points are 5 centimeters from a line and 3 centimeters from a point P on the line?

11. Two points, A and B, are 8 inches apart. Find the number of points that are equidistant from A and B and 3 inches from the line passing through A and B.

12. How many points are equidistant from two given parallel lines and also equidistant from 2 different points in one of the lines?

13. What is the locus of the center of a bicycle wheel as the bicycle travels along a straight path?

14. A bicycle path is to be constructed between two parallel highways that are 20 kilometers apart. What path would keep the bikers as far from traffic as possible?

15. At the Veterans Day Parade, a veteran carrying the American flag walked the straight parade route. A photographer who wanted to take special pictures of the veteran was told that she must remain at least 7 feet away from the parade route. Show in this drawing the possible locations where the photographer can stand.

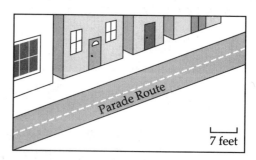

16. What properties of parallel lines are used in the construction of a parallel line?

350 Chapter 9: Transformation, Locus, and Construction

5. Equidistant From Intersecting Lines: The Angle Bisector

Given $\overleftrightarrow{AB}$ intersecting $\overleftrightarrow{CD}$, find all points equidistant from these two intersecting lines.

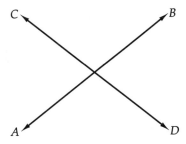

Locus: A pair of perpendicular lines, $\overleftrightarrow{WX}$ and $\overleftrightarrow{YZ}$, that bisect the angles formed by the intersecting lines.

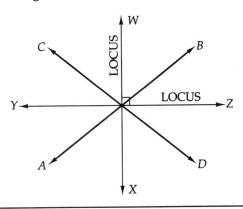

An alternative solution: the slope of one of the lines of the locus is the average of the slopes of the two given lines.

To find an angle bisector in the coordinate plane, identify the line of symmetry. The lines of symmetry are the angle bisectors for the figure.

Note: Most problems of this type will involve horizontal or vertical lines.

MODEL PROBLEM

Find the points equidistant from the lines $y = 3x + 2$ and $y = -3x + 2$.

SOLUTION

Graph the lines. The lines of symmetry, $x = 0$ and $y = 2$, are the locus required.

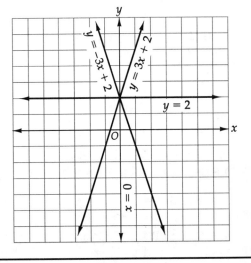

To Construct the Bisector of ∠ABC

- With B as the center, draw an arc that intersects $\overrightarrow{BA}$ at D and $\overrightarrow{BC}$ at E.
- With D and E as centers, draw intersecting arcs. The radii of these two arcs must be the same. Be certain that the radius is large enough to allow an intersection.
- Label the intersection F.
- Draw $\overrightarrow{BF}$, the angle bisector.

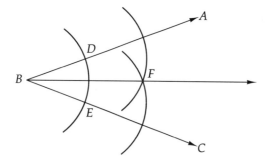

 Practice

1. How many points are equidistant from two intersecting lines and also 5 units from the point of intersection?
   - (1) 0
   - (2) 2
   - (3) 4
   - (4) 5

2. Which is an equation for the locus of points equidistant from the lines $y = x + 3$ and $y = -x + 3$?
   - (1) $y = 3$
   - (2) $y = 0$
   - (3) $y = 3$ and $x = 0$
   - (4) $x = 3$ and $y = 0$

3. Which are equations for the locus of points equidistant from the lines $y = 3$ and $x = 2$?
   - (1) $x = 3$ and $y = 2$
   - (2) $y = x$ and $y = -x$
   - (3) $y = x + 3$ and $y = -x + 3$
   - (4) $y = x + 1$ and $y = -x + 5$

4. ∠ABC is bisected by $\overrightarrow{BD}$. Points R on $\overrightarrow{BA}$ and S on $\overrightarrow{BC}$ are equidistant from B. If point X is a point on BD, then which of the following is *always* true?
   - (1) △SRX is an equilateral triangle.
   - (2) X lies on the locus of points equidistant from R and S.
   - (3) X is closer to S than to B.
   - (4) △SBX is an isosceles triangle.

5. Write equations for the locus of points equidistant from the lines $y = 2 - 3x$ and $y = 3x - 2$.

6. Write equations for the locus of points equidistant from the lines $y = 0$ and $x = 3$.

7. Draw an angle and construct the bisector.

8. Points A and B are on line *l*. Line *m* is parallel to line *l*. How many points are equidistant from points A and B and also equidistant from lines *l* and *m*?

9 An elementary school is situated at the intersection of 2 busy roads. The property is triangular, as shown in the accompanying diagram. The PTA wants to build a new playground in front of the school that is the same distance from each of the two roads. The PTA has a scale drawing similar to the diagram. How can the PTA determine possible locations?

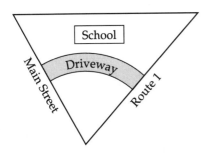

10 Draw a triangle and construct a circle that is inscribed inside the triangle. (Hint: The bisectors of each of the angles in any triangle meet at a common point. This point will have the same perpendicular distance from each of the sides.)

CHAPTER REVIEW

1 After the mapping $M:(x, y) \to (-x, y)$, the image of a point that lies in the second quadrant would now lie in quadrant:
   (1) I
   (2) II
   (3) III
   (4) IV

2 Which word has vertical line symmetry?
   (1) WOW
   (2) EVE
   (3) DAD
   (4) BOB

3 If point (0, 6) is mapped onto point (2, 6) by some line reflection, how would you describe the line of reflection?
   (1) y-axis
   (2) parallel to the x-axis
   (3) parallel to the y-axis
   (4) line $y = x$

4 The locus of the midpoints of all radii of two concentric circles is
   (1) a circle
   (2) 2 circles
   (3) a line
   (4) the empty set

5 Point P is 3 inches from line k. How many points are 5 inches from point P and also 2 inches from line k?
   (1) 1
   (2) 2
   (3) 3
   (4) 4

6 Points A and B lie on line m. Line k intersects line m at A. The number of points equidistant from points A and B and also equidistant from lines m and k is:
   (1) 0
   (2) 1
   (3) 2
   (4) 4

7 If M is the transformation $(x, y) \to (-x, y - 3)$, then what is the image of $(-1, 4)$ after the transformation M?

8 What is the image of point $(-3, 9)$ under the mapping $M:(x, y) \to \left(\frac{2}{3}x, \frac{2}{3}y\right)$?

9 If translation T is defined as $(x, y) \to (x + 2, y - 1)$, then what is the image of $(-5, 1)$ under translation T?

10. Find the coordinates of the image of $N(-1, 4)$ after a reflection over the line $y = -x$.

11. Write an equation of the line of reflection that maps $A(1, 8)$ onto $A'(8, 1)$.

12. What is the image of point $(9, -4)$ under the rotation $R_{90°}$ about the origin?

13. Write an equation for the locus of points whose distance from point $(4, -2)$ is 1.

14. Write an equation for the locus of points equidistant from $A(0, 0)$ and $B(4, 4)$.

15. Write equations for the locus of points equidistant from the lines $y = 2$ and $x = 0$.

16. What is the locus of points 4 units from the y-axis and 3 units from the origin?

17. What is the locus of points 2 units from the origin and 2 units from $P(0, -4)$?

18. Frances tells you that from where she lives, it is the same distance to the two nearest bus stops. Draw an X for each bus stop and sketch the possible places where Francis might live.

19. An architect is deciding where to install a fountain in a long rectangular courtyard. The fountain should be equidistant from the north and south sides of the courtyard. It should also be 30 feet away from the entrance on the north side. The courtyard is 60 feet across. Create a map showing where the fountain should be installed.

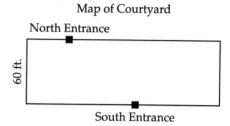

Map of Courtyard

20. Given: pentagon $FRANK$ with coordinates $F(2, 2)$, $R(4, 0)$, $A(2, -2)$, $N(-2, -2)$, $K(-4, 0)$. Write the coordinates of the images of the given point after the transformation described.

   a. the image of point F after a reflection in the line $y = -x$

   b. the image of point R after a reflection in the y-axis

   c. the image of point A after a reflection in the origin

   d. the image of point N after a rotation of $-90°$ about the origin

   e. the image of point K after the transformation $(x, y) \rightarrow (x + 7, y - 2)$

Now turn to page 422 and take the Cumulative Review test for Chapters 1–9. This will help you monitor your progress and keep all your skills sharp.

Chapter 10:

Statistics

Statistics is the study of numerical **data**. These data are collections of related numerical information, such as sports scores and records, election results, real estate sales, and grades on Regents examinations.

Once data are collected, they must be organized in a meaningful way so that conclusions can be drawn from them.

Sampling and Collecting Data

If a study requires information about every member of a population—such as the birthplace of every student in a particular high school—a **census** is conducted. A census includes every person or every situation. However, a census is often impractical or even impossible. For instance, to determine the life of 40-watt bulbs produced in a certain factory, only a small **sample**, or number, of the bulbs would be tested, because the owners of the factory would not want to burn out all of their inventory. An election poll does not include all voters; it counts only a sample of voters. Samples must be chosen carefully to avoid bias. **Bias** is a tendency to favor the selection of certain members of a population.

Selecting a Sample

- *A sample must fairly represent the entire population being studied.* The sample for a study of residents' satisfaction with a local park should represent all people in the neighborhood, not just those who use the park or come on weekends. People dissatisfied with the park may not visit it at all, and their opinions should also be included.

- *The number of data values must be significant.* The study of statistics includes determining the minimum amount of data needed to make a study meaningful.

- *The sample should be random or selected in an organized way that avoids bias.* Picking names out of a hat is an example of random selection. Studying every tenth person who enters a supermarket is an example of organized selection.

 **Practice**

1. A survey that includes every member of a population is called a
   (1) poll
   (2) census
   (3) random sample
   (4) biased study

2. A well-chosen sample
   (1) represents an entire population
   (2) is large enough to provide meaningful data
   (3) is unbiased
   (4) all of the above

3. Statistics is the study of
   (1) lightbulbs
   (2) random samples
   (3) numerical data
   (4) election polls

4. Sampling is used to choose
   (1) an entire population
   (2) a fair representation of a group
   (3) a small number of people
   (4) sports statistics

Exercises 5–14: A sample must be selected to determine the favorite summer activities of people in a particular town. Is each of the following samples representative or not? Why?

5. tenth-grade students

6. women

7. people at a beach

8. residents of a nursing home

9. shoppers in a mall

10. people in the produce aisle of a supermarket

11. fans at a baseball game

12. every twentieth name in the town's telephone directory

13. concert musicians

14. millionaires

Organizing Data

Stem-and-Leaf Plots

A **stem-and-leaf plot** is a display that shows each data value.

 MODEL PROBLEM

In September, each of the 24 students in a math class reported the number of days he or she worked that summer:

25, 17, 15, 28, 24, 13, 16, 28, 19, 25, 24, 36, 33, 18, 24, 38, 28, 25, 27, 14, 37, 28, 35, 43

Represent these data in a stem-and-leaf plot, with the values in the *tens* place as the stem and the corresponding values in the *units* place as the leaves.

SOLUTION

Step 1. Find the smallest and largest values in the data.

Step 2. Draw a vertical line. To the left of the line, write each consecutive digit from the smallest stem value to the largest stem value.

Step 3: Insert the leaves.
Step 4: For each stem, reorder the leaves from smallest to largest.
Step 5: Include a key.

For the data above, the smallest value is 13 and the largest value is 43. The smallest tens value is 1 and the largest tens value is 4. After step 3, the plot looks like this:

```
1 | 7 5 3 6 9 8 4
2 | 5 8 4 8 5 4 4 8 5 7 8
3 | 6 3 8 7 5
4 | 3
```

After Step 4—ordering the leaves—the plot looks like this:

```
1 | 3 4 5 6 7 8 9
2 | 4 4 4 5 5 5 7 8 8 8 8
3 | 3 5 6 7 8
4 | 3
```

The key (Step 5) can show any value, such as 2|4 = 24.

From this plot, we find that most students worked from 24 to 28 days.

A **back-to-back stem-and-leaf plot** can be used to compare two sets of data. The stem is drawn down the middle of the chart, and the leaves extend to the left for one data set and to the right for the other data set. The leaves are in increasing order moving away from the stems.

MODEL PROBLEM

Two brands of soup are advertised as "low-fat." A study compares grams of fat per serving for 11 kinds of soup of each brand.

Brand A:
1.5, 1.6, 3.0, 4.2, 1.2, 3.1, 4.6, 2.4, 3.6, 2.4, 1.8

Brand B:
1.3, 3.5, 4.2, 3.3, 5.2, 4.1, 3.2, 5.1, 4.8, 4.1, 4.4

Display these results in a back-to-back stem-and-leaf plot.

SOLUTION

The stem for these data is the units digit. The leaves show the numbers in the tenths places.

Brand A		Brand B
8 6 5 2	1	3
4 4	2	
6 1 0	3	2 3 5
6 2	4	1 1 2 4 8
	5	1 2
Key: 6|3 = 3.6		Key: 3|2 = 3.2

This plot shows that brand A soups generally have less fat than brand B soups.

Organizing Data

Tally and Frequency Tables

Suppose you want to know how many hours members of your class spend on the Internet each week. You conduct a census and get these results: 0, 0, 7, 5, 0, 5, 0, 1, 5, 5, 3, 5, 0, 0, 5, 5, 7, 1, 0, 5.

Data presented in an unorganized format such as this are difficult to use. Questions like "What is the most common amount of time spent on the Internet?" and "How many students spend more than 4 hours a week on the Internet?" are answered more easily if the data are organized. A **tally** or **frequency table** is one form of organization.

MODEL PROBLEM

Organize the above results in a tally or frequency table.

SOLUTION

Step 1. Label the columns: in this case "Hours on Internet," "Tally," and "Frequency."

Step 2. Start with the largest number in the data, in this case 7. Keep subtracting 1, entering the numbers down the first column. Stop at 0.

Step 3. For each piece of data, enter a mark (|) in the appropriate row under "Tally." Every fifth mark is a diagonal through the preceding four marks (||||).

Step 4. Under "Frequency," complete each row by counting the tally marks and recording the number.

Step 5. Add all the frequency values to get total frequency.

> Sometimes, you need only tally marks or only frequencies, not both.

Hours on Internet	Tally	Frequency
7	||	2
6		0
5	|||| |||	8
4		0
3	|	1
2		0
1	||	2
0	|||| ||	7
Total frequency: 20		

We can see from the frequency table that 10 out of 20 or 50% of the students spent more than 4 hours weekly on the Internet.

In the model problem above, only a few pieces of data were reported. Sometimes, however, the number of values is so large that a table showing each value would be very long. In these cases, we group the data into equal-sized **intervals**. The resulting frequency chart looks the same except for the first column, which shows intervals.

For example, suppose that a guidance counselor wants to make a chart of the Math SAT I scores for all her students. She groups scores into six intervals: 201–300, 301–400, 401–500, 501–600, 601–700, 701–800. (Each interval has a length of 100.) She then looks at each student's score and enters a tally mark in the appropriate interval. Then she enters the total for each interval in the frequency column.

Math SAT I Scores

Interval	Tally	Frequency
701–800	|||| |||	8
601–700	|||| |||| |||| ||||	19
501–600	|||| |||| |||| |||| ||||	25
401–500	|||| |||| |||| |||| |||	23
301–400	|||| |||| ||||		16
201–300	|||	3
	Total frequency:	94

To Group Data for a Tally or Frequency Table

- Make all your intervals equal in size. Try to use a length that will give a total of 5 to 10 intervals.

- Make sure that the intervals include the entire range of values in the data set.

- Construct the intervals from the bottom up. The highest value in each interval should be one less than the lowest value in the next interval. The intervals should not overlap.

Note: A grouped frequency table does not show individual data values.

Frequency Histograms

Often, it is useful to display grouped data in a **frequency histogram**, similar to a vertical bar graph. The widths of all the bars are equal and represent equal intervals. The frequency for each interval is represented by the height of its bar. Just as there are no gaps between intervals, there should be no gaps between bars.

Other kinds of bar graphs are reviewed on the next three pages.

To Construct a Frequency Histogram

- Construct a frequency table.

- Draw a horizontal axis and mark off the intervals. Label the horizontal axis. If the first interval does not start at 0, use a "break" symbol (⌇) on the axis.

- Draw a vertical axis and identify a scale for the frequencies. Label the vertical axis. Often, the vertical axis is "Frequency."

- Draw bars with heights corresponding to the frequency values in the table.

- Give the graph an appropriate title.

Organizing Data

MODEL PROBLEM

Construct a frequency histogram for these test grades: 100, 93, 71, 74, 85, 56, 62, 68, 70, 100, 99, 85, 77, 85, 48, 51, 79, 25, 86, 93, 67, 88, 70, 100, 26

SOLUTION

First construct the frequency table. Intervals of 10 are used here.

Test Scores	Frequency
91–100	6
81–90	5
71–80	4
61–70	5
51–60	2
41–50	1
31–40	0
21–30	2

Next draw the frequency histogram.

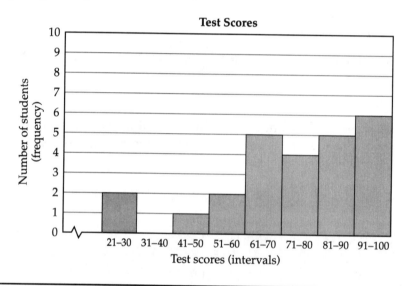

A histogram can be constructed on a graphing calculator. On a TI-83 calculator, use the [STAT] menu to enter a list. Use [WINDOW] to set up the axes. Use [STAT PLOT] to select the histogram, then press [GRAPH]. See the Appendix for more information on constructing graphs.

Bar Graphs

A **bar graph** looks much like a histogram. However, the height or length of each bar represents an amount, or value, rather than a frequency. The greater the value, the longer or higher the bar. A bar graph can be vertical or horizontal.

360 Chapter 10: Statistics

For example, the bar graph below compares types of TV programs watched by a family during prime time over 1 week. Each bar represents a number of hours.

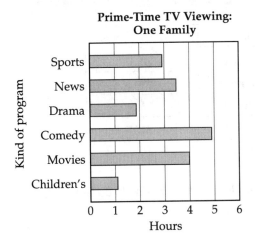

A **double bar graph** can be used to compare two quantities, as in the example below:

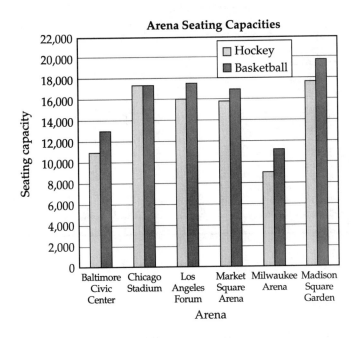

Organizing Data 361

MODEL PROBLEM

Construct a bar graph for these data on weekly TV watching by young people:

Group	Weekly hours
Teenage girls	20.3
Teenage boys	22
Children 6–11	21.5
Children 2–5	24.7

SOLUTION

Step 1. Determine and label the numerical scale, which must include the range of data values. Here, the label is "Hours."

Step 2. On the other scale, provide a label for each bar.

Step 3. Plot the data on the numerical scale to get the length of each bar. Then fill in the bars.

Step 4. Give the bar graph a title.

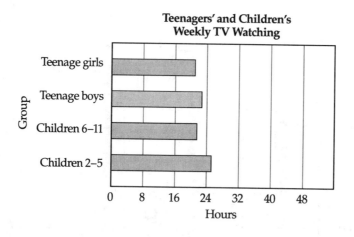

Broken-Line Graphs

A **line graph**, or **broken-line graph**, is used to describe a trend, or changes over time, for one particular item. The horizontal axis is a number line showing time. The vertical axis shows numeric amounts.

To Construct a Broken-Line Graph

- Construct the horizontal axis. Label the axis (for instance, days, weeks, months, or years). Use equal time intervals.

- Construct and label a vertical axis, using a scale appropriate for the values. Remember to use a "break" symbol if the values do not start at 0.

- Plot the data points. Connect the points with lines.

- Create a title for the graph.

MODEL PROBLEMS

A student civics club is raising money for a charity. The graph below shows amounts raised over 8 weeks.

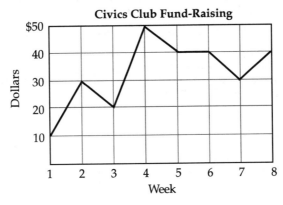

Answer each question:

1. When did the greatest increase in dollars raised occur?

2. When did no change occur?

3. When did the greatest percent increase occur?

SOLUTIONS

From week 3 to week 4. (The graph is steepest here.)

From week 5 to 6. (The graph is flat.)

From week 1 to week 2.
$$\left(\frac{30-10}{10} = \frac{20}{10} = 2 = 200\%\right)$$

Scatter Plots

A **scatter plot** relates two sets of data. Usually, both the vertical axis and the horizontal axis show numeric amounts.

To Construct a Scatter Plot

- Construct and label both axes, using reasonable scales for the values. (Use a "break" symbol if necessary.)
- Plot the data points.
- Give the scatter plot a title.

For example, the graph below shows heights and shoe sizes for a group of males.

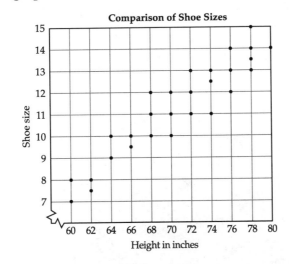

Organizing Data

Here, the plotted points appear to cluster around a straight line. This graph is an example of **positive correlation**. The two sets of data increase together. The plotted points go from lower left to upper right.

By contrast, in a graph showing **negative correlation**, one data set increases while the other decreases. The graph would go from upper left to lower right.

Sometimes there is no obvious relationship between data sets. In such cases, we say that there is **no correlation**.

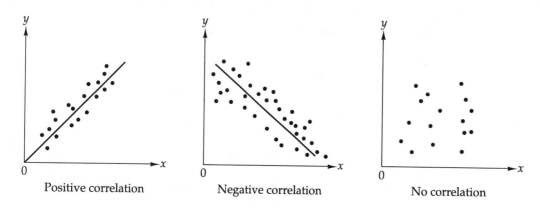

Positive correlation · Negative correlation · No correlation

Note: Scatter plots can be drawn with a graphing calculator. See the Appendix.

Circle Graphs

A **circle graph** shows how percents or parts of a whole are distributed. The whole circle always represents 100% of the data.

The area of a section is proportional to the fraction or percentage it represents. The central angle of the sector is equal to the given percentage of 360° (the whole circle). For example, a sector representing 25% of the data would have a central angle equal to 25% × 360° = 0.25 × 360° = 90°; a sector representing 12% would have a central angle equal to 12% of 360° = 0.12 × 360° = 43.2°.

To Construct a Circle Graph

- Find the number that represents the whole if it is not given.

- Find what part of the whole each category represents. It is usually most convenient to express this part as a decimal.

- Calculate the central angle for each sector: multiply 360° by each part of the whole found in step 2.

- Draw a circle. Use a protractor to construct the central angle for each sector. Label each sector.

MODEL PROBLEMS

1 In a survey, 400 people named their favorite TV sport: baseball, 115; football, 129; basketball, 99; tennis, 57. Draw a circle graph representing the results of the survey.

SOLUTION

The total number, 400, is given.

baseball: $\frac{115}{400} = 0.2875$ $0.2875 \times 360° = 103.5°$

football: $\frac{129}{400} = 0.3225$ $0.3225 \times 360° = 116.1°$

basketball: $\frac{99}{400} = 0.2475$ $0.2475 \times 360° = 89.1°$

tennis: $\frac{57}{400} = 0.1425$ $0.1425 \times 360° = 51.3°$

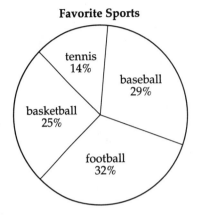

Check: $103.5° + 116.1° + 89.1° + 51.3° = 360°$

2 This circle graph shows votes in a student council election. If 1,200 students voted, how many voted for each candidate?

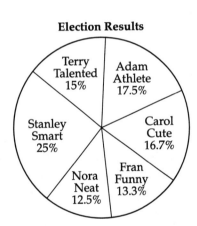

SOLUTION

Use your calculator.

Adam Athlete: 17.5% of 1,200 = 210
Carol Cute: 16.7% of 1,200 = 200.4 (round to 200)
Fran Funny: 13.3% of 1,200 = 159.6 (round to 160)
Nora Neat: 12.5% of 1,200 = 150
Stanley Smart: 25% of 1,200 = 300
Terry Talented: 15% of 1,200 = 180

Check: $210 + 200 + 160 + 150 + 300 + 180 = 1,200$

1. The key for a stem-and-leaf plot is 4|7 = 47. The value 63 would be plotted as:
   (1) 3|6
   (2) 60|3
   (3) 6|3
   (4) 63

2. The frequency table shows favorite types of television programs for a group of 50 students. What percent of the students chose comedy or drama?

   Favorite TV Programs

Program	Frequency
News	7
Sports	11
Drama	8
Talk	5
Comedy	14
Science	3
Cooking	2

   (1) 14%
   (2) 22%
   (3) 28%
   (4) 44%

3. The heights of a group of 36 people were measured to the nearest inch. The heights ranged from 47 inches to 76 inches. To construct a histogram showing 10 intervals, the length of each interval should be:
   (1) 2 inches
   (2) 3 inches
   (3) 4 inches
   (4) 5 inches

4. Two quantities for which a positive correlation could be expected are:
   (1) years of school and annual salary
   (2) age and minutes spent watching cartoons
   (3) outdoor temperature and math test grades
   (4) shoe size and number of books read

5. Arthur's monthly income is $3,000. His rent is $850 per month. The measure of the central angle for the sector "Rent" on a circle graph for Arthur's budget would be approximately:
   (1) 28°   (3) 102°
   (2) 85°   (4) 127°

6. Sajan, a camp counselor, recorded the high temperature at camp each day. Construct a back-to-back stem-and-leaf plot comparing temperatures in July with temperatures in August. State one fact about the temperatures you are able to learn from the plot.

   July (°F): 78, 79, 80, 86, 82, 90, 91, 91, 89, 94, 89, 95, 94, 91, 88, 86, 91, 87, 90, 88, 92, 90, 94, 93

   August (°F): 92, 89, 89, 90, 86, 87, 90, 85, 84, 79, 90, 90, 89, 88, 90, 91, 98, 91, 87, 79, 82, 81, 86, 82

Exercises 7 and 8: Construct a frequency table for the data in each set.

7. Heights, in inches, of 30 students:
   65, 66, 64, 70, 71, 68, 65, 70, 69, 67, 66, 65, 68, 67, 70, 64, 68, 67, 66, 70, 64, 65, 70, 65, 64, 71, 71, 65, 69, 68

8. Grades on a 10-point quiz for 25 students:
   10, 5, 4, 8, 6, 3, 9, 10, 7, 8, 3, 7, 5, 5, 8, 8, 6, 7, 9, 10, 7, 5, 8, 3, 9

9. Following are test grades for 18 students:
   72, 86, 95, 75, 100, 85, 87, 100, 81, 86, 78, 94, 96, 80, 100, 98, 96, 91

   a Complete the tally and frequency table:

Interval	Tally	Frequency
96–100		
91–95		
86–90		
81–85		
76–80		
71–75		
Total frequency:		

   b Draw a frequency histogram.
   c How many students had grades less than 81?

10 Following are weights, in pounds, of 30 students:

174, 126, 115, 150, 180, 185, 130, 105, 200, 176, 151, 194, 146, 180, 120, 118, 116, 131, 136, 190, 177, 149, 145, 165, 182, 156, 143, 199, 103, 130

a Complete the tally and frequency table.

Interval	Tally	Frequency
191–200		
181–190		
171–180		
161–170		
151–160		
141–150		
131–140		
121–130		
111–120		
101–110		
Total frequency:		

b Draw a frequency histogram.
c Which intervals have the greatest number of students?

11 Following are numbers of students registered for Advanced Placement U.S. History in North High School each year since 1995.

Year	1995	1996	1997	1998	1999
Number	30	32	35	43	45

a Graph these data on a line graph.
b Which year had the largest percent increase in students taking the course?

12 A class of 24 students was surveyed to find favorite toppings on pizza. The results were as follows.

Sausage 9
Bacon 4
Pepperoni 3
Mushrooms 6
Peppers 2

In a circle graph of this survey, what would be the measure of the central angle for each sector of the graph?

Sausage _____°
Bacon _____°
Pepperoni _____°
Mushrooms _____°
Peppers _____°

13 The following circle graph represents 34 people's favorite ice cream flavor. How many people preferred each flavor?

Favorite Ice Cream Flavors

14 Describe a real-world example for which you would predict a negative correlation.

15 Describe a real-world example for which you would predict no correlation.

Analyzing Data

Measures of Central Tendency: Mean, Median, and Mode

Mean (average), median, and mode are **measures of central tendency**. They are selected values used to describe all the numbers in a set.

The **mean** (or average) of a set of numbers is found by adding the numbers and then dividing that total by the number of data items in the set. For example, if Matthew's grades are 80, 92, 85, 91, 95, and 88, the mean is:

$$\frac{80 + 92 + 85 + 91 + 95 + 88}{6} = \frac{531}{6} = 88.5$$

The **median** is the middle value when a set of numbers is arranged in order from least to greatest (or from greatest to least):

- If there is an odd number of items, the median is the middle number. For the numbers 2, 5, 7, 19, 20 the median is 7.

- If there is an even number of items, the median is the average of the two middle numbers. For the numbers 3, 4, 6, 7, 8, 9 the median is $\frac{6+7}{2} = \frac{13}{2} = 6.5$.

The median is not necessarily a number in the data set.

Both mean and median can be found on a graphing calculator.

The **mode** is the value or values that appear most often in a set. A set of values may have **no mode** (if each value appears the same number of times), **one mode**, or **more than one mode**. For the numbers 3, 7, 5, 3, 2, 5, 8 the modes are 3 and 5. This set of numbers is said to be **bimodal** because it has two modes.

MODEL PROBLEMS

1 The average grade for the 26 students in Mr. Rottman's U.S. history class is 82. The average for the 24 students in Mr. Spence's U.S. history class is 86. What is the mean grade for all of the students in both classes?

SOLUTION

Mean = $\frac{\text{sum}}{\text{number of items}}$. Thus:

$82 = \frac{\text{sum of Mr. Rottman's grades}}{26}$

Sum of Mr. Rottman's grades = $82 \times 26 = 2{,}132$

$86 = \frac{\text{sum of Mr. Spence's grades}}{24}$

Sum of Mr. Spence's grades = $86 \times 24 = 2{,}064$

Average = $\frac{\text{sum of all scores}}{\text{total number of students}}$

$= \frac{2{,}064 + 2{,}132}{24 + 26} = \frac{4{,}196}{50} = 83.92$

2 On Memorial Day, the Community Chest float had one student from each grade, 3 to 12, except grade 6. What was the median grade of the students on the float?

SOLUTION

Grades on the float were 3, 4, 5, 7, 8, 9, 10, 11, 12. The middle grade is 8.

Answer: Grade 8

3 Allan's test scores are 80, 90, 92, and 88. He needs a 90 average to make the principal's honor roll. What is the lowest score he can get on his next test to raise his average to at least 90?

SOLUTION

Let x = Allan's next test score. Then:

$\frac{80 + 90 + 92 + 88 + x}{5} \geq 90$

$\frac{350 + x}{5} \geq 90$

$350 + x \geq 450$

$x \geq 100$

Answer: To make the honor roll, Allan must get 100 on his next test. The answer actually comes out as ≥ 100, but since 100 is the highest possible score, we discard *greater than*.

4 The prices of beaded bracelets at a crafts fair were $8 for amethyst, $10 for jade, $10 for onyx, $9 for hematite, and $10 for topaz. What is the mode (in dollars) for these prices?

SOLUTION

8, 9, 10, 10, 10
The most common price is $10.

Answer: $10

Practice

1. What is the median of the set 80, 50, 67, 55, 70, 65, 75, 50?
   (1) 50
   (2) 64
   (3) 66
   (4) 67

2. The mean of a set of test scores is 83. If a score of 85 is added to the set and a new mean is calculated:
   (1) the new mean is greater than 83
   (2) the new mean is less than 83
   (3) the new mean is 83
   (4) the effect on the new mean cannot be determined

3. Which *must* be a value in the data set?
   (1) mean
   (2) median
   (3) mode
   (4) none of the above

4. Find the mean of these shoe sizes.

Size	Frequency
5	2
6	4
7	5
8	6
9	3

   (1) 7.0
   (2) 7.2
   (3) 7.8
   (4) 8.4

5. Find the median of this stem-and-leaf plot.

   ```
   3 | 3 5
   4 | 4 4 6 8
   5 | 0 1 3 7 9 9
   6 | 2 2 5 8
   ```
   Key: 3|3 = 33

   (1) 50
   (2) 52
   (3) 52.25
   (4) 59

6. The average of A and B is 6. The average of X, Y, and Z is 16. What is the average of A, B, X, Y, and Z?

7. The average grade on a test taken by 20 students was 75. When one more student took the test after school, the class average became 76. What score did that student receive?

8. Find the mean, median, and mode of this data set:
   $$-2, 5, 10, -6, 7, 5, -2, 5, 10, -15, 14$$

9. On New Year's Day in Watertown, the following temperatures were recorded. What is their average (mean)?

Time	°F
6 A.M.	−10
10 A.M.	2
2 P.M.	8
6 P.M.	4

10. The average of four different positive integers is 9. What is the greatest value for one of the integers?

11. Five students took a makeup quiz for Mrs. Hald. Three students picked up their papers after school. Their grades were 75, 73, and 86. Mrs. Hald told them that the average of the five grades was 80. What was the average of the other two scores?

12. If $j > 0$, what is the average of $2j$, $4j - 3$, and 6?

13. If the sum of an odd number of consecutive integers is zero, what is the median?

14. Create a data set of seven integers in which the median is greater than the mode.

15. The average and the median of three fractions are each $\frac{1}{3}$. The largest fraction is $\frac{1}{2}$. What is the smallest fraction?

Range, Quartiles, and Box-and-Whisker Plots

The **range** of a set of data is the difference between the greatest value and the least value. To find the range, it is often helpful to rank the members of the set from greatest to least (or vice versa). For example, when 84, 60, 89, 95, 63, and 74 are ranked in order, the range is obvious:

95, 89, 84, 74, 63, 60

Range = 95 − 60 = 35

The concept of range is important in quartiles and box-and-whisker plots.

The median of a data set divides the set into two equal parts—that is, two parts with the same number of members. **Quartiles** are values that separate a data set into four parts, each containing one-fourth or 25% of the members. For example:

```
 53, 60, 61, 63, 64,   65, 65, 65, 65, 66,    66, 67, 67, 68, 69,    70, 70, 71, 71, 73
        64.5                   66                     69.5
                              Median
     First quartile      Second quartile       Third quartile
```

- The **first** or **lower quartile** is the center of the lower half.
- The **second quartile** is the median.
- The **third** or **upper quartile** is the center of the upper half.
- The **interquartile range** is the difference between the third quartile and the first quartile. In the example above, interquartile range = 69.5 − 64.5 = 5.

MODEL PROBLEM

Find the quartiles and the interquartile range:

42, 25, 55, 58, 60, 75, 80, 85, 65, 55, 19, 72, 77, 50

SOLUTION

Step 1. Rank the members in order:
19, 25, 42, 50, 55, 55, 58, 60, 65, 72, 75, 77, 80, 85

Step 2. Find the median. This is the second quartile:
19, 25, 42, 50, 55, 55, <u>58, 60</u>, 65, 72, 75, 77, 80, 85
Median = (58 + 60) ÷ 2 = 59

Step 3. First quartile is 50, the median of the seven values to the left of 59:
19, 25, 42, <u>50</u>, 55, 55, 58

Step 4. Third quartile is 75, the median of the seven values to the right of 59:
60, 65, 72, <u>75</u>, 77, 80, 85

Step 5. To find the interquartile range, subtract the first quartile from the third:
75 − 50 = 25

Answer: The quartiles are 50, 59, and 75. The interquartile range is 25.

A **box-and-whisker** plot is a graph that describes data using the quartiles and the highest and lowest values (the **extreme** values) in the data. This plot is useful for comparing two or more data sets. The box-and-whisker plots show how the data for each set are distributed and what the extreme values are.

To Construct a Box-and-Whisker Plot

- Draw a number line to include the lowest value and the highest value in the data set.
- Above the number line, mark the quartiles and the extreme values.
- Draw a box above the number line, with vertical sides passing through the lower and upper quartiles. Draw a vertical line in the box through the median (second quartile).
- Draw the "whiskers," horizontal lines extending from the vertical sides of the rectangle to the extreme values.

Box-and-whisker plots can also be constructed with a graphing calculator. See the Appendix.

Note: **Outliers** are values much lower or much higher than most of the data. In a box-and-whisker plot, outliers are data that fall more than 1.5 times the interquartile range from the quartiles. Do *not* extend whiskers to any outliers.

MODEL PROBLEM

Draw a box-and-whisker plot for these data:

20 27 28 29 30 31 33 33 37 39 55

SOLUTION

Draw the box: The median is 31. The lower quartile is 28. The upper quartile is 37. Plot those values above a number line and draw the rectangle.

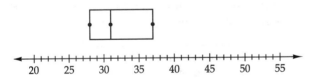

Draw the whiskers: The interquartile range is $37 - 28 = 9$. Data more than $1.5(9) = 13.5$ from the quartiles are outliers.

Left whisker:

$28 - 13.5 = 14.5$

No data are smaller than 14.5, so there are no low outliers. The left whisker will extend from the box to 20.

Right whisker:

$37 + 13.5 = 50.5$

One value, 55, is more than 50.5. The right whisker will therefore extend only to the next highest value, 39.

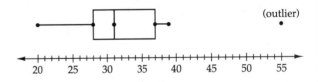

Practice

1 Identify the first quartile:

32, 24, 38, 26, 38, 36, 37, 39, 23, 40, 21, 31

(1) 21
(2) 25
(3) 32
(4) 34

2 The interquartile range of a data set is 18. The first quartile is 52. Which value could be the median?

(1) 25
(2) 34
(3) 61
(4) 97

Analyzing Data

3 This is a box-and-whisker plot for 60 test scores.

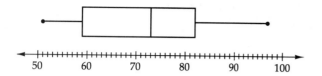

The upper quartile is:

(1) 59
(2) 73
(3) 82
(4) 97

4 Find the interquartile range for this set of children's heights (in centimeters):

147, 130, 160, 150, 152, 120, 121, 125, 128, 121, 140, 142, 134, 126

(1) 7
(2) 22
(3) 33
(4) 40

5 These box-and-whisker plots show test scores for two classes:

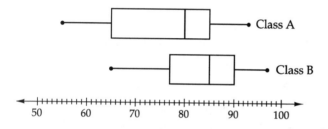

Which statement is FALSE?

(1) The interquartile range is greater for class A than for class B.
(2) The second quartile is higher for class B than for class A.
(3) The lowest score was in class A.
(4) Class A did better than class B.

Exercises 6 and 7: Use this data set.

60, 62, 62, 64, 68, 69, 71, 72, 74, 76, 78, 81, 83, 87

6 Find the quartiles and interquartile range.

7 Construct a box-and-whisker plot.

Exercises 8 and 9: use this data set.

30, 67, 67, 68, 68, 69, 71, 71, 71, 72, 73, 74, 74, 75, 80

8 Find the quartiles and interquartile range.

9 Construct a box-and-whisker plot.

10 In a box-and-whisker plot, what percentage of the scores are represented by the box?

11 In a box-and-whisker plot, what percentage of the scores are represented by each whisker?

12 In a box-and-whisker plot, what characteristic of the data would position the median in the middle of the box?

13 Is it possible to have a box-and-whisker plot with only one whisker? Explain your answer.

14 Two box-and-whisker plots for the same number of data points have the same extremes and the same median. However, the box is twice as long for the first plot as for the second. What differences would you expect to find if you compared the data?

15 A class decided to compare a supermarket-brand chocolate chip cookie with a famous name brand. The students broke apart nine cookies of each brand and recorded the number of chips. Make a box-and-whisker plot for each brand and compare the data, including quartiles, interquartile ranges, and the length of the whiskers.

Supermarket brand	Name brand
4	7
4	7
5	8
5	9
6	9
7	9
10	10
12	10
18	11

Percentiles and Cumulative Frequency Histograms

The percentile rank of a member of a data set tells us the percent of values in the set less than or equal to that member. If more than one value matches the member, include half of the matching values.

 MODEL PROBLEM

Find the percentile rank of 80 in the following 30-item data set:

63, 63, 63, 64, 65, 66, 66, 68, 70, 70, 74, 75, 80, 80, 80, 83, 83, 84, 87, 87, 88, 88, 89, 89, 89, 90, 90, 90, 91, 91

SOLUTION

12 numbers are less than 80

3 numbers are 80

$\frac{1}{2} \times 3 = 1.5$

$12 + 1.5 = 13.5$

$\frac{13.5}{30} = 0.45 = 45\%$

Answer: The number 80 is at the 45th percentile.

We have already reviewed frequency tables and frequency histograms for grouped data. Suppose we collected data on test scores and made this table:

Interval (Test Scores)	Frequency (Number)
91–100	50
81–90	60
71–80	60
61–70	20
51–60	10

By adding one more column to the table, **cumulative frequency**, we can record the number of scores that are:

100 or less
90 or less
80 or less
70 or less
60 or less

Starting with the top row, the cumulative frequency column is calculated by adding all the frequencies from that row down to the bottom of the table. Or, starting from the bottom, each new cumulative frequency would be found by adding the frequency of that row to the cumulative frequency recorded on the next lower row.

Analyzing Data

For this example, the cumulative frequency table is:

Interval (Test Scores)	Frequency (Number)	Cumulative Frequency
91–100	50	200
81–90	60	150
71–80	60	90
61–70	20	30
51–60	10	10

The graph of a cumulative frequency table is called a **cumulative frequency histogram**. It is constructed in exactly the same way as a frequency histogram. Notice that as we look at a cumulative frequency histogram from left to right, the bars increase in height.

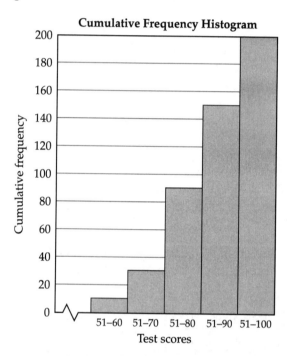

In the cumulative frequency histogram above, the frequency scale goes from 0 to 200. A different vertical scale, showing the cumulative frequency in percents, can also be used. We would write 0% instead of 0 and 100% in place of 200. The other markers usually indicate **quartiles** (0%, 25%, 50%, 75%, 100%) or **deciles** (multiples of 10%). For example, if we display quartiles:

0% is at 0

25% is at 25% of 200, or 50

50% is at 50% of 200, or 100

75% is at 75% of 200, or 150

100% is at 100% of 200, or 200

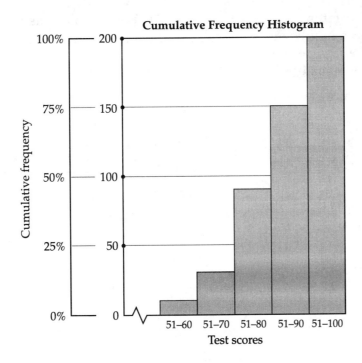

From this version of the cumulative frequency histogram, we can see that 75% of the students scored 90 or less. Thus, 90 is the value of the third quartile.

Often, we use a cumulative frequency histogram to find approximate quartile values. If we want to know the number of scores at or below an interval upper value, we can approximate from this chart. For instance, in this example about 45% of the scores are at or below 80, and about 5% of the scores are at or below 60.

 Practice

1. If Bradley scored at the 90th percentile on a test, which statement *must* be true?

   (1) Bradley answered 90% of the questions correctly.
   (2) About 90% of the students who took the test had a score greater than or equal to Bradley's.
   (3) 90% of the students who took the test received the same score as Bradley.
   (4) About 90% of the students who took the test had a score less than or equal to Bradley's.

2. For a set of data consisting of basketball players' heights in inches, the 50th percentile is 76. Which statement could be *false*?

   (1) Fifty percent of the heights are 76 inches or less.
   (2) Half of the heights are 76 inches or more.
   (3) The mean height is 76 inches.
   (4) The median height is 76 inches.

Analyzing Data   **375**

3 For a set of data recording temperatures (degrees Fahrenheit) in Binghamton in July, the 25th percentile was 68° and the 75th percentile was 80°. Which statement *must* be true?

(1) Fifty percent of the temperatures were between 68 and 80 degrees.
(2) The median temperature was 74 degrees.
(3) Temperatures were recorded below 68 degrees.
(4) The mean temperature was 74 degrees.

4 What is the percentile rank of 90 in the following data set?

65 67 68 75 77 77 79 82 83 84 84 85 85 86 88 88 90 90 92 93 93 94 95 97 99

(1) 64
(2) 68
(3) 72
(4) 90

5 In the table below, what is the cumulative frequency for the salary range $25,000–$29,999?

Salary	Frequency	Cumulative Frequency
$40,000–44,999	6	60
$35,000–39,999	9	54
$30,000–34,999	18	45
$25,000–29,999	0	?
$20,000–24,999	12	27
$15,000–19,999	10	15
$10,000–14,999	5	5

(1) 0
(2) 12
(3) 27
(4) 39

Exercises 6–10: Refer to the table below.

Interval	Frequency	Cumulative Frequency
91–100	8	
81–90	10	
71–80	10	
61–70	8	
51–60	4	

6 Complete the table.

7 Draw a cumulative frequency histogram with a percent scale by quartile.

8 Find the interval that contains the median.

9 Find the interval that contains the 25th percentile.

10 Estimate what percent of the scores are 70 or less.

Exercises 11–15: Refer to the table below, which shows ungrouped data. The interval length is 1.

Score	Frequency	Cumulative Frequency
10	6	
9	7	
8	7	
7	6	
6	3	

11 Complete the table.

12 Draw a cumulative frequency histogram with a percent scale by decile.

13 Find the score that contains the median.

14 Find the score that contains the 10th percentile.

15 Estimate what percent of the scores are 6 or less.

CHAPTER REVIEW

1. For a set of data showing the number of books read by Ms. Corson's AP English class, the median was 8 and the average was 12. Which statement *must* be true?

   (1) Fifty percent of the students read 8 books.
   (2) Fifty percent of the students read 12 books.
   (3) At least one student read more than 12 books.
   (4) Most students read 8 books.

2. This table shows the ages of 30 tennis players in a teen tournament. Which statement is true?

Age	Frequency
19	6
18	6
17	12
16	4
15	2

   (1) median > mode
   (2) median = mode
   (3) median < mode
   (4) median > mean

3. For seven weeks, the number of TVs Mark Stone sold were 2, 5, 0, 7, 4, 6, and 4. What is the minimum number of TVs Mark must sell next week to have a weekly sales average greater than 4?

   (1) 3
   (2) 4
   (3) 5
   (6) 6

4. What is the mean of the following data?

   $3 + \sqrt{3}, \ 4 + \sqrt{3}, \ 5 + \sqrt{3}, \ 6 + \sqrt{3}, \ 7 + \sqrt{3}$

   (1) 5
   (2) 25
   (3) $5 + \sqrt{3}$
   (4) $5 + 5\sqrt{3}$

5. A box-and-whisker plot was constructed for a set of 100 test scores. Approximately how many scores are between the lower quartile and the highest score?

   (1) 75
   (2) 50
   (3) 25
   (4) cannot be determined

6. Which statements are true?

   I   The interquartile range is half the distance between the lower quartile and the upper quartile.
   II The interquartile range is not affected by outliers.
   III The median represents the 50th percentile.

   (1) I and II
   (2) II only
   (3) II and III
   (4) I and III

7. Bruce's class counted the raisins in individual-portion boxes of Raisin Toasties and Rocky Raisins. Create a back-to-back stem-and-leaf plot comparing the number of raisins recorded for the two brands. State one fact you are able to learn from the plot.

   Raisin Toasties:
   18, 19, 12, 15, 11, 20, 21, 21, 18, 24, 19

   Rocky Raisins:
   19, 18, 20, 21, 18, 21, 19, 10, 12, 11, 16

8. Following are test grades of 18 students:
   80, 80, 95, 90, 90, 85, 83, 97, 86, 86, 80, 94, 100, 80, 100, 98, 96, 91

   a. Complete the table:

Interval	Tally	Frequency
96–100		
91–95		
86–90		
81–85		
76–80		
71–75		
Total frequency:		

   b. Draw a frequency histogram.

   c. How many students received grades less than 81?

9 Following are weights in pounds of 30 students:

174, 126, 115, 150, 180, 185, 130, 105, 200, 176, 151, 194, 146, 180, 120, 118, 116, 131, 136, 190, 177, 149, 145, 165, 182, 156, 143, 199, 103, 130

a Complete the table:

Interval	Tally	Frequency
191–200		
181–190		
171–180		
161–170		
151–160		
141–150		
131–140		
121–130		
111–120		
101–110		
Total frequency:		

b Draw a frequency histogram.

10 Shown below are the numbers of students hurt in car accidents in a suburban town yearly since 1994.

Year	Number
1994	15
1995	14
1996	10
1997	12
1998	16
1999	20

a Display these data on a line graph.

b Which year had the largest percent increase in injuries?

11 A class of 30 students was surveyed to find favorite subjects. The results were as follows. On a circle graph for this survey, what is the measure of the central angle, in degrees, for each sector?

Subject	Number of Students	Measure of Central Angle
English	8	
Social studies	9	
Mathematics	6	
Science	6	
Foreign language	1	

12 Inez has totaled her expenses for the last year and represented the results in this circle graph. If her total expenses were $18,000, find the amount spent on each category.

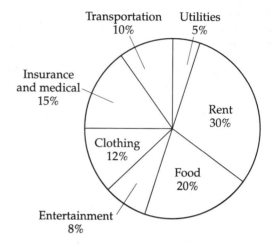

13 The average grade on a test taken by 15 students was 70. When one more student took the test after school, the class average became 71. What score did that student receive?

14 If $k > 0$, what is the average of $3k - 1$, $4k - 3$, and $1 - 10k$?

15 Give an example of a data set for each of the following situations.

a mean = median = mode

b mean < median = mode

c mean = median < mode

Exercises 16 and 17: Use this data set.

51, 75, 75, 77, 81, 82, 84, 86, 87, 90, 91, 92, 96, 99, 100

16 Find the quartiles and interquartile range.

17 Construct a box-and-whisker plot.

18 Find the percentile rank of 189 in the following set of heights (in centimeters):

148 162 166 167 168 175 175 177 179 180 182
182 183 188 189 189 189 189 190 190 191 191
191 193 194 197 198 200 200 202

19 A class average on a science test was 28.5 correct out of 40 questions. The 19 girls in the class scored 539 correct. How many correct did the 11 boys score?

378 Chapter 10: Statistics

20 Use this table:

Interval	Frequency	Cumulative Frequency
91–100	6	
81–90	12	
71–80	8	
61–70	8	
51–60	6	

a Complete the table.
b Draw a cumulative frequency histogram with a percent scale by quartile.
c Find the interval that contains the median.
d Find the interval that contains the 25th percentile.

Now turn to page 425 and take the Cumulative Review test for Chapters 1–10. This will help you monitor your progress and keep all your skills sharp.

Chapter 11:
Probability

Probability is a branch of mathematics in which a numerical value is used to express the likelihood of an event, or outcome. Probability has to do with certainty, uncertainty, and prediction—that is, with chance. The probability of an event is the chance that it will occur, expressed as a ratio of a specific event to all possible events:

$$\text{Probability} = \frac{\text{number of actual events}}{\text{number of possible events}}$$

This ratio can be written as a fraction (such as $\frac{1}{2}$), as a percent (50%), or as a decimal (.5). In probability it is customary to omit the zero before a decimal point.

Principles of Probability

Theoretical and Empirical Probability

If you flip a fair coin, two results are equally likely: heads or tails. **Theoretical probability** deals with sets of equally likely possibilities such as these.

In real life, outcomes are often *not* equally likely. For instance, if you toss a paper cup it can land facing up, facing down, or on its side. You might predict that it is most likely to land on its side; but to determine the probability of each outcome, you would need to experiment and observe the results. **Empirical probability** deals with sets of possibilities based on experimentation; thus it is sometimes called **experimental probability**.

Rules of Probability

The following definitions and rules are generally applicable to situations involving probability, but we will refer to a specific example of theoretical probability: rolling a single fair die, or number cube, one time.

380

- **Outcome.** The result of some activity. For the die, the possible outcomes are:
  1, 2, 3, 4, 5, 6
- **Sample space.** Set of all possible outcomes. For the die:
  sample space $S = \{1, 2, 3, 4, 5, 6\}$
- **Event.** Any subset of the sample space. The event of rolling a 6 is:
  $E = \{6\}$
- A **formula for probability** P of a single event E is:
  $$P(E) = \frac{n(E)}{n(S)}$$
  where $n(E)$ is the number of ways E can occur, and $n(S)$ is the number of elements in sample space S. For example, the probability of rolling a 6 is:
  $$P(6) = \frac{n(6)}{n(S)} = \frac{1}{6} \text{ or } 16.\overline{6}\% \text{ or } .1\overline{6}$$
- The **probability of an impossible event is 0**, because $n(E) = 0$. For example, the probability of rolling a 7 is:
  $$P(7) = \frac{n(7)}{n(S)} = \frac{0}{6} = 0$$
- The **probability of a certain event is 1**. For example, the probability of rolling a positive number is:
  $$P(+) = \frac{n(+)}{n(S)} = \frac{6}{6} = 1$$
- For any event E, the range of probabilities is from 0 to 1:
  $$0 \leq P(E) \leq 1$$
- The **probability of a single event with two conditions** is the number of ways both conditions can be satisfied, divided by $n(S)$. For example, to find the probability $P(A \text{ and } B)$ of (A) rolling an even number and (B) rolling a number greater than 3, we first find $n(A \text{ and } B)$, in this case $n(\text{even}, >3)$:

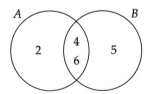

  There are two elements, 4 and 6, in the intersection, so $n(\text{even}, >3) = 2$. Then:
  $$P(\text{even}, >3) = \frac{2}{n(S)} = \frac{2}{6} = \frac{1}{3}$$
- The **probability that two different events will both occur**, simultaneously or in succession, is the product of their individual probabilities:
  $$P(A \text{ and } B) = P(A) \times P(B)$$
  For example, if you roll the die twice, the probability of getting a 5 both times is:
  $$P(5 \text{ and } 5) = \frac{1}{6} \times \frac{1}{6} = \frac{1}{36}$$

Principles of Probability

- The **probability of a single event that must satisfy one condition *or* another** is found with this formula:

  $P(A \text{ or } B) = P(A) + P(B) - P(A \text{ and } B)$

  For example, to find the probability of rolling (A) an odd number *or* (B) a number greater than 2, we find the set that satisfies A: {1, 3, 5}. Then we find the set that satisfies B: {3, 4, 5, 6}. Two numbers, 3 and 5, are members of both sets, so $n(A \text{ and } B) = 2$. Therefore:

  $P(A \text{ or } B) = \dfrac{3}{6} + \dfrac{4}{6} - \dfrac{2}{6} = \dfrac{5}{6}$

- Two events are **mutually exclusive** if they cannot occur at the same time. Thus, $P(A \text{ and } B) = 0$ and the **probability of two mutually exclusive events** is:

  $P(A \text{ or } B) = P(A) + P(B) - 0 = P(A) + P(B)$

  For example, you cannot roll a 2 and a 6 at the same time, so the probability of rolling (A) a 2 *or* (B) a 6 is:

  $P(2 \text{ or } 6) = \dfrac{1}{6} + \dfrac{1}{6} = \dfrac{2}{6} = \dfrac{1}{3}$

- The condition "not A" is called the **complement** of A and can be written $\overline{A}$ or $\sim A$. The probability of this condition is:

  $P(\text{not } A) = 1 - P(A) \quad \text{or} \quad P(\overline{A}) = 1 - P(A) \quad \text{or} \quad P(\sim A) = 1 - P(A)$

  For example, the probability of rolling any number that is not 3 is:

  $P(\text{not } 3) = 1 - P(3) = 1 - \dfrac{1}{6} = \dfrac{5}{6}$

- In calculating an **empirical probability**, $n(S)$ is the total number of past occurrences, such as observations or trials, and $n(E)$ is the number of times the particular event occurred:

  $P(E) = \dfrac{\text{number of occurrences of } E}{\text{total number of observations}}$

  For instance, if we toss a cup 25 times and observe that it lands up 3 times, down 2 times, and sideways 20 times, the empirical probabilities are:

  $P(\text{up}) = \dfrac{3}{25} \qquad P(\text{down}) = \dfrac{2}{25} \qquad P(\text{side}) = \dfrac{20}{25} = \dfrac{4}{5}$

Random Variables and Probability Distributions

Often each outcome of a probability experiment can be represented by a real number. If x represents the different numbers corresponding to the possible outcomes of some probability situation, then x is called a **random variable**.

For example, suppose a travel agent knows that $\dfrac{1}{5}$ of her customers book 2-day trips, $\dfrac{1}{4}$ book 3-day trips, $\dfrac{2}{5}$ book 4-day trips, and $\dfrac{3}{20}$ book 5-day trips. If x represents number of days, then x is a random variable that takes the value 2 with probability $\dfrac{1}{5}$, the value 3 with probability $\dfrac{1}{4}$, the value 4 with probability $\dfrac{2}{5}$, and the value 5 with probability $\dfrac{3}{20}$.

A **probability distribution** for a random variable is a list or formula that gives the probability for each value of the random variable. The sum of the probabilities must equal 1:

x	P(x)
2	$\frac{1}{5}$
3	$\frac{1}{4}$
4	$\frac{2}{5}$
5	$\frac{3}{20}$

$$\frac{1}{5} + \frac{1}{4} + \frac{2}{5} + \frac{3}{20} = \frac{4}{20} + \frac{5}{20} + \frac{8}{20} + \frac{3}{20} = \frac{20}{20} = 1$$

MODEL PROBLEMS

1 Two dice are rolled. Find P(dice show the same number).

SOLUTION

We need to determine how many outcomes are possible (sample space). We can list the possible outcomes in a table as ordered pairs.

		Second Die					
		1	2	3	4	5	6
First Die	1	(1, 1)	(1, 2)	(1, 3)	(1, 4)	(1, 5)	(1, 6)
	2	(2, 1)	(2, 2)	(2, 3)	(2, 4)	(2, 5)	(2, 6)
	3	(3, 1)	(3, 2)	(3, 3)	(3, 4)	(3, 5)	(3, 6)
	4	(4, 1)	(4, 2)	(4, 3)	(4, 4)	(4, 5)	(4, 6)
	5	(5, 1)	(5, 2)	(5, 3)	(5, 4)	(5, 5)	(5, 6)
	6	(6, 1)	(6, 2)	(6, 3)	(6, 4)	(6, 5)	(6, 6)

There are 36 possible outcomes in the sample space. Of these, there are 6 outcomes where the dice show the same number. So:

$$P(\text{dice show the same number}) = \frac{n(E)}{n(S)} = \frac{6}{36} = \frac{1}{6}$$

Answer: $\frac{1}{6}$

2 Two dice are rolled. Find P(sum of 11).

SOLUTION

To get a sum of 11, one die must show 5 and the other die must show 6. Referring to the table, we find 2 possible outcomes with a sum of 11. So:

$$P(\text{sum of 11}) = \frac{n(E)}{n(S)} = \frac{2}{36} = \frac{1}{18}$$

Answer: $\frac{1}{18}$

Principles of Probability

1 The probability that the Talbot Agency will win a new advertising contract is .3. What is the probability that it will *not* win the contract?

(1) .5
(2) .7
(3) .9
(4) 1.3

2 Dana, Ed, and Laurie are the only candidates for class president. Their respective probabilities of winning are $\frac{1}{2}$, $\frac{3}{10}$, and $\frac{1}{5}$. What is the probability that the winner will be either Dana or Laurie?

(1) $\frac{2}{5}$
(2) $\frac{1}{2}$
(3) $\frac{7}{10}$
(4) $\frac{4}{5}$

3 A die was tossed 100 times. The outcomes are shown below:

Number	Frequency
1	14
2	18
3	20
4	4
5	22
6	22

On the basis of this experiment, $P(4 \text{ or } 5)$ is:

(1) $\frac{1}{25}$
(2) $\frac{1}{3}$
(3) $\frac{7}{10}$
(4) $\frac{13}{50}$

4 Which of the following could NOT represent a probability distribution for the random variable x?

(1)

x	$P(x)$
1	.2
2	.2
3	.3
4	.4

(2)

x	$P(x)$
0	$\frac{1}{8}$
2	$\frac{3}{8}$
5	$\frac{3}{8}$
8	$\frac{1}{8}$

(3)

x	$P(x)$
1	$\frac{1}{6}$
2	$\frac{1}{6}$
3	$\frac{1}{6}$
4	$\frac{1}{6}$
5	$\frac{1}{6}$
6	$\frac{1}{6}$

(4)

x	$P(x)$
10	.1
11	.2
12	.1
13	.55
14	.05

5 Two dice are rolled. What is P(sum is at least 2)?

(1) 0

(2) $\frac{1}{6}$

(3) $\frac{3}{4}$

(4) 1

6 The table shows the number of juices sold in the cafeteria one morning.

Juice	Number
apple	13
orange	21
grapefruit	11
tomato	5
cranberry	6
mango	4

On the basis of these data, P(orange or grapefruit) is:

(1) $\frac{8}{15}$

(2) $\frac{11}{21}$

(3) $\frac{16}{25}$

(4) $\frac{7}{20}$

Exercises 7–12: A spinner with eight equal sectors numbered 1 to 8 is spun. Find each probability.

7 P(even)

8 P(greater than 2 or less than 7)

9 P(less than 1)

10 P(less than 10)

11 P(multiple of 3 and odd)

12 P(not a multiple of 4)

Exercises 13–18: Two dice are rolled. Find the probability of each outcome:

13 At least one die shows a 5.

14 Neither die shows a 2.

15 The dice show different numbers.

16 Exactly one die shows a 1.

17 The sum of the numbers shown is 7 or 11.

18 The sum of the numbers shown is greater than 9.

19 There are 5 flashlights in a box. Each flashlight can be either good or defective. Define a random variable x to represent the number of defective flashlights. What are the possible values of x?

20 Of 500 sweaters examined, 6 have irregular sleeves, 19 are discolored, and 133 have other flaws. The rest have no defects.

   a What is the probability of getting a sweater with other flaws?

   b What is the probability of getting a sweater with no defects?

   c Which kind of probability does this situation involve: theoretical or empirical? Why?

Calculating Sample Spaces

As reviewed above, the **sample space** S is the number of ways an event can occur.

In some cases, the sample space is obvious: for instance, if a spinner has four equal sectors, $n(S) = 4$ for a single spin. And in some cases the sample space is a familiar array, such as the table of possible outcomes of rolling two dice. However, in many problems involving a series of possibilities, it is necessary to calculate the sample space. This section reviews several calculation methods.

Tree Diagrams

Using a **tree diagram** is one way to show a sample space and count the number of possible outcomes.

MODEL PROBLEM

If a family has 2 children, how many different arrangements of male (M) and female (F) are there? Make a tree diagram to show the sample space.

SOLUTION

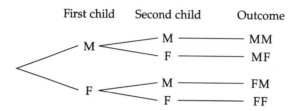

The tree diagram shows that there are 4 possible outcomes.

Answer: Four different arrangements are possible.

The Counting Principle

The **counting principle** states that:

- For a sequence of possibilities, the total number of possible outcomes is the product of the number of outcomes for each part of the sequence.

Thus if there are m choices for event A and n choices for event B, then there are mn choices for event A followed by event B.

In this situation, two concepts are important:

- **With replacement** means that an outcome can be included more than once. That is, the first event does *not* affect the possibilities for the second event. We call such events **independent**.

- **Without replacement** means that an outcome *cannot* be included more than once. In this case, the first event *does* affect the possibilities for the second event. These events are **dependent**.

For instance, consider a bag containing 2 apples and 2 oranges. Suppose that:

- (1) We take out one piece of fruit at random. It is an apple. We replace it. Then we draw another piece at random.

Or:

- (2) We take out one piece at random. It is an apple. We do *not* replace it. Then we draw another piece at random.

The events in situation 1—with *replacement*—are *independent*. Since the first apple drawn is replaced, the probability of getting an apple on the second draw does not change. In contrast, the events in situation 2—*without* replacement—are *dependent*. Since the first apple drawn is *not* replaced, there are fewer apples left for the second draw, and so the probability of drawing an apple changes.

To take another case, tossing a coin is a situation involving independent events, since the outcome of one toss does not affect the outcome of the next.

MODEL PROBLEMS

1. A pizza can be ordered in 3 sizes: small (S), medium (M), or large (L). One of 3 extra toppings can be ordered: pepperoni (P), vegetables (V), or cheese (C). How many different ways can a pizza be ordered?

SOLUTION

Use a tree diagram:

Size	Topping	Outcome
S	P	SP
	V	SV
	C	SC
M	P	MP
	V	MV
	C	MC
L	P	LP
	V	LV
	C	LC

Use the counting principle:

Choice of sizes is event A, with $m = 3$ possibilities.

Choice of toppings is event B, with $n = 3$ possibilities.

Then:

$$\text{total possible outcomes} = mn = 3 \times 3 = 9$$

Answer: A pizza can be ordered 9 different ways.

2. A department store has 6 entrances. There are 3 ways of getting from floor to floor: stairs, elevator, or escalator. How many ways are there to enter the store and go from the first floor to the second floor? Use the counting principle.

SOLUTION

6 entrances × 3 ways to the second floor = 18 different ways

3. How many ways can a red flag, a white flag, and a blue flag be lined up in a row?

SOLUTION

This is a problem *without replacement*, since the same flag cannot appear twice. There are 3 choices for the first flag. After the choice is made, there are 2 choices left for the second flag. After that choice is made, there is one choice for the third flag:

$3 \times 2 \times 1 = 6$

Answer: 6 arrangements

4. How many different 3-digit numbers can be made from the 10 digits if each digit may be used only once and the 3-digit number must be greater than 600?

SOLUTION

Since each digit is used only once, this is a problem *without replacement*. The number could have any of these forms:

6__ __   7__ __   8__ __   9__ __

There are 4 choices for the hundreds digit: 6, 7, 8, or 9.

After the hundreds digit is chosen, 9 choices are left for the tens digit. After the tens digit is chosen, 8 choices are left for the units digit.

Then:

$4 \times 9 \times 8 = 288$

Answer: 288 numbers

Calculating Sample Spaces

Sets of Ordered Pairs

Another way to count a sample space is to write **sets of ordered pairs**. For example, in a toss of two coins, the possible outcomes are the ordered pairs:

{(H, H), (H, T), (T, H), (T, T)}

Note: We can also write **sets of ordered triples**. For example, the outcomes of tossing 3 coins would be:

{(H, H, H), (H, H, T), (H, T, T), (T, T, T), (T, T, H), (T, H, H), (H, T, H), (T, H, T)}

MODEL PROBLEM

Pairs of black (B), white (W), and red (R) socks are in a drawer. You take out 1 sock at random and then take out another sock. Use ordered pairs to find the number of possible outcomes.

SOLUTION

The set of possible outcomes is:

{(B, B), (B, W), (B, R), (W, B), (W, W), (W, R), (R, B), (R, W), (R, R)}

Answer: There are 9 possible outcomes.

 Practice

1. Three children are born in a family. How many of the possible outcomes include exactly 2 females?
   (1) 2
   (2) 3
   (3) 4
   (4) 5

2. For his vacation, Leon will travel to New York City and then to Paris. He can travel to New York by bus, train, car, or plane. He can travel from New York to Paris by plane or ship. How many different ways can Leon plan his trip?
   (1) 4
   (2) 6
   (3) 7
   (4) 8

3. How many different ways can the letters in the word MONEY be arranged?
   (1) 10
   (2) 120
   (3) 3,125
   (4) 1,953,125

4. How many different 3-digit numbers are possible if the first digit cannot be 0 and the second digit must be 0 or 1?
   (1) 90
   (2) 162
   (3) 180
   (4) 729

5. A true-false test has 4 questions. How many different sets of answers are possible?
   (1) 4
   (2) 8
   (3) 16
   (4) 256

6 Ellen has 4 pairs of slacks, 3 shirts, and 3 scarves that all match. How many outfits that include slacks, a shirt, and a scarf can she put together? Make a tree diagram.

7 A delicatessen serves sandwiches on rye bread, white bread, or rolls. There are eight different sandwich fillings. Customers have a choice of lettuce or no lettuce on the sandwich. How many different sandwiches can be assembled?

8 A spinner is divided into 4 equal sectors, each a different color. If the spinner is spun twice, how many different results can occur?

9 There are 4 entrances to a building, 3 escalators up, and 2 escalators down. How many ways can a person enter from outside, go to the second floor, return to the first floor, and then leave the building?

10 A six-sided die and a fair coin are tossed together. How many outcomes are in the sample space?

11 A six-sided die and a fair coin are tossed together. How many outcomes that do NOT include both an even number and tails are in the sample space?

12 How many 7-digit telephone numbers can be created if the first digit cannot be 0?

Permutations

In some situations, the order of outcomes, or events, is important. In other situations, the order of outcomes is *not* important. This difference can affect the size of a sample space. A **permutation** (reviewed here) is an ordered arrangement of events. A **combination** (reviewed next) is a selection of events in which the order does not matter.

Two kinds of sample spaces involving permutations are:

- Number of permutations of a set taken as a whole.
- Number of permutations when we select from a set.

To illustrate **permutations of a set taken as a whole**, consider the earlier example of finding how many arrangements can be made if 3 different flags are lined up. Using the counting principle, we found that the number of arrangements is $3 \times 2 \times 1 = 6$. Two standard ways to write $3 \times 2 \times 1$ are 3! (3 **factorial**) and $_3P_3$ (the permutation of 3 items taken 3 at a time.)

In general, $n! = {_nP_n} = n \times (n-1) \times (n-2) \times \ldots \times 1$.

0! is defined as 1. 1! is also defined as 1.

MODEL PROBLEM

How many different ways can 5 students be seated in a row of 5 chairs?

SOLUTION

$_5P_5 = 5! = 5(5-1)(5-2)(5-3)(5-4) = 5 \times 4 \times 3 \times 2 \times 1 = 120$

Answer: 120 ways

To illustrate **permutations of selected members of a set**, consider the model problem below, which involves 3 items out of a set of 10. In a situation like this, the arrangement involves some number r that is less than the number n of all the items. The number of arrangements is $10 \times 9 \times 8$ or $_{10}P_3$ (the permutation of 10 items taken 3 at a time.)

In general, $_nP_r = \dfrac{n!}{(n-r)!} = n \times (n-1) \times (n-2) \times \ldots \times (n-r+1)$.

Calculating Sample Spaces **389**

MODEL PROBLEM

How many signals can be made by lining up 3 flags selected from 10 different colored flags?

SOLUTION

Here, $n = 10$ and $r = 3$.

There are 10 choices for the first spot, 9 choices for the second spot, and 8 choices for the third spot. Then:

$$_{10}P_3 = 10 \times 9 \times 8 = 720$$

Answer: 720 ways

When we find $n!$ or $_nP_n$ each item must be unique. For example, think of the letters BOB and suppose we treated the two B's as different items, B_1 and B_2. Then B_1OB_2 and B_2OB_1 would appear, misleadingly, as different arrangements.

Thus, if a set includes repetition, there are less than $n!$ arrangements of n items. This is a **special case of permutation**:

- Number of permutations of n items taken n at a time with r items identical $= \dfrac{n!}{r!}$

If there is more than one set of identical items, the denominator shows each of these sets as a factorial product.

 MODEL PROBLEM

How many different arrangements can be made from the letters in MISSISSIPPI?

SOLUTION

MISSISSIPPI has 11 letters, with 3 sets the same: 4 I's, 4 S's, and 2 P's.

Use the special case formula $\dfrac{n!}{r!}$ with $n = 11$. Show each of the repeated sets as a factorial in the denominator:

$$\frac{11!}{4! \cdot 4! \cdot 2!} = \frac{39{,}916{,}800}{24 \cdot 24 \cdot 2} = \frac{39{,}916{,}800}{1{,}152} = 34{,}650$$

Answer: 34,650 arrangements

In summary, here are the **formulas** for numbers of permutations.

- For a set of n events taken n at a time: $_nP_n = n!$

- For a set of n events taken r events at a time: $_nP_r = \dfrac{n!}{(n-r)!}$

- For n events taken n at a time with r items identical, permutations $= \dfrac{n!}{r!}$

Scientific and graphing calculators have keys or menu choices for n! (or x!) and $_nP_r$. See the appendix.

Practice

1. The expression $_8P_4$ is equal to:
   (1) 12
   (2) 32
   (3) 1,680
   (4) 4,096

2. How many different ways can 7 runners be assigned to 7 lanes at the start of a race?
   (1) 14
   (2) 49
   (3) 210
   (4) 5,040

3. Steve has 6 coins in his pocket: 1 penny, 1 nickel, 1 dime, and 3 quarters. In how many different orders can Steve take the 6 coins out of his pocket?
   (1) 36
   (2) 120
   (3) 720
   (4) 46,656

4. There are 12 students in a club. In how many different ways can a president, vice president, treasurer, and secretary be elected from the club members?
   (1) 16
   (2) 48
   (3) 1,320
   (4) 11,880

5. The expression $_{10}P_1$ is equal to:
   (1) 1
   (2) 10
   (3) 11
   (4) 100

6. How many ways can all of 6 different flags be lined up?

7. How many ways can 4 of 10 flags be lined up?

8. How many ways can a 3-digit number be created if the digits are all different and the number is at least 500?

9. How many 4-digit numbers can be created if all the digits are odd?

10. How many different ways can you arrange the letters in BUFFALO?

11. How many different ways can you arrange the letters in BUFFALO if the F's must remain together?

12. How many ways can you arrange the letters in BUFFALO if the first letter must be B?

13. How many different ways can you arrange the letters in CALIFORNIA?

14. How many different ways can you arrange the letters in MINISINK?

15. How many different ways can you arrange 2 of 6 minivans, followed by 4 of 5 convertibles?

Combinations

A selection of events in which order does *not* matter is called a **combination**. For example, if a committee of 3 students is being chosen from a group of 10, the order of the students does not matter. The committee {Bob, Carol, Sam} is the same as {Carol, Bob, Sam} or {Sam, Carol, Bob}, and so on.

Two kinds of sample spaces involving combinations are:

- Number of **combinations of a set of *n* events taken *n* events at a time**. This number is always 1. In other words, a set of *n* events taken as a whole, $_nC_n$, has only one combination:

  $_nC_n = 1$

Calculating Sample Spaces   **391**

- Number of **combinations of a set of *n* events taken *r* events at a time**. This situation can be written as $_nC_r$. The **formula** is:

$$_nC_r = \frac{_nP_r}{r!} \quad \text{or} \quad _nC_r = \frac{n!}{r! \cdot (n-r)!}$$

Remember the standard expressions:

Notation	Meaning
$_nC_n$	Combinations of a set of *n* events taken *n* events at a time (always = 1)
$_nC_r$	Combinations of a set of *n* events taken *r* events at a time

Graphing calculators have a key or menu item for $_nC_r$.

MODEL PROBLEMS

1 Ms. Simpson is faculty advisor to the school's National Honor Society chapter. There are 50 members. How many ways can she choose 5 members to attend a conference?

SOLUTION

This situation does *not* involve a permutation: order does *not* matter. Therefore, we want the number of *combinations* of 50 items taken 5 at a time. Find $_{50}C_5$ on your calculator:

$$_{50}C_5 = 2{,}118{,}760$$

As a check, find $\frac{_{50}P_5}{5!}$ on your calculator. You should get the same result:

$$\frac{_{50}P_5}{5!} = \frac{254{,}251{,}200}{120} = 2{,}118{,}760$$

Answer: 2,118,760 ways

2 Ms. Leonardi is forming a committee of students to work on a school project. If 3 freshmen, 2 sophomores, 4 juniors, and 4 seniors volunteer, how many ways can Ms. Leonardi choose a committee of 8 students, with 2 from each grade level?

SOLUTION

Ms. Leonardi needs 2 freshmen out of 3, 2 sophomores out of 2, 2 juniors out of 4, and 2 seniors out of 4.

Calculate:

$$_3C_2 \cdot {_2C_2} \cdot {_4C_2} \cdot {_4C_2} = 3 \cdot 1 \cdot 6 \cdot 6 = 108$$

Answer: 108 ways

Note: For the $_nC_r$'s in this problem, the sum of the *n*'s is the number of elements in the sample space and the sum of the *r*'s is the total number of elements being chosen.

Here are some more facts about combinations.

- Just as there is only 1 combination of a set as a whole, or only 1 way to choose all the items ($_nC_n = 1$), there is only 1 way to choose *none* of the items:

$$_nC_0 = 1$$

- Choosing (including) *r* out of *n* items is the same as *not* choosing (excluding) $n - r$ items. This is reasonable, since selecting 2 of 5 is the same as excluding or leaving out 3 of 5. Thus if $r \le n$, then:

$$_nC_r = {_nC_{n-r}}$$

Note: Here are some guidelines for determining whether a problem involves permutations or combinations.

- Ask whether *order* is important (a permutation) or unimportant (a combination).

- Look for *key words*:
  Words associated with permutations include *order*, *ordering*, *arrange*, and *arrangement*.
  Words associated with combinations include *group*, *gather*, and *assemble*.

- Recognize *typical situations*. For example:
  A problem that involves choosing people to fill specific roles—such as club president, secretary, and treasurer—is a permutation problem.
  A problem that involves choosing people for a group with no specific roles—such as members of a committee—is a combination problem.

Practice

1. $_8C_3 =$
   (1) 24
   (2) 56
   (3) 336
   (4) 1,680

2. $_{14}C_{14} =$
   (1) 0
   (2) 1
   (3) 14
   (4) 196

3. If $_nC_6 = {_nC_4}$, then $n =$
   (1) 2
   (2) 8
   (3) 10
   (4) 24

4. $_{15}C_8 =$
   (1) $_8C_7$
   (2) $_{15}C_7$
   (3) $_7C_{15}$
   (4) $_{23}C_{15}$

5. Acme Company wants to open accounts at 3 different local banks. If 10 banks are available, how many ways can Acme choose the 3 banks?
   (1) 30
   (2) 120
   (3) 720
   (4) 1,000

6. Leslie bought 5 flavors of ice cream and 3 toppings. She will choose 2 flavors and 1 topping for a sundae. How many different sundaes can she make?

7. If 4 Jets, 4 Giants, and 4 Bills volunteer for a Big Brother project, how many ways can 2 men from each team be chosen?

8. A family of 5 received 3 free tickets to a concert. The baby will not go. How many ways can the tickets be distributed?

Exercises 9–12: Kyle goes to a video rental store with enough money to rent 3 videos. He finds 3 mysteries, 4 dramas, and 2 comedies that he wants to see.

9. How many selections of 3 videos would be all mysteries?

10. How many selections of 3 videos would be all dramas?

11. How many selections of 3 videos would be all comedies?

12. How many selections of 3 videos would be one of each kind?

Exercises 13–16: Ten girls and 8 boys are interested in visiting a job fair off campus. The school district has a minibus that can hold 14 students. Susan and Peter must go to the job fair.

13 How many ways can students be chosen for the bus?

14 How many ways can students be chosen if all of the girls must go?

15 How many ways can the students be chosen if the same number of girls and boys must go?

16 How many ways can the students be chosen if Marta must go?

17 How many ways can a committee of 4 juniors and 4 seniors be selected from 10 juniors and 12 seniors?

18 A box of 12 cell phones has 3 defective ones. How many ways can 4 phones be selected so that 2 are good and 2 are defective?

Probability Problems Involving Permutations and Combinations

Probability problems, including real-world situations, often involve permutations and combinations. The model problems below are typical examples.

MODEL PROBLEMS

1 From a standard 52-card deck, you draw 1 card and put it aside. Then you draw another card. What is the probability that the first card is an ace (A) and the second is a king (K)?

SOLUTION

Since order is important, this is a permutation problem. Note that the events are dependent, and that there are 4 aces and 4 kings in a standard deck. Three solution methods are shown.

Method 1: Counting Principle	Method 2: $P(A \text{ and } B) = P(A) \cdot P(B)$	Method 3: $_nP_r$
Possibilities for 1st card: $4 \text{ out of } 52 = \frac{4}{52}$ Possibilities for 2nd card: $4 \text{ out of } 51 = \frac{4}{51}$ Multiply: $\frac{4}{52} \cdot \frac{4}{51} =$ $\frac{16}{2,652} = \frac{4}{633} \approx .006$	Event A is an ace (A) and event B is a king (K). For 1st card: $P(A) = \frac{n(E)}{n(S)} = \frac{n(A)}{52} = \frac{4}{52}$ For 2nd card: $P(K) = \frac{n(E)}{n(S)} = \frac{n(K)}{51} = \frac{4}{51}$ Then: $P(A \text{ and } K) = P(A) \cdot P(K)$ $= \frac{4}{52} \cdot \frac{4}{51} = \frac{16}{2,652} \approx .006$	Ways to get 1 ace of $4 = {_4P_1}$ Ways to get 1 king of $4 = {_4P_1}$ So for ace followed by king: $n(E) = {_4P_1} \cdot {_4P_1}$ Ways to draw 2 cards in order from $52 = {_{52}P_2}$, so: $n(S) = {_{52}P_2}$ Calculate: $\frac{n(E)}{n(S)} = \frac{{_4P_1} \cdot {_4P_1}}{{_{52}P_2}}$ $= \frac{4 \cdot 4}{52 \cdot 51} = \frac{16}{2,652} \approx .006$

Answer: The probability is approximately .006.

2 At the Pet of the Month awards ceremony, 9 collies and 6 poodles are available for a news photo. The photographer chooses 3 dogs at random. What is the probability that 2 collies and 1 poodle will be chosen?

SOLUTION

Order is not important, so this is a combination problem. The events are dependent, since any dog can be chosen only once. Here is one solution method.

Calculate:

$n(E) = {_9}C_2 \cdot {_6}C_1 = 36 \cdot 6 = 216$

$n(S) = {_{15}}C_3 = 455$

$P = \dfrac{n(E)}{n(S)} = \dfrac{{_9}C_2 \cdot {_6}C_1}{{_{15}}C_3} = \dfrac{216}{455} \approx .47$

Answer: The probability is approximately .47.

Practice

1 Two cards are drawn without replacement from a standard 52-card deck. The probability that both cards are diamonds is:

(1) $\dfrac{1}{17}$

(2) $\dfrac{169}{2,652}$

(3) $\dfrac{1}{221}$

(4) $\dfrac{1}{4}$

2 A box contains 3 red pens, 2 green pens, and 5 blacks pens. Two pens are picked at random (without replacement of the first pen before the second is picked). What is the probability of getting a red pen, then a green pen?

(1) $\dfrac{3}{50}$

(2) $\dfrac{1}{15}$

(3) $\dfrac{1}{6}$

(4) $\dfrac{2}{27}$

3 Cindy has forgotten her friend's phone number. She knows it starts with 555 and the other four numbers consist of one 3, one 7, and two 9's. If she picks a combination and dials, what is the chance she will dial the correct number?

(1) 1 in 4
(2) 1 in 12
(3) 1 in 24
(4) 1 in 420

4 Denzel picked six numbers out of 20 to play on an internet lottery. Which expression is equal to the probability that his six numbers will match the six winning numbers? (The order of the numbers does not matter.)

(1) $\dfrac{14!}{20! \times 6!}$

(2) $\dfrac{14! \times 6!}{20!}$

(3) $\dfrac{20!}{14!}$

(4) $\dfrac{6!}{20!}$

Probability Problems Involving Permutations and Combinations

5. A coin bank contains 9 nickels, 4 dimes, and 5 quarters. If you shake out 3 coins from the bank, without replacement, what is the probability that you will get a nickel, a quarter, and a nickel in that order?

   (1) $\dfrac{1}{6}$

   (2) $\dfrac{3}{34}$

   (3) $\dfrac{5}{68}$

   (4) $\dfrac{5}{136}$

6. A box of 12 batteries has 3 defective ones. If 2 batteries are taken from the box without replacement, what is the probability that 1 battery is defective?

   (1) $\dfrac{1}{22}$

   (2) $\dfrac{1}{16}$

   (3) $\dfrac{9}{44}$

   (4) $\dfrac{9}{22}$

7. Asa, Bill, Cathy, Dino, and Eve are randomly seated in a row of 5 seats. Find the probability that Asa will sit next to Bill.

   (1) $\dfrac{1}{60}$

   (2) $\dfrac{2}{5}$

   (3) $\dfrac{1}{10}$

   (4) $\dfrac{4}{15}$

8. A librarian is arranging a display of new books. She has room for 8 books. She has 3 mysteries, 4 computer books, 3 biographies, and 2 books on family health. If her choices are random, what is the probability that the display will contain 2 of each type of book?

9. Ten students are applying for 3 positions on a team. The students include 4 boys (Adam, Alex, Anthony, and Arnold) and 6 girls (Abbey, Aurora, Agnes, Alice, Amanda, and Anna). All the students have an equal chance of being selected. Find the probability that the students selected will include:

   a 3 girls

   b 1 boy and 2 girls

   c at most 1 girl

   d Adam, Anthony, and Alice

   e Agnes and 2 other students

10. If the letters of the word MATHEMATICS are arranged at random, find the probability that the first letter is:

    a M

    b a vowel

11. Three cards are drawn at random from a standard 52-card deck, without replacement. Find the probability of selecting:

    a any 3 hearts

    b any 3 black cards

    c any 3 red picture cards

12. A box of identically shaped candies contains 3 chocolate, 4 butterscotch, and 2 berry. Claire closes her eyes and picks 3 candies. Find the probability that she will get:

    a 3 of the same kind

    b 1 of each kind

    c 2 butterscotch and 1 chocolate

CHAPTER REVIEW

1 A spinner has 8 sectors, each a different color. If it is spun twice, how many different results are possible?
 (1) 8
 (2) 16
 (3) 32
 (4) 64

2 A store has 3 entrances. Shoppers can go between the first floor and the second floor by a stairway, an elevator, or an escalator. How many ways can a shopper enter from outside, go to the second floor, return to the first floor, and leave the building?
 (1) 12
 (2) 27
 (3) 81
 (4) 243

3 A travel club has 200 members. If 50 members are picked at random each year to go on a trip, what is the probability that a member is picked 3 years in a row?
 (1) $\frac{1}{4}$
 (2) $\frac{1}{50}$
 (3) $\frac{1}{64}$
 (4) $\frac{1}{150}$

4 A 6-sided die is rolled and a coin is tossed. (Both are fair.) How many outcomes that include an even number on the die are in the sample space?
 (1) 2
 (2) 4
 (3) 6
 (4) 8

5 Meg buys 2 kinds of lettuce, 2 kinds of tomatoes, and 3 kinds of dressing. To make a salad, she will use 1 kind of lettuce, 1 kind of tomato, and 1 kind of dressing. How many salads can she make?
 (1) 3
 (2) 6
 (3) 7
 (4) 12

Exercises 6–7: A spinner has 6 equal sectors numbered 1 to 6. It is spun once. Find:

6 P(not greater than 1)

7 P(less than 3)

Exercises 8–9: Two dice are rolled. Find each probability:

8 Both dice show 1.

9 The sum is 3 or 10.

10 Mr. Goldman has 3 posters, 1 calculator chart, and 3 banners. If he arranges them all from left to right across a bulletin board, how many different arrangements can he make?

11 How many ways can a 3-digit number be created if the digits are all different and odd?

12 How many 3-digit numbers can be created if each number is a multiple of 10?

13 How many ways can you arrange the letters in CHESAPEAKE if the first letter is C?

14 How many ways can you arrange 2 marching bands out of 4, followed by 4 kick-lines out of 6?

Exercises 15–18: Six girls and 8 boys are interested in an internship program that can take 10 students.

15 How many ways can students be chosen for the internship?

16 How many ways can all the boys be selected?

17 How many ways can the same number of girls and boys be selected?

18 What is the probability all of the boys will be selected?

19 Five students are applying for 2 student manager positions for the football team. The students include 4 boys and 1 girl. All the students have an equal chance of being selected. Find the probability that the 2 managers will include:
 a 1 girl
 b 2 boys

20 Four cards are drawn at random from a 52-card deck, without replacement. Find the probability of selecting:
 a 1 card from each suit
 b all 4 queens

Now turn to page 428 and take the Cumulative Review test for Chapters 1–11. This will help you monitor your progress and keep all your skills sharp.

Appendix:

Using Your Calculator

by Kim E. Genzer and David Vintinner

Many high school mathematics examinations allow you to have access to a scientific or graphing calculator for the duration of the test. These calculators have a multitude of functions to aid you in finding answers and checking your calculations.

Using a Scientific Calculator

A scientific calculator has several rows of function keys above the standard 0–9 and operations keys. There are usually additional functions printed above each key. These are accessed by pressing [SHIFT] or [2nd], followed by the appropriate key.

Calculators vary widely in which functions are *on* the keys and which are *above*. We will show procedures as if all functions were *on* the key. If your calculator has a function above a key, you will need to press [SHIFT] or [2nd] first.

Basic Operations

The four basic operation keys are: [+] [−] [×] [÷]. They work as you would expect for a single calculation. *Example*: To calculate twenty-seven divided by nine, you would press 27 [÷] 9 [=] and get a display reading [3.].

Order of Operations

All scientific calculators should follow the correct order of operations. To check this, enter 6 [+] 3 [×] 5. If you get 21, then your calculator was correctly programmed to perform multiplication before addition. If, however, you get 45, then your calculator doesn't know the rules and you will need to be careful about how you enter expressions. In this case, you would need to enter the expression following the correct order of operations to get the correct answer: 3 [×] 5 [+] 6.

Sign Change

This key changes the sign of the number in the display. *Example*: To enter -7, press 7 [+/−]. A minus sign will appear in front of the seven: [-7.] or [- 7.]. To calculate 7×-5, press 7 [×] 5 [+/−] [=]. You will get [-35.].

Note: Often this key is marked [(−)] instead of [+/−]. If this is the case with your calculator, press [(−)] before the number you wish to make negative. To calculate $7 \times (-5)$, press 7 [×] [(−)] 5 [=].

If you do not have a sign change key, you may be able to get a negative number by pressing the minus key [−]. If you want to calculate $-5 - 6$, you can press [−] 5 [−] 6 [=] to get -11. However, to calculate $-5 - (-6)$, you must first simplify the expression to get $-5 + 6$. Then press [−] 5 [+] 6 [=] to get 1.

[(] [)] Parentheses

Parenthesis keys can be used to define the order in which your calculator performs operations. They are pressed in the same order as they would be written. *Example*: To calculate $(6 + 3) \times 5$, press [(] 6 [+] 3 [)] [×] 5 [=]. You will get [45.].

[x^2] Square a Number

This function is frequently paired with the square root key. It gives you the square of the number in your display. *Example*: 9 [x^2] [=] gives you [81.].

Note: With functions like [x^2], calculators treat negative signs differently. Enter 7 [+/−] [x^2] or [(−)] 7 [x^2]. If you get 49, then your calculator includes the negative sign in the function: $(-7)^2$. If you get -49, then your calculator ignores the negative when it squares: $-(7)^2$. Use parentheses to be sure the calculator does what you want it to do.

[x^y] Raise a Number to a Power

To use this key, first type the base (the number corresponding to x) and press [x^y]. Then type the exponent (the number corresponding to y) and press [=]. *Example*: To find 3^4, press 3 [x^y] 4 [=] . Your display should read [81.].

Note: Occasionally this key is marked [y^x] or [^] instead of [x^y]. It works the same way in either case. First type the base, then [y^x] or [^], and then the exponent.

[√] Square Root

This key is used to find the square root of whatever follows. *Examples*: To find the square root of 3, press [√] 3 [=] to get [1.732058], rounded to the number of digits your calculator can handle. To find $\sqrt{3^2 + 4^2}$, press [√] [(] 3 [x^2] [+] 4 [x^2] [)] [=] to get [5.].

Note: On a few calculators, [√] finds the square root of the number already in your display. Press 3 [√] if this is the case.

[1/x] Inverse

This key is useful for finding the reciprocal of a number in decimal form. Press the number and then press [1/x]. *Example*: To find the decimal value of $\frac{1}{8}$, press 8 [1/x] [=] to get [0.125].

Note: Many calculators have [x⁻¹] instead of [1/x].

[π] Pi

Pressing this key will display an approximation of π to as many decimal places as the calculator can show. *Example*: To find the approximate circumference of a circle with a diameter of 2, you can press 2 [×] [π] [=] to get [6.283185], which can then be rounded to as many places as you need.

[EE] Scientific Notation

Most scientific calculators have a key that allows you to enter numbers in scientific notation. In general, the base 10 is not shown. *Example*: To find the value of 5.8×10^3, press 5 [.] 8 [EE] 3 [=] to get [5800.]. This key may also appear as [EXP].

Note: Some calculators have a key that lets you toggle to and from scientific notation. It may be [EE], [EXP], or [Sci]. Experiment with your calculator or refer to its instructions to find out.

[x!] Factorial

Pressing this key gives you the factorial of the number displayed. *Example*: To find 7! (or 7 • 6 • 5 • 4 • 3 • 2 • 1), simply press 7 [x!] [=] and the display will read [5040.].

[nPr] Permutations

This key calculates permutations—the number of ways to arrange n items taken r items at a time: $\frac{n!}{(n-r)!}$. First press your value for n, then [nPr], then r, and finally [=]. *Example*: To find the number of ways 3 students can be seated in a row when chosen from a group of 5 students, press 5 [nPr] 3 [=] to get [60.].

[nCr] Combinations

This key calculates combinations—the number of groups with r members that can be selected from a group with n members: $\frac{n!}{r! \cdot (n-r)!}$. First press your value for n, then press [nCr], then r, then [=]. *Example*: To find the number of ways 3 people can be chosen for a project out of a group of 5 people, press 5 [nCr] 3 [=] to get [10.].

[%] Percent

Calculators vary in how they handle percents. For example, to find 25 percent of 80, on some calculators you will press 80 [×] 25 [%] [=] and the display will read [20.]. On others you press 25 [%], which will give you [0.25], and then you press [×] 80 [=] to get [20.]. Experiment with your calculator to figure out how it works.

Note: Many calculators allow you to enter percentage increase or decrease problems quickly with the percent key. *Example*: To find the value of an $80 charge with a 20 percent discount, press 80 [−] 20 [%] [=]. If you get [64.], your calculator automatically found that 20 percent of 80 is 16.

[F↔D] [SIMP] [a] [b/c] Fractions

Some specific calculators have a fraction feature which may allow you to convert between fractions and decimals, simplify fractions, and perform calculations on fractions and mixed numbers.

- To enter a fraction, type the numerator, then press [b/c], and then type the denominator. You may then proceed with your desired calculation.
- To enter a mixed number, type the whole number portion, press [a], type the numerator, press [b/c], and then type the denominator.
- If a decimal is shown on the display, pressing [F↔D] will change it to a fraction. When there is a fraction in the display, pressing this key converts it to a decimal.
- The [SIMP] key will simplify a fraction shown in the display. It may be necessary to press this key more than once to reduce the fraction to lowest terms.

Memory

Most scientific calculators can store numbers in memory. The exact keys and their use vary somewhat from one calculator to another. Two common systems are explained below, but you may need to experiment with your calculator or refer to its instructions to find out how it works.

[M+] [M−] [MR] [MC]

- If there is nothing in memory, press [M+] to store the value that is in your display.
- In general, the [M+] key adds the number in the display to any number already in memory.
- The [M−] key subtracts the number in the display from whatever is in memory. If you have not stored anything, this key subtracts the displayed number from zero.
- [MR] stands for Memory Recall. Pressing this key will show you the number currently in memory. Pressing it does not erase the value from memory.
- To clear the memory, you may need to press [AC], [MC], [MR] [MR], or some other key, depending on your calculator.

[STO] [ALPHA]

- To store a value with [STO], enter it into the display, press [STO] and [ALPHA], then choose any letter by pressing the key beneath it on the keypad.
- To retrieve the value in a subsequent calculation, use [ALPHA] to enter the letter as you would a variable.

The Quadratic Formula

You can calculate the solutions to a quadratic equation through careful use of the parenthesis and special function keys. Suppose you wanted to find the solutions of the equation $6x^2 + 5x - 25 = 0$. Remember that the quadratic formula is $x = \dfrac{-b \pm \sqrt{b^2 - 4ac}}{2a}$, where a, b, and c are the coefficients of the quadratic equation, in order. So our calculation is $x = \dfrac{-5 \pm \sqrt{(5)^2 - 4(6)(-25)}}{2(6)}$.

You find the values of x by first finding the value of the **determinant** (the part under the radical) and storing it in memory. You then substitute this value into the quadratic formula:

[( ] [(−)] 5 [+] [√] [MR] [ )] [+] [( ] 2 [×] 6 [ )] [=]

Repeat this calculation with [−] in the place of [+].

You should obtain 1.6666667 and −2.5. You can check these answers by substituting them in the original equation.

Trigonometric Functions

[DRG] **or** [DEG] This key sets the calculator's measurement system to degrees, radians, or gradians. For the Mathematics A Examination, you should be sure your calculator is set on degrees. Look for a small "D" or "deg" somewhere on the display. If you find an "R" or a "G" instead, press [DRG] until you see "D."

[SIN] [COS] [TAN] These keys will give you the sine, cosine, and tangent, respectively, of the degree measure you enter. Enter the angle (in degrees) whose sine, cosine, or tangent you want to know. Press the appropriate trig key, and the display will give you the value. Do not press [=]. *Example*: To find the cosine of 30°, press 30 [COS] to get [0.866025].

Note: Many calculators operate in the opposite order. Press the appropriate trig key *before* entering the angle. *Example*: To find the cosine of 30°, press [COS] 30 [=] to get [0.866025].

[INV] This key will allow you to inverse the trigonometric function and find the measure of an angle when you know the value of the ratio. Press this key before pressing the appropriate trig key. *Example*: To find the angle with a sine ratio of 0.25, enter 0.25 [INV] [SIN] or [INV] [SIN] 0.25 [=] to get [14.477512].

Note: Rather than [INV], your calculator may have keys labeled [SIN⁻¹] [COS⁻¹] and [TAN⁻¹].

MODEL PROBLEMS

Evaluate using a scientific calculator.

SOLUTIONS

1. $16 + 4 \times 10 - 8 \div 2 =$ _____
2. $\sqrt{25} + (-5) =$ _____
3. 15% of 90 = _____
4. $-1 + (-2) + (-3) - (-4) =$ _____
5. $(3 \times 9) + (81 \div 3) + (-6 + 8) - (4 - 7) =$ _____ (Use the memory feature.)
6. $3^2 \times 3! + 46 =$ _____
7. $(-5 + 3)^5 =$ _____
8. How many committees of four people can be formed from a club with six members? _____
9. What is the sale price of a $59 sweater during a 35% off sale? _____
10. What angle has a tangent ratio of $\frac{\sqrt{3}}{1}$? _____
11. What is the sine of an 80° angle, rounded to the nearest thousandth? _____
12. $\frac{1}{\pi} \approx$ _____
13. $3.3 \times 10^{-2} =$ _____
14. What are the roots of $36x^2 - 229x + 25$? _____

1. 52
2. 0
3. 13.5
4. -2
5. 59

6. 100
7. -32
8. 15

9. $38.35

10. 60°

11. 0.985

12. 0.3183098862
13. 0.033
14. $0.\overline{1}$ and 6.25 or $\frac{1}{9}$ and $6\frac{1}{4}$

Using a Graphing Calculator

Graphing calculators vary widely in format and capability. As their name implies, these calculators can graph equations and draw bar charts, scatterplots, and other statistical graphs. They have several rows of function keys like scientific calculators, but many of their special functions can only be accessed by selecting menu choices on the display screen. For example, to change from radians to degrees, you will need to press [MODE] to access a menu. Then you can highlight the word *Degree* (if it is not already highlighted) and press [ENTER].

Calculators vary widely in the location of function keys and menu choices. We will show general guidelines for the procedures you are most likely to use on a high school math exam, but there are many more powerful operations open to you. Take the time to explore your calculator's menus and find the functions *you* will need the most.

One important difference you should notice is that the graphing calculator does not have an equals key: [=]. Instead, it has [ENTER]. Other types of calculators perform some operations as you go along. However, a graphing calculator does not calculate anything until you press [ENTER]. It then follows the rules for order of operations to find the answer. One nice feature is that the operation you put in is generally still visible on the screen above the answer that comes out. Then if you don't get the answer you expected, you can see what you typed to determine if you made an error in how you entered it.

Another difference: it may not be necessary to press × in order to do multiplication. Some calculators let you type the same way as you would write an expression. *Example*: 3(5 + 2) or 6 [TAN] 30 could be entered exactly as just shown, without having to insert a times sign.

Note: Some graphing calculators can also solve linear and quadratic equations, draw lines of best fit, calculate with matrices, and perform various other high-level mathematical calculations. Since the procedures for such activities are not consistent from one calculator model to the next, we will not cover them here. You will need to experiment with your calculator, ask your teacher, or consult the calculator's manual.

[▲] [▼] [◄] [►] Moving the Cursor

These four arrow keys allow you to move around in equations, menus, and graphs.

[WINDOW] Defining the Screen Layout

This key lets you access a menu that defines the area of a graph. The horizontal axis will extend from *Xmin* to *Xmax* with markings at an interval equal to *Xscl*. Similarly, the vertical axis will extend from *Ymin* to *Ymax* with markings at an interval of *Yscl*.

Graphing Equations

[Y=] This key lets you enter equations to be graphed. It is sometimes possible to enter more than one equation. The equation should be solved for y in terms of x (which must be entered with the special variable key [X, T, θ, n]). Use [CLEAR] to erase equations you no longer need.

To see a table of values for the equation(s) you enter in the [Y=] screen, press [TABLE]. To change the increment by which the x-value changes, or to enter the values individually, press [TBLSET] and adjust the settings as you wish.

[GRAPH] This key allows you to see the graph defined by the [WINDOW] menu with the equations from [Y=] graphed on it. To find the coordinates of specific points on the graph, such as x-intercepts, press [TRACE] and use the arrows to move the cursor to the point of interest. You may also be able to enter x-values with the keypad.

You can alter the type of line drawn on a graph or add shading to a graph by choosing one of several options. In the [Y=] screen, press the left arrow key until the cursor is on the diagonal slash to the left of the Y for the equation whose graph you wish to alter. Press [ENTER] several times to see the various options for shading and line weight and style. When you have chosen the one you want, press [GRAPH] to see the result.

Graphing Data

[STAT] Pressing this key allows you to access statistical data. Choosing EDIT allows you to enter lists of data you wish to graph. Choosing CALC allows you to calculate values (like the median of your data, $\bar{x}$) or find equations to approximate your data (like the linear regression and median-median lines of best fit.)

[STAT PLOT] Pressing this key allows you to select and turn on different types of graphs of your data: broken-line graphs, histograms, scatterplots, and box-and-whisker plots are among the choices.

Depending on which graph you choose, options such as Xlist, Ylist, and Mark may appear to allow you to tell the calculator where the lists of data are stored (e.g. L1), and what type of point you would like the graph to display. Once you have entered this information, press [GRAPH] to see the display. You can use [TRACE] and the arrow keys to identify points on the graph.

Note: The [WINDOW] menu also controls the appearance of statistical graphs. The *Xscl* value controls the range of values for each bar in the histogram.

MODEL PROBLEMS

SOLUTIONS

1. Graph the line $y = -0.068x + 0.34$ and use [TRACE] to find where it crosses the *x*-axis. You may need to change your window settings to zoom in on the solution.

   1. It crosses at $x = 5$.

2. Graph the lines $y = 10x + 15$ and $y = \frac{1}{2}(x + 11)$ and use [TRACE] to find where they intersect.

   2. They intersect at $(-1, 5)$.

3. Graph the parabola $y = x^2 + \frac{41}{18}x - 10$ and find where the graph crosses the *x*-axis. (Hint: One of the roots is a repeating decimal.)

   3. The *x*-intercepts are -4.5 and $2.\overline{2}$.

4. Enter the following quiz grades into a [STAT] list:

   60, 65, 67, 70, 70, 86, 86, 89, 89, 90, 90, 95, 97, 99, 100

   a. Find the mean grade ($\bar{x}$) with the CALC menu.

   b. Plot the grades as a histogram with settings $Xmin = 60$, $Xmax = 100$, and $Xscl = 4$ and use [TRACE] to find the bar with the most scores.

   c. Plot the grades as a box-and-whisker diagram and use [TRACE] to find the median grade and the upper and lower quartiles.

   4. a $\bar{x} = 83.5\overline{3}$

   b There are 4 quiz scores in the range 88 to 91.

   c The median score is 89. The lower and upper quartile scores are 70 and 95.

Assessment:
Cumulative Reviews and Practice Examinations

CUMULATIVE REVIEW
CHAPTERS 1–2

Part I

All questions in this part will receive 2 credits. No partial credit will be allowed.

1. Order these values from least to greatest.
   (a) $-|5|$ (b) $-|7|$ (c) $-3 - 3^2$
   (d) $|-8|$ (e) $-(-3)^2$

   (1) a, b, e, c, d
   (2) b, c, a, d, e
   (3) c, e, d, b, a
   (4) c, e, b, a, d

2. If $n - 1$ is an odd integer, what is the next largest even integer?
   (1) $n - 2$
   (2) n
   (3) $n + 1$
   (4) $n + 2$

3. Before redistricting, a small town in Ohio had 1,200 elementary school pupils. After redistricting, it had 300. What was the percent of decrease?
   (1) 25%
   (2) 30%
   (3) $33\frac{1}{3}\%$
   (4) 75%

4. What is the smallest number that has 4 different prime factors?
   (1) 16
   (2) 30
   (3) 210
   (4) 945

5. Which expression is equivalent to 7.001×10^3?
   (1) 0.00701
   (2) 0.000701
   (3) 700.1×10^5
   (4) 7001×10^0

6. If $\frac{2}{5} < \frac{x}{20} < \frac{1}{2}$, what is the integer value of x?
   (1) 10
   (2) 9
   (3) 8
   (4) 7

7. On Thursday at 8 A.M., the temperature is 25°F. By noon it rises 11 degrees, by 6 P.M. it drops 4 degrees, and by 8 A.M. on Friday it drops another 10 degrees. What is the change in temperature from 8 A.M. on Thursday to 8 A.M. on Friday?
   (1) −14 degrees
   (2) −3 degrees
   (3) 5 degrees
   (4) 22 degrees

8. $2 \times 3^3 \times 3^2 =$
   (1) $6^3 \times 3^2$
   (2) 2×9^5
   (3) 2×3^5
   (4) 2×3^6

9 $\left(4\frac{5}{8} - 2\frac{3}{4}\right) \div \frac{5}{4} =$

(1) $\frac{2}{3}$ (3) $2\frac{11}{32}$

(2) $1\frac{1}{2}$ (4) $5\frac{1}{5}$

10 If $\frac{4}{5} \cdot \frac{5}{6} \cdot \frac{6}{7} \cdot \frac{7}{8} \cdot \frac{8}{9} \cdot \frac{9}{10} \cdot \frac{10}{11} \cdot \frac{11}{12} = \frac{1}{n}$, then $n =$

(1) 3   (3) 12

(2) 4   (4) 48

Part II

Each correct answer will receive 2 credits. Clearly indicate the necessary steps, including appropriate formula substitutions, diagrams, charts, etc. Correct numerical answers without work shown will receive only 1 credit.

11 Is the set of irrational numbers closed under multiplication? Explain why or why not, using at least one example.

12 List the numbers in the set {natural numbers between -4 and 8} and graph them on a number line.

13 Is the set of real numbers an infinite set? Explain your reasoning.

14 For the local minor league baseball team, the ratio of games won to games played is 8 : 11. Write the ratio of games lost to games played.

Part III

Each correct answer will receive 3 credits. Partial credit will be allowed. Clearly indicate the necessary steps, including appropriate formula substitutions, diagrams, charts, etc. Correct numerical answers without work shown will receive only 1 credit.

15 In each equation, find the value of k.

    a $3^3 + 3^4 = k$

    b $(a^k)^k = a^9$

    c $2^k \cdot 2^4 = 2^7$

16 The product of two consecutive positive integers is always divisible by an integer greater than 1. What is that number? Why is this true?

17 Which number below is irrational? Why?

    $\frac{\sqrt{4}}{3}$ $\sqrt{8}$ $\sqrt{169}$

Part IV

Each correct answer will receive 4 credits. Partial credit will be allowed. Clearly indicate the necessary steps, including appropriate formula substitutions, diagrams, charts, etc. Correct numerical answers without work shown will receive only 1 credit.

18 Match each item in Column I with one item in Column II

Column I	Column II
1 ∅	(a) {Numbers that equal their own square}
2 {0, 1}	(b) {Squares of even numbers}
3 {0, 4, 16, 36, . . .}	(c) {Integers greater than 0 and less than 1}
4 {1, 9, 25, 49, . . .}	(d) {Squares of prime numbers}
5 {4, 9, 25, 49, 121, . . .}	(e) {Squares of odd numbers}

19 Neil and Jackie are describing the same number. Neil says, "The number is a negative integer greater than -12." Jackie says, "The number is divisible by 3." If Neil's statement is true and Jackie's statement is false, what are all the possible numbers?

20 An antique dealer paid \$300 for a desk. She wants to put a price tag on it so that she can offer her customers a discount of 10% of the price marked on the tag and yet still make a profit of 20% of the amount she paid. What price should she put on the tag?

CUMULATIVE REVIEW
CHAPTERS 1–3

Part I

All questions in this part will receive 2 credits. No partial credit will be allowed.

1. Order these real numbers from smallest to greatest: $0.5, \sqrt{0.5}, 0.\overline{5}$.
   (1) $0.5 < \sqrt{0.5} < 0.\overline{5}$
   (2) $\sqrt{0.5} < 0.\overline{5} < 0.5$
   (3) $0.5 < 0.\overline{5} < \sqrt{0.5}$
   (4) $\sqrt{0.5} < 0.5 < 0.\overline{5}$

2. The distance from the planet Pluto to the sun is approximately 3,700,000,000 miles. Which of the following represents that distance in scientific notation?
   (1) 3.7×10^9
   (2) 3.7×10^{10}
   (3) 0.37×10^{10}
   (4) 37×10^{10}

3. A verbal description of the set $\{8, 12, 16, 20, \ldots\}$ is:
   (1) the set of multiples of 8
   (2) the set of multiples of 4
   (3) the set of multiples of 4 greater than 4
   (4) the set of multiples of 4 greater than 4 and less than 24

4. If a, b, and c are real numbers, the commutative property of addition states that:
   (1) $(a + b) + c = c + (a + b)$
   (2) $a(b + c) = ab + ac$
   (3) $a(bc) = (ab)c$
   (4) $(a + b) + c = a + (b + c)$

5. If the temperature was $-39.5°$ yesterday and $+6.7°$ today, by how much did the temperature rise?
   (1) $3.28°$
   (2) $4.62°$
   (3) $32.8°$
   (4) $46.2°$

6. Which number below is rational?
   (1) π
   (2) $\sqrt{0.9}$
   (3) $\sqrt{0.09}$
   (4) $\sqrt{-9}$

7. Which number line is the proper representation of the set of all numbers with an absolute value less than 2?

8. Which monomial is equivalent to $2x^5y^2 \cdot 5xy^3z^2 \cdot 2yz$?
   (1) $10x^{15}y^6z^2$
   (2) $20x^6y^6z^3$
   (3) $20x^5y^6z^2$
   (4) $20x^5y^3z^3$

9. If $m - 3 = n$, then what is the value of $(n - m)^3$?
   (1) -27
   (2) -9
   (3) 9
   (4) 27

10. Express $\dfrac{1}{4x} + \dfrac{3}{x^2}$ as a single fraction.
    (1) $\dfrac{x + 12}{4x^2}$
    (2) $\dfrac{x^2 + 12}{4x^3}$
    (3) $\dfrac{13}{4x}$
    (4) $\dfrac{1 + 12x}{4x^2}$

Part II

Each correct answer will receive 2 credits. Clearly indicate the necessary steps, including appropriate formula substitutions, diagrams, charts, etc. Correct numerical answers without work shown will receive only 1 credit.

11 What are the common elements of the set of even whole numbers less than 4 and the set of even integers less than 4 and greater than -4?

12 Simplify: $-\dfrac{7}{16} \div \left(-\dfrac{21}{32}\right)$

13 Simplify: $\sqrt{12} - \sqrt{3}$

14 Maria exercised on her bicycle for 40 minutes. Michael exercised $\dfrac{3}{8}$ as long. How many more minutes did Maria exercise than Michael?

Part III

Each correct answer will receive 3 credits. Partial credit will be allowed. Clearly indicate the necessary steps, including appropriate formula substitutions, diagrams, charts, etc. Correct numerical answers without work shown will receive only 1 credit.

15 Do $(-1)^{27}$ and $(-1)^{28}$ represent the same number? Why or why not?

16 Pam saved money to buy Jean special golf clubs for her birthday. She forgot that she needed money for the 8% sales tax. The tax was $29.70. What was the price of the golf clubs?

17 Graph the solution set of $2(5 - x) < 16$ on a number line.

Part IV

Each correct answer will receive 4 credits. Partial credit will be allowed. Clearly indicate the necessary steps, including appropriate formula substitutions, diagrams, charts, etc. Correct numerical answers without work shown will receive only 1 credit.

18 Natasha is a finalist in a spelling bee. She wants to learn at least 250 new words in time for the competition in 10 days. Each day, she will learn 2 more words than she learned the day before. What is the *least* number of words she must learn on the first day?

19 If $2 : k = k : 16$, find a positive real number that could equal k.

20 A 1-hour driver's test has 100 questions. Mike spends an average of 30 seconds each on the first 60 questions. To finish the test on time, what is the average number of seconds he can spend on each of the remaining questions?

CUMULATIVE REVIEW
CHAPTERS 1–4

Part I

All questions in this part will receive 2 credits. No partial credit will be allowed.

1. Which radical represents a rational number?
   (1) $\sqrt{10}$ (3) $\sqrt{14}$
   (2) $\sqrt{12}$ (4) $\sqrt{16}$

2. Which of the following rational numbers is between -2.1 and -2.2?
   (1) -2.25 (3) -2.04
   (2) -2.15 (4) 2.14

3. Which choice represents the scientific notation for 0.0000801?
   (1) 8.01×10^4 (3) 80.1×10^6
   (2) 80.1×10^{-5} (4) 8.01×10^{-5}

4. If a soccer team played 40 games in a season and won 24 of them, what percent of the games did the team win?
   (1) 16%   (3) 40%
   (2) 24%   (4) 60%

5. Write the ratio of the number of males to the number of students if there are 45 male students and 60 female students.
   (1) $105:45$ (3) $3:7$
   (2) $45:60$ (4) $3:4$

6. If a is an odd integer and b is an even integer, which of the following is an even integer?
   (1) $a + 2b$ (3) $a^2 + b$
   (2) $3a + 3$ (4) $a^2 + 2b$

7. What is the sum of $m^2n + 2mn + 7n$ and $3m^2n - 5mn + 2m$?
   (1) $4m^4n^2 - 3m^2n^2 + 7n + 2m$
   (2) $4m^2n - 3mn + 9mn$
   (3) $2m^2n + 3mn + 2m + 7n$
   (4) $4m^2n - 3mn + 7n + 2m$

8. If $A = \frac{1}{2}h(b + c)$, what is the value of b when $A = 50$, $h = 4$, and $c = 11$?
   (1) 14   (3) 36
   (2) 24   (4) 56

9. Solve for the positive value of x: $5x^2 - 80 = 0$
   (1) 2   (3) $4\sqrt{5}$
   (2) 4   (4) 8

10. Which is the correct factorization of $x^2 - 100$?
    (1) $-(x + 10)^2$ (3) $(x + 10)(x - 10)$
    (2) $(x + 10)^2$ (4) $(x - 10)^2$

Part II

Each correct answer will receive 2 credits. Clearly indicate the necessary steps, including appropriate formula substitutions, diagrams, charts, etc. Correct numerical answers without work shown will receive only 1 credit.

11. Susan told Bill that every time an even number is divided by an even number, the result is an even number. Bill tried to think of an example to show Susan that she was incorrect. Find an example for Bill.

12. Simplify $\dfrac{7a^3 + 14a}{7a}$.

13. The formula for simple interest is $I = prt$.
    a. Solve for p.
    b. Find p if the interest is \$45, the rate is 5%, and $t = \dfrac{1}{2}$ year.

14. Factor $8x^2 - 24x + 18$ completely.

Part III

Each correct answer will receive 3 credits. Partial credit will be allowed. Clearly indicate the necessary steps, including appropriate formula substitutions, diagrams, charts, etc. Correct numerical answers without work shown will receive only 1 credit.

15 State whether each of the following sets is closed or not closed under the given operation. If it is *not* closed, provide an illustration to support your answer.

   a {0, 1, 2} addition

   b {1, 3, 5} subtraction

   c {−1, 0, 1} multiplication

16 Solve for x: $x^2 - 4x - 45 = 0$

17 In the tenth grade, 52 students made phone calls on a blood drive and 16 helped register blood donors. If 120 of the 180 students in the tenth grade did not participate in the blood drive, how many students both phoned and registered?

Part IV

Each correct answer will receive 4 credits. Partial credit will be allowed. Clearly indicate the necessary steps, including appropriate formula substitutions, diagrams, charts, etc. Correct numerical answers without work shown will receive only 1 credit.

18 In a science lab, a lever positioned on a fulcrum is used to balance weights. The positions of the weights with respect to the fulcrum can be represented with a proportion: $\dfrac{F_1}{F_2} = \dfrac{D_2}{D_1}$, where F_1 and F_2 are the weights (forces) and D_1 and D_2 are the distances from the fulcrum.

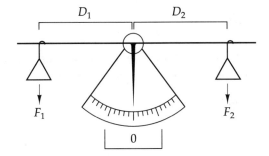

Solve each of the following:

   a A movable weight is 2 pounds and hangs 3 inches from the fulcrum. If a gold deposit weighs 0.5 pound, how far from the fulcrum must it hang to create a balance?

   b An 8-pound weight hangs 8 inches from the fulcrum. A quartz stone hanging 5 inches from the fulcrum balances the scale. How much does the stone weigh?

19 Popcorn is sold at a carnival for $2.50 per bag and candy apples are sold for $1.50 per apple. An inventory for last week showed that the number of bags of popcorn sold was four times the number of apples. The total amount collected for the two snacks was at least $500. What was the least possible number of bags of popcorn sold?

20 Solve for x and check: $\dfrac{x-5}{3} = 1 - \dfrac{5}{x}$

CUMULATIVE REVIEW
CHAPTERS 1–5

Part I

All questions in this part will receive 2 credits. No partial credit will be allowed.

1. Which equation illustrates the additive inverse property?
   (1) $a + (-a) = 0$
   (2) $a + 0 = a$
   (3) $a + (-a) = 1$
   (4) $a + \dfrac{1}{a} = 1$

2. The expression $\dfrac{4^3}{2^3}$ is equivalent to:
   (1) 2
   (2) 2^2
   (3) 2^3
   (4) 2^6

3. The smallest positive integer is:
   (1) 2
   (2) 1
   (3) 0.00001
   (4) 0

4. Which fraction represents $\dfrac{1}{8} \div \dfrac{1}{4}$?
   (1) $\dfrac{1}{32}$
   (2) $\dfrac{1}{4}$
   (3) $\dfrac{1}{2}$
   (4) $\dfrac{2}{1}$

5. Expressed as a single fraction, what is $\dfrac{2}{x+2} - \dfrac{1}{x}$?
   (1) $\dfrac{1}{x}$
   (2) $\dfrac{1}{2}$
   (3) $\dfrac{1}{x^2 + 2x}$
   (4) $\dfrac{x-2}{x^2 + 2x}$

6. Solve the following system of equations.
   $x + 2y = 10$ and $3x - 2y = 6$
   (1) (4, 2)
   (2) (4, 3)
   (3) (1, −4.5)
   (4) (1, 4.5)

7. The larger root of $2x^2 - 2 = 0$ is:
   (1) −1
   (2) 0
   (3) 1
   (4) 2

8. Which expression is the reduced form of $\dfrac{x^2 - 3x - 4}{16 - x^2}$?
   (1) $-\dfrac{x-1}{x-4}$
   (2) $-\dfrac{x+1}{x+4}$
   (3) $\dfrac{x+1}{x+4}$
   (4) $\dfrac{x-1}{x+4}$

9. If two points on a line are (−1, 4) and (2, −2), what is the slope of the line?
   (1) $-\dfrac{2}{3}$
   (2) −2
   (3) 2
   (4) 6

10. If the line $y = 6$ is graphed in the coordinate plane, which point would *not* be on the line?
    (1) (−6, 6)
    (2) (6, 0)
    (3) (0, 6)
    (4) (6, 6)

Part II

Each correct answer will receive 2 credits. Clearly indicate the necessary steps, including appropriate formula substitutions, diagrams, charts, etc. Correct numerical answers without work shown will receive only 1 credit.

11. Evaluate: $2^3 + 100\left(\dfrac{1}{5}\right)^2$

12. Translate the following sentence into an algebraic statement and solve: If one-third of a number is 7 more than one-half of the number, what is that number?

13. Simplify: $\dfrac{6x^2 + 9xy}{4x^2 - 9y^2}$

14. Write an equation that describes y in terms of x in the table below.

x	2	3	4	5
y	$3.50	$5.25	$7.00	$8.75

Part III

Each correct answer will receive 3 credits. Partial credit will be allowed. Clearly indicate the necessary steps, including appropriate formula substitutions, diagrams, charts, etc. Correct numerical answers without work shown will receive only 1 credit.

15 Lunch at Washington High School costs $2 for the basic lunch, 25¢ extra for a king-size drink, and 40¢ extra for a bag of potato chips. Twelve members of the golf team ate lunch together. Ten members ordered a king-size drink. The total bill for lunch for the golf team was $29.30. How many bags of potato chips were purchased?

16 If 5 is a root of $x^2 - 3x + k = 0$, then (a) what is the value of k, and (b) what is the other root?

17 Line l contains points $(0, -3)$ and $(1, 0)$. Show that point $(-10, -27)$ lies on line l, or show that it does *not* lie on line l.

Part IV

Each correct answer will receive 4 credits. Partial credit will be allowed. Clearly indicate the necessary steps, including appropriate formula substitutions, diagrams, charts, etc. Correct numerical answers without work shown will receive only 1 credit.

18 Shelley and Michael are describing the same number. Shelley says, "The number is between 10 and 30 and is a multiple of 4." Michael says, "The number is between 10 and 30 and is a multiple of 3."

  a If Shelley's statement is true and Michael's statement is true, what are all the possible numbers?

  b If Shelley's statement is true and Michael's statement is false, what are all the possible numbers?

19 A gardener is paid $8 per hour for an eight-hour workday. His boss wants to give him a 10% raise in his hourly salary, but will cut his time to 7 hours per day. What will be his change in weekly salary if he works 5 days each week?

20 Solve the following system of quadratic-linear equations using an algebraic method.

   $y = -x^2 + 4x + 7$ and $y = -2x + 15$

CUMULATIVE REVIEW
CHAPTERS 1–6

Part I

All questions in this part will receive 2 credits. No partial credit will be allowed.

1. If $\sqrt{k} > k > k^2$, then k could be
   (1) 0
   (2) $\frac{1}{9}$
   (3) 1
   (4) 9

2. The expression $\sqrt{80}$ can be simplified to
   (1) $4\sqrt{5}$
   (2) $4\sqrt{10}$
   (3) $10\sqrt{2}$
   (4) $8\sqrt{5}$

3. The product of $4x^2y^2$ and xy^3 is
   (1) $4x^2y^5$
   (2) $4x^2y^6$
   (3) $4x^3y^5$
   (4) $4x^3y^6$

4. The formula $S = 3F - 24$ can be used to find a man's shoe size (S) for a given foot length measured in inches (F). What is the foot length of a man who wears a size 9 shoe?
   (1) 3 inches
   (2) 5 inches
   (3) 9 inches
   (4) 11 inches

5. For what value or values of x is the fraction $\dfrac{x+4}{x^2+3x-4}$ undefined?
   (1) {−4}
   (2) {4}
   (3) {1, −4}
   (4) {−1, 4}

6. Factor the binomial $4xy^2 - 6xy^3$ completely.
   (1) $2xy(2y - 3y^2)$
   (2) $2xy^2(2 - 3y)$
   (3) $xy^2(4 - 6y)$
   (4) $2xy^2(2xy - 3y)$

7. If y varies directly as x and $y = 2$ when $x = 3$, find y when $x = 18$.
   (1) $\frac{1}{3}$
   (2) 3
   (3) 12
   (4) 27

8. What is the slope of the line whose equation is $4x - 5y - 20 = 0$?
   (1) $\frac{5}{4}$
   (2) $\frac{4}{5}$
   (3) 4
   (4) −5

9. The direct distance between the main entrance to a high school and the entrance to the football field is $\frac{1}{2}$ mile. The direct distance between the main entrance to a middle school and the entrance to the football field is $1\frac{1}{2}$ miles. Which could *not* be the direct distance between the main entrances of the high school and the middle school?
   (1) $\frac{3}{4}$ mile
   (2) 1 mile
   (3) $1\frac{1}{4}$ miles
   (4) $1\frac{1}{2}$ miles

10. In rectangle $ABCD$, $BC = 24$ and $BA = 26$. Find CD.
    (1) 5
    (2) 10
    (3) 24
    (4) 26

Part II

Each correct answer will receive 2 credits. Clearly indicate the necessary steps, including appropriate formula substitutions, diagrams, charts, etc. Correct numerical answers without work shown will receive only 1 credit.

11 Which number below is irrational? Why?

$0.\overline{77}$ $\sqrt{50}$ $\sqrt{\dfrac{81}{121}}$

12 Solve for x: $3(x-2) = 2.4 - x$

13 On a coordinate plane, draw a line with slope $\dfrac{3}{2}$ that passes through point $(3, -2)$.

14 The sum of the measures of the interior angles of a polygon is $1{,}080°$. Name the polygon.

Part III

Each correct answer will receive 3 credits. Partial credit will be allowed. Clearly indicate the necessary steps, including appropriate formula substitutions, diagrams, charts, etc. Correct numerical answers without work shown will receive only 1 credit.

15 A commission was divided among three salespeople so that the most experienced person, Brendan, received 50% of the money and the least experienced person, Chad, received $\dfrac{1}{3}$ the amount Brendan received. What fraction of the commission did the third salesperson, Donna, receive?

16 The original retail price for a pair of earrings was $30. Becky used a 15% off coupon to help pay for the earrings. If the tax on the sale price was 8%, what was her change from $30?

17 Solve for all possible values of x:

$$\dfrac{x-5}{3} = \dfrac{-3}{x+5}$$

Part IV

Each correct answer will receive 4 credits. Partial credit will be allowed. Clearly indicate the necessary steps, including appropriate formula substitutions, diagrams, charts, etc. Correct numerical answers without work shown will receive only 1 credit.

18 The profits of a family business are divided so that the father receives half of the profits and the three children receive profits in the ratio of 4 to 3 to 3. The profit for the year was $238,000. Determine the amount of profit each family member received.

19 Solve the following system of equations algebraically or graphically for x and y.

$y = 2x^2 - 4x + 1$ and $y = 2x + 9$

20 Lorna is playing a video game. In the game, she can drive through red gates or green gates. The first time she played, she drove through 3 red gates and 2 green gates and got three bonus points. The second time she played, she drove through 2 red gates and 4 green gates and got 10 bonus points.

a How many points is a green gate worth?

b How many points is a red gate worth?

CUMULATIVE REVIEW
CHAPTERS 1–7

Part I

All questions in this part will receive 2 credits. No partial credit will be allowed.

1. The statement "If x is divisible by 6 then x is divisible by 4" is false if x equals:
   - (1) 0
   - (2) 12
   - (3) 18
   - (4) 24

2. The sum of $2x^2 - 4 + x^2 + x - 4$ can be expressed as:
   - (1) $2x^2 - 8$
   - (2) $3x^2 + x - 8$
   - (3) $2x^2 + x - 8$
   - (4) $3x^2 + x$

3. What is the integer root of the equation $(x + 1)(2x - 3) = 0$?
   - (1) -1
   - (2) 1
   - (3) -3
   - (4) 3

4. Expressed as a single fraction, what is $\dfrac{1}{2x} + \dfrac{1}{3x}$?
   - (1) $\dfrac{1}{5x}$
   - (2) $\dfrac{1}{6x}$
   - (3) $\dfrac{2}{5x}$
   - (4) $\dfrac{5}{6x}$

5. If $x = 4$ and $y = -2$, which point on the graph represents $(x, -y)$?

   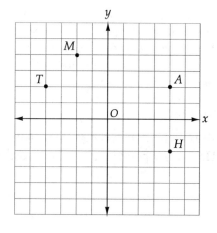

   - (1) M
   - (2) A
   - (3) T
   - (4) H

6. What is a solution for the system of equations $x + y = 0$ and $y = \dfrac{1}{2}x + 3$?
   - (1) $(-1, 1)$
   - (2) $(-2, 2)$
   - (3) $(2, -2)$
   - (4) $(-3, 3)$

7. Two angles of a triangle measure 68° and 52°. Which is the measure of an exterior angle of the triangle?
   - (1) 28°
   - (2) 60°
   - (3) 120°
   - (4) 138°

8. What is true about the statement "If a figure is a trapezoid, then the figure is a quadrilateral" and its converse?
   - (1) The statement is always true but its converse is sometimes false.
   - (2) The statement is always false but its converse is sometimes true.
   - (3) Both the statement and its converse are always true.
   - (4) Both the statement and its converse are sometimes false.

9. Which equation is an illustration of the commutative property of multiplication?
   - (1) $3 + 4 = 4 + 3$
   - (2) $\dfrac{3}{4} \cdot \dfrac{4}{3} = 1$
   - (3) $4 \cdot 3 = 3 \cdot 4$
   - (4) $\dfrac{4}{3} \cdot 1 = \dfrac{4}{3}$

10. In parallelogram $ABCD$, $m\angle A = 14x + 2$ and $m\angle B = 16x - 2$. Find $m\angle D$.
    - (1) 6
    - (2) 30
    - (3) 86
    - (4) 94

Part II

Each correct answer will receive 2 credits. Clearly indicate the necessary steps, including appropriate formula substitutions, diagrams, charts, etc. Correct numerical answers without work shown will receive only 1 credit.

11 Find three consecutive even integers such that the sum of 6 times the first number and 2 times the third number equals 200.

12 Simplify: $\dfrac{20xy - 4y^2}{25x^2 - y^2}$

13 The equation of a line is $2x + y - 4 = 0$. Find the equation of a line parallel to this line, with no points in the first quadrant.

14 Rewrite the sentence "I am happy when I succeed" as an if-then statement. Write the contrapositive of this statement.

Part III

Each correct answer will receive 3 credits. Partial credit will be allowed. Clearly indicate the necessary steps, including appropriate formula substitutions, diagrams, charts, etc. Correct numerical answers without work shown will receive only 1 credit.

15 Members of a track team must run at least 25 laps each school week. Vinnie wants to build his endurance. He will increase the number of laps he runs by one every day for the five-day week. What is the least number of laps Vinnie must complete on Monday?

16 As shown in the graph, Jack and Susan participated in the New York City Marathon. They each maintained a constant pace for the first hour. Their times and distances for the first 40 minutes are shown in the accompanying graph. Find, in miles per hour, how much faster Jack was running than Susan.

17 A flagpole casts a shadow of 42 feet while nearby a boy 5 feet tall casts a shadow of 7 feet. What is the height of the flagpole?

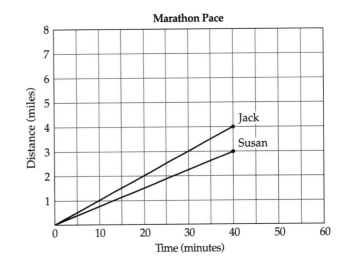

Part IV

Each correct answer will receive 4 credits. Partial credit will be allowed. Clearly indicate the necessary steps, including appropriate formula substitutions, diagrams, charts, etc. Correct numerical answers without work shown will receive only 1 credit.

18 A sale of property is to be divided by three sisters in the ratio of 4 to 3 to 5. The selling price of the property was $600,132. Determine the number of dollars each sister is to receive.

19 Solve the following system of equations algebraically or graphically for x and y.

$y = x^2 - 2x - 4$ and $y = 2x - 4$

20. Sal wants to earn money by washing cars. Speedy Car Wash and Quick Car Wash offer different salary plans. The graph below represents the salary schedule for each car wash.

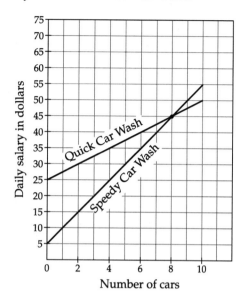

a. If the daily salary includes base pay and a commission based upon the number of cars washed, what is the difference in base pay between Speedy Car Wash and Quick Car Wash?

b. What is the rate of commission per car paid by Speedy Car Wash?

c. How many cars would Sal have to wash in order to make the same salary in both car washes?

d. How much salary would Sal earn at the work level where both salary plans pay the same?

CUMULATIVE REVIEW
CHAPTERS 1–8

Part I

All questions in this part will receive 2 credits. No partial credit will be allowed.

1. The Eagle Scouts organized a party for the Cub Scouts. The heights of the Eagle Scouts ranged from 54 inches to 74 inches. The heights of the Cub Scouts ranged from 40 inches to 48 inches. Pictures of the scouts were taken in pairs, with one Eagle Scout and one Cub Scout. The difference between the heights of the boys in any picture was between:

   (1) 13 inches and 27 inches
   (2) 5 inches and 27 inches
   (3) 5 inches and 35 inches
   (4) 13 inches and 35 inches

2. The roots of the equation $x^2 + 4x - 12 = 0$ are:

   (1) $\{2, -6\}$ (3) $\{4, -3\}$
   (2) $\{3, -4\}$ (4) $\{6, -2\}$

3. Which is true of the graph of $y = -3$?

   (1) It is parallel to the y-axis.
   (2) It contains point $(-3, 3)$.
   (3) It has a slope of -3.
   (4) Its slope is zero.

4. The shaded half-plane in the accompanying figure is a graph of which inequality?

   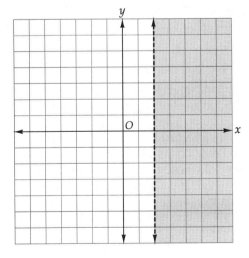

   (1) $x > 2$ (3) $x < 2$
   (2) $y > 2$ (4) $y \geq 2$

5. For which point in the given figure is it true that $|x| + |y| = 5$?

   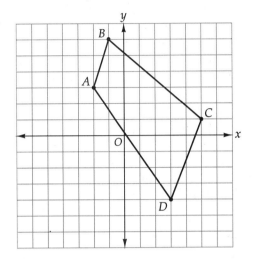

   (1) A (3) C
   (2) B (4) D

6. A road sign is shaped like a regular hexagon. If the length of one side is represented by $k + 8$, its perimeter would be represented by

   (1) $6k + 8$ (3) $6k + 48$
   (2) $8k + 8$ (4) $8k + 64$

7. In the accompanying diagram of triangle ABC, segment BC is extended through C to D. If m$\angle ABC = 4x - 6$, m$\angle BAC = 8x$, and m$\angle ACD = 10x + 10$, find m$\angle ACB$.

   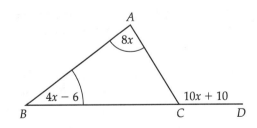

   (1) 26 (3) 90
   (2) 60 (4) 120

8 The statement "If x is a perfect square, then it is an even integer" is false if x is:

(1) 15 (3) 21
(2) 16 (4) 25

9 Which expression is equivalent to $\dfrac{1}{a^2 - 1} + \dfrac{1}{a - 1}$?

(1) $\dfrac{a + 2}{a^2 - 1}$ (3) $\dfrac{2}{a^2 - 2}$
(2) $\dfrac{a + 1}{a^2 - 1}$ (4) $\dfrac{a + 2}{a^2 - 2}$

10 What is the height of an equilateral triangle whose perimeter is 18?

(1) 3
(2) $3\sqrt{2}$
(3) $3\sqrt{3}$
(4) 4

Part II

Each correct answer will receive 2 credits. Clearly indicate the necessary steps, including appropriate formula substitutions, diagrams, charts, etc. Correct numerical answers without work shown will receive only 1 credit.

11 If $7 \times 10^2 + 3 \times 10^4 = N \times 10^4$, then what is the value of N?

12 Simplify: $\dfrac{3x^2 - 27}{x^2 - x - 12}$

13 If the measure of an interior angle of a regular polygon is 135°, and a side of the polygon is 4 inches, find the perimeter of the polygon.

14 Falynn charges baby-sitting rates to the Brownes, who have triplets, as follows:

Before midnight: $6 per hour
After midnight: $8 per hour

If the Brownes came home at 2 A.M. and owed Falynn $40, what time did the job start?

Part III

Each correct answer will receive 3 credits. Partial credit will be allowed. Clearly indicate the necessary steps, including appropriate formula substitutions, diagrams, charts, etc. Correct numerical answers without work shown will receive only 1 credit.

15 The legs of a right triangle are represented by x and $x + 2$, and the hypotenuse is 10. Find the area of the triangle.

16 A parabola whose equation is $y = x^2 - 2x + k$ has an x-intercept at $x = 5$. Find the other x-intercept.

17 At a local high school football game, $1,350 was collected for hot dogs, hamburgers, tortillas, and soft drinks. Three times as many hot dogs were sold as hamburgers and twice as many tortillas were sold as hamburgers. As many soft drinks were sold as hot dogs, hamburgers, and tortillas combined. If each item sold for $1.50, how many soft drinks were sold?

Part IV

Each correct answer will receive 4 credits. Partial credit will be allowed. Clearly indicate the necessary steps, including appropriate formula substitutions, diagrams, charts, etc. Correct numerical answers without work shown will receive only 1 credit.

18 In the accompanying diagram of triangle *ABC*, $\overline{DE} \parallel \overline{AB}$, $DE = 2x$, $AB = 5x - 1$, $CE = 3x$, and $BE = 1$. Find the length of $\overline{AB}$.

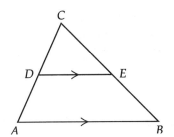

19 In the accompanying diagram of rhombus *GIFT*, m∠*TGF* = 24 and *TI* = 12.

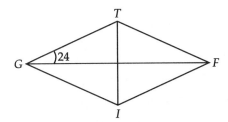

Find, *to the nearest tenth*:

a length of *GF*

b perimeter of *GIFT*

20 A ladder is leaning against a building as shown in the accompanying diagram. The ladder is 24 feet long and touches the building 8 feet from the roof. The angle of elevation to the point where the ladder touches the building is 64°. Find the height, *h*, of the building, to the *nearest foot*.

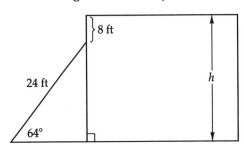

CUMULATIVE REVIEW
CHAPTERS 1–9

Part I

All questions in this part will receive 2 credits. No partial credit will be allowed.

1. How many integer values of x are there so that x, 6, and 10 could be sides of a triangle?
   (1) 10
   (2) 11
   (3) 12
   (4) 13

2. What is the distance between points $(-3, 2)$ and $(-2, 3)$?
   (1) 1
   (2) $\sqrt{2}$
   (3) $\sqrt{5}$
   (4) $\sqrt{10}$

3. In the accompanying diagram, $\overline{AB}$ is parallel to $\overline{CD}$. Transversal $\overline{EF}$ intersects $\overline{AB}$ and $\overline{CD}$ at G and H respectively. If m$\angle CHE = 4x - 4$ and m$\angle EGB = 3x + 2$, find m$\angle DHG$.

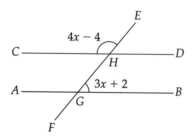

   (1) 20
   (2) 26
   (3) 80
   (4) 100

4. The distance between points A and B is 10 units. How many points are 5 units from A and 15 units from B?
   (1) 0
   (2) 1
   (3) 2
   (4) 4

5. The transformation of $\triangle ABC$ to $\triangle A'B'C'$ is shown in the accompanying diagram.

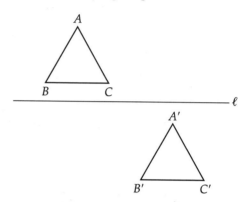

   This transformation is an example of:
   (1) reflection over line ℓ
   (2) point rotation about point A
   (3) dilation
   (4) translation

6. The ratio of the surface areas of two cubes is 3 to 4. What is the ratio of the volume of the smaller cube to the volume of the larger cube?
   (1) 3 to 8
   (2) $\sqrt{3}$ to 2
   (3) 3 to 4
   (4) $3\sqrt{3}$ to 8

7. What is the equation of a line that has a y-intercept of -3 and is parallel to the line whose equation is $4y = 3x + 12$?
   (1) $y = \frac{3}{4}x + 3$
   (2) $y = \frac{4}{3}x - 3$
   (3) $y = \frac{3}{4}x - 3$
   (4) $y = -\frac{4}{3}x - 3$

8. What is the sum of $\frac{c}{2z}$ and $\frac{d}{3z}$?
   (1) $\frac{3c + 2d}{6z}$
   (2) $\frac{c + d}{5z}$
   (3) $\frac{c + d}{6z}$
   (4) $\frac{cd}{6z}$

9. Melissa saved money by working over the summer. Before school started, she spent 25% of the money on jewelry and then spent $\frac{2}{3}$ of the remaining money on a digital camera. She still had 213 dollars left. How much money did Melissa have before she started shopping?

   (1) $568 (3) $852
   (2) $639 (4) $2,556

10. What is the equation of the axis of symmetry for the graph of the equation $y = \frac{1}{2}x^2 + 2x + 3$?

    (1) $x = -1$
    (2) $x = -2$
    (3) $x = -4$
    (4) $x = 4$

Part II

Each correct answer will receive 2 credits. Clearly indicate the necessary steps, including appropriate formula substitutions, diagrams, charts, etc. Correct numerical answers without work shown will receive only 1 credit.

11. Find the number of square units in the area of quadrilateral *ABCD*.

    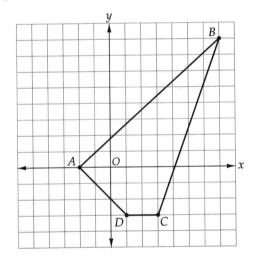

12. Solve for the positive value of y:

    $$\frac{2+y}{2y} = \frac{y-1}{y}$$

13. Solve this system of equations for the value of y:

    $2x + y = 7$ and $x + y = 3$

14. If the lengths of two sides of an equilateral triangle are represented by $2x + 7$ and $3x - 5$, what is the perimeter of the triangle?

Part III

Each correct answer will receive 3 credits. Partial credit will be allowed. Clearly indicate the necessary steps, including appropriate formula substitutions, diagrams, charts, etc. Correct numerical answers without work shown will receive only 1 credit.

15. Mr. Fox bought a car for d dollars. After 5 years, the value of the car was half the original value. The following year, the value decreased another $1,500.

    a Write an expression in terms of d to represent the value of the car after six years.

    b If the price of the car when it was new was $24,000, find the value of the car when it was 6 years old.

16. A tree 15 feet tall casts a shadow 8 feet long. Find, to the *nearest degree*, the angle of elevation of the sun.

17. The angles in a triangle measure $7x - 1$, $18x + 2$, and $5x + 10$. Determine whether the triangle is acute, obtuse, or right. State your reason clearly.

Part IV

Each correct answer will receive 4 credits. Partial credit will be allowed. Clearly indicate the necessary steps, including appropriate formula substitutions, diagrams, charts, etc. Correct numerical answers without work shown will receive only 1 credit.

18 A survey showed that the ratio of Democrats to Republicans to Independents in Center City was 7 to 4 to 1. If 6,000 people were surveyed, determine how many identified with each political party.

19 Solve the following system of equations algebraically or graphically.
$$y = x^2 - 6x - 7 \quad \text{and} \quad y = -x + 7$$

20 Graph these inequalities on the same set of axes.
$$y < 2x + 3 \quad \text{and} \quad 2x + y < 1$$

CUMULATIVE REVIEW
CHAPTERS 1–10

Part I

All questions in this part will receive 2 credits. No partial credit will be allowed.

1. A package of chocolate is shown below:

   The shape of the package is best described as a:
   (1) rectangular solid
   (2) prism
   (3) pyramid
   (4) tetrahedron

2. The expression $\sqrt{48}$ can be simplified to:
   (1) $2\sqrt{12}$ (3) $4\sqrt{12}$
   (2) $4\sqrt{3}$ (4) $16\sqrt{3}$

3. What is the image of point $(-2, -4)$ after a dilation of $-\frac{1}{2}$ whose center of dilation is the origin?
   (1) $(1, 2)$ (3) $(2, 1)$
   (2) $(-1, -2)$ (4) $(-2, -1)$

4. What is true about the statement "If two angles are vertical angles, the angles have equal measures" and its inverse "If two angles are not vertical angles, they do not have equal measures"?
   (1) The statement is always true but its inverse is sometimes false.
   (2) The statement is always false but its inverse is sometimes true.
   (3) Both the statement and its inverse are always true.
   (4) Both the statement and its inverse are sometimes false.

5. What is the slope of the line whose equation is $x + 2y - 4 = 0$?
   (1) $-\frac{1}{2}$ (3) -2
   (2) $\frac{1}{2}$ (4) 4

6. If two hamburgers and one soda cost $6 and one hamburger and two sodas cost $4.50, find the cost of one hamburger and one soda.
   (1) $2.50   (3) $3.50
   (2) $3.00   (4) $4.00

7. The negative root of the equation $(2x + 1)(2x - 3) = 0$ is:
   (1) -3 (3) -1
   (2) $-\frac{3}{2}$ (4) $-\frac{1}{2}$

8. Which of the following expressions could be used to show that integers are not closed under division?
   (1) $0 \div 4$ (3) $-8 \div 4$
   (2) $3 \div 4$ (4) $4.5 \div 1.5$

9. Which of the following inequalities is represented by the shaded region?

   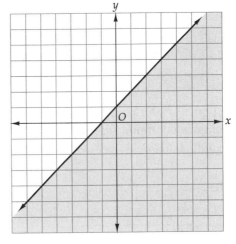

   (1) $y \leq x + 1$ (3) $x + y \leq 1$
   (2) $y \geq x + 1$ (4) $y + 1 \leq x$

10. If x and $2x$ have the same average as 11, 12, and 13, what is the value of x?
    (1) 3   (3) 8
    (2) 4   (4) 12

Chapters 1–10 **425**

Part II

Each correct answer will receive 2 credits. Clearly indicate the necessary steps, including appropriate formula substitutions, diagrams, charts, etc. Correct numerical answers without work shown will receive only 1 credit.

11 Point P is on line l. What is the total number of points 3 units from line l and 2 units from point P?

12 A highway rest area is planned so that the entry and exit paths form a right angle. The entry path is completed. It is 80 meters long and forms a 40-degree angle with the highway. How long will the exit path be to the nearest meter?

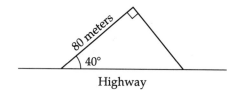

13 Two railroad lines intersect as shown in the diagram. The electric company wants to route poles for electric lines in a straight line so that they cross the intersection and remain equidistant from each railroad path. Draw a sketch to show a possible path for the poles. How many paths are possible?

14 Battery life for two brands of batteries was tested under identical situations. The results were recorded in the box-and-whisker plots shown below.

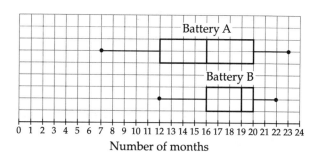

If both batteries cost the same, which would you buy? Explain why.

Part III

Each correct answer will receive 3 credits. Partial credit will be allowed. Clearly indicate the necessary steps, including appropriate formula substitutions, diagrams, charts, etc. Correct numerical answers without work shown will receive only 1 credit.

15 A student wrote the following four numbers in a table.

$\sqrt{\frac{4}{9}}$	$0.\overline{123}$
$0.\overline{3}$	$\sqrt{12}$

Which one of the above numbers is different from the others? Give a mathematical justification for your choice.

16 Solve for x: $(x + 1)^2 + 3(x - 3) - 4 = x(x - 1)$

17 A cardboard square, 2 feet on each side, has 3 inches square cut out of each corner so that an open box can be constructed. Find the volume of the box in cubic inches.

Part IV

Each correct answer will receive 4 credits. Partial credit will be allowed. Clearly indicate the necessary steps, including appropriate formula substitutions, diagrams, charts, etc. Correct numerical answers without work shown will receive only 1 credit.

18. Line l contains points $A(2, -6)$ and $B(-3, 6)$. Points A' and B' are the images of points A and B after a reflection over the y-axis.

    a What are the coordinates of the points A' and B'? What is the slope of $\overline{A'B'}$?

    b What is the midpoint of $\overline{A'B'}$?

    c What is the length of $\overline{A'B'}$?

19. A rectangular garden, 60 feet by 80 feet, is surrounded by a walk of uniform width as indicated in the diagram. The area of the walk itself (the shaded region) is 7,200 square feet. Find the width, x, of the walk.

    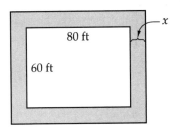

20. A candy store sells 32-ounce bags of a mixture of chocolate and raisins. The price of the 32-ounce bag depends upon the amount of chocolate. If c ounces of chocolate are in the bag, the price p can be expressed as $p = 0.2c + 0.1(32 - c)$.

    a Find the price of a bag containing 24 ounces of chocolate.

    b Find the price of a bag containing 24 ounces of raisins.

    c Kate spent $4.20 on her purchase. Describe the mixture she bought.

CUMULATIVE REVIEW
CHAPTERS 1–11

Part I

All questions in this part will receive 2 credits. No partial credit will be allowed.

1. An expression equivalent to 7.003×10^{-3} is:
   (1) 0.7003
   (2) 0.07003
   (3) 0.007003
   (4) 0.0007003

2. The statement "If x is divisible by 3, then it is divisible by 6" is false if x equals:
   (1) 12
   (2) 15
   (3) 25
   (4) 30

3. The original price of a radio was $84. The sale ticket said "25% off original price." If Evan bought the radio on sale, what did he pay for it?
   (1) $21
   (2) $59
   (3) $63
   (4) $72

4. In right triangle *PET*, with $\angle T = 90°$, $\angle P = 44°$, and side $PT = 10$, what is the value, to the nearest hundredth, of side *ET*?
   (1) 6.94
   (2) 7.19
   (3) 9.66
   (4) 10.36

5. The ratio of the areas of two similar circles is 1 to 2. what is the ratio of the radii of the circles?
   (1) $1 : \sqrt{2}$
   (2) $1 : 2$
   (3) $1 : 2\pi$
   (4) $1 : 4$

6. The direct distance between Washington High School and Jefferson High School is 10 miles. The direct distance between Jefferson High School and Adams High School is 23 miles. Which could *not* be the direct distance between Washington High School and Adams High School?
   (1) 10 miles
   (2) 15 miles
   (3) 20 miles
   (4) 25 miles

7. Which equation represents the graph of a circle?
   (1) $y = x$
   (2) $y = x^2$
   (3) $x^2 + 4y^2 = 1$
   (4) $x^2 + y^2 = 4$

8. What is the image of point $(-3, 6)$ under a reflection in the *x*-axis?
   (1) $(3, 6)$
   (2) $(3, -6)$
   (3) $(-3, -6)$
   (4) $(6, -3)$

9. Which shape has the greatest number of lines of symmetry?
   (1) square
   (2) equilateral triangle
   (3) rhombus
   (4) isosceles trapezoid

10. How many unique ways can the letters in the word ORCHARD be arranged?
    (1) 5,040
    (2) 2,520
    (3) 1,260
    (4) 720

428 Cumulative Review

Part II

Each correct answer will receive 2 credits. Clearly indicate the necessary steps, including appropriate formula substitutions, diagrams, charts, etc. Correct numerical answers without work shown will receive only 1 credit.

11 Solve for x: $\dfrac{1}{x} + \dfrac{2}{3} = \dfrac{7}{3x}$

12 An emergency fuel container holds 6 quarts of gas. Jerry keeps the container in his trailer when he goes on wilderness hunting trips. If Jerry's trailer uses fuel at the rate of 12 miles per gallon, how far can Jerry travel using his reserve supply?

13 A coin is tossed three times.
   a Draw a tree diagram or list a sample space to show all the possible arrangements of heads and tails in the three tosses.
   b Using your information from part **a**, what is the probability that there will be only one tail tossed?

14 The accompanying graph shows the closing price per share of a certain stock over a period of 7 days. (Note that a price such as $1\dfrac{1}{2}$ dollars is $1.50.)

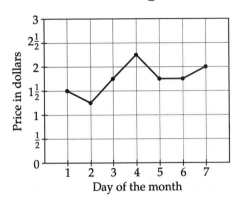

Between what two days did the price decrease most sharply? How much did the price decrease over those two days?

Part III

Each correct answer will receive 3 credits. Partial credit will be allowed. Clearly indicate the necessary steps, including appropriate formula substitutions, diagrams, charts, etc. Correct numerical answers without work shown will receive only 1 credit.

15 The grades on a 12-item quiz were scaled so that a perfect quiz received a grade of 100 by the following formula: $G = 8N + 4$, where G represents the final grade and N represents the number of correct answers.
   a Solve the grading formula for N.
   b For Abdul to receive a grade of 76, how many correct answers must he have?

16 The measures of the angles of a quadrilateral are in the ratio 3 to 4 to 5 to 6. Determine the number of degrees in each angle of the quadrilateral.

17 Quadrilateral $ABCD$ has vertices $A(-6, -1)$, $B(-1, -1)$, $C(2, 3)$, and $D(-3, 3)$.
   a On graph paper, construct and label the quadrilateral.
   b Using coordinate geometry, prove $ABCD$ is a rhombus.

Part IV

Each correct answer will receive 4 credits. Partial credit will be allowed. Clearly indicate the necessary steps, including appropriate formula substitutions, diagrams, charts, etc. Correct numerical answers without work shown will receive only 1 credit.

18 Solve the following system of equations algebraically or graphically for x and y.

$y = x^2 - 4x$ and $y = -x + 4$

19 Two bridge clubs offer different membership plans. The graph below represents the total cost of belonging to club A for one year.

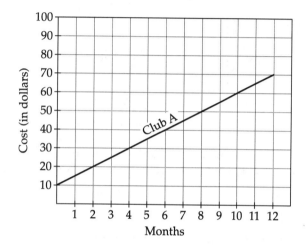

a On the same set of axes, draw the graph of the cost of belonging to club B for a year, if there is no initial membership fee and the monthly fee is $10.

b What is the initial membership fee for club A?

c What is the number of the month when the total cost is the same for both clubs?

d What is the monthly charge for club A?

20 Eunice is sending a package from Maine to California on Saturday by Speedy Mail Service. When she asked when the package would arrive in California, the clerk showed her the circle graph below.

Delivery Dates for Saturday Packages

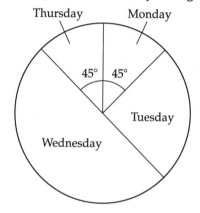

a What is the probability that her package will arrive on Tuesday?

b What is the probability that her package will arrive on or before Wednesday?

c Construct a cumulative frequency histogram to show the probability that the package will arrive on or before each weekday.

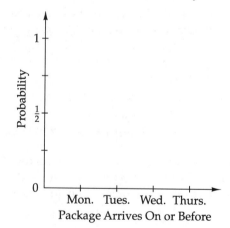

PRACTICE EXAMINATION A

Part I

Answer all questions in this part. Each correct answer will receive 2 credits. No partial credit will be allowed. Record your answers in the spaces provided on the separate answer sheet. [40]

1. A fair coin is thrown in the air four times. If the coin lands with the head up on the first three tosses, what is the probability that the coin will land with the head up on the fourth toss?

   (1) 0   (2) $\frac{1}{16}$ (3) $\frac{1}{8}$ (4) $\frac{1}{2}$

2. The statement "If x is divisible by 8, then it is divisible by 6" is false if x equals
   (1) 6   (2) 14   (3) 32   (4) 48

3. What is the image of point (2, 5) under the translation that shifts (x, y) to $(x + 3, y - 2)$?
   (1) (0, 3)   (2) (0, 8)   (3) (5, 3)   (4) (5, 8)

4. The sum of $3x^2 + x + 8$ and $x^2 - 9$ can be expressed as
   (1) $4x^2 + x - 1$ (2) $4x^2 + x - 17$ (3) $4x^4 + x - 1$ (4) $3x^4 + x - 1$

5. The direct distance between city A and city B is 200 miles. The direct distance between city B and city C is 300 miles. Which could be the direct distance between city C and city A?
   (1) 50 miles   (2) 350 miles   (3) 550 miles   (4) 650 miles

6. Expressed as a single fraction, what is $\frac{1}{x+1} + \frac{1}{x}, x \neq 0, -1$?

   (1) $\frac{2x+3}{x^2+x}$ (2) $\frac{2x+1}{x^2+x}$ (3) $\frac{2}{2x+1}$ (4) $\frac{3}{x^2}$

7. How many different three-member teams can be formed from six students?
   (1) 20   (2) 120   (3) 216   (4) 720

8. If $x = -3$ and $y = 2$, which point on the accompanying graph represents $(-x, -y)$?
   (1) P (2) Q (3) R (4) S

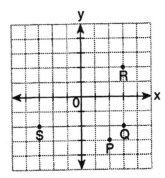

9. The larger root of the equation $(x + 4)(x - 3) = 0$ is
   (1) −4 (2) −3 (3) 3 (4) 4

10. Linda paid $48 for a jacket that was on sale for 25% of the original price. What was the original price of the jacket?
    (1) $60 (2) $72 (3) $96 (4) $192

11. The expression $2^3 \cdot 4^2$ is equivalent to
    (1) 2^7 (2) 2^{12} (3) 8^5 (4) 8^6

12. In the accompanying diagram of $\triangle ABC$, $\overline{AB}$ is extended to D, exterior angle CBD measures 145°, and m$\angle C$ = 75. What is m$\angle CAB$?
    (1) 35 (2) 70 (3) 110 (4) 220

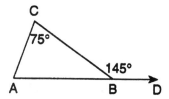

13. A total of $450 is divided into equal shares. If Kate receives four shares, Kevin receives three shares, and Anna receives the remaining two shares, how much money did Kevin receive?
    (1) $100 (2) $150 (3) $200 (4) $250

14. What is the diameter of a circle whose circumference is 5?
    (1) $\dfrac{2.5}{\pi^2}$ (2) $\dfrac{2.5}{\pi}$ (3) $\dfrac{5}{\pi^2}$ (4) $\dfrac{5}{\pi}$

15. During a recent winter, the ratio of deer to foxes was 7 to 3 in one county of New York State. If there were 210 foxes in the county, what was the number of deer in the county?
    (1) 90 (2) 147 (3) 280 (4) 490

16. In the accompanying figure, $ACDH$ and $BCEF$ are rectangles, $AH = 2$, $GH = 3$, $GF = 4$, and $FE = 5$. What is the area of $BCDG$?
    (1) 6 (2) 8 (3) 10 (4) 20

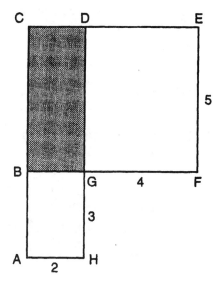

17. If $t^2 < t < \sqrt{t}$, then t could be
    (1) $-\frac{1}{4}$ (2) 0 (3) $\frac{1}{4}$ (4) 4

18. What is the slope of line ℓ shown in the accompanying diagram?
    (1) $\frac{4}{3}$ (2) $\frac{3}{4}$ (3) $-\frac{3}{4}$ (4) $-\frac{4}{3}$

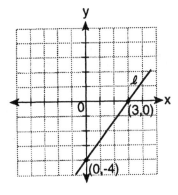

19. In a class of 50 students, 18 take music, 26 take art, and 2 take both art and music. How many students in the class are not enrolled in either music or art?
    (1) 6 (2) 8 (3) 16 (4) 24

20. The expression $\sqrt{27} + \sqrt{12}$ is equivalent to
    (1) $5\sqrt{3}$ (2) $13\sqrt{3}$ (3) $5\sqrt{6}$ (4) $\sqrt{39}$

Part II

Answer all questions in this part. Each correct answer will receive 2 credits. Clearly indicate the necessary steps, including appropriate formula substitutions, diagrams, graphs, charts, etc. For all questions in this part, a correct numerical answer with no work shown will receive only 1 credit. [10]

21. Draw all the symmetry lines on the accompanying figure.

22. Shoe sizes and foot length are related by the formula $S = 3F - 24$, where S represents the shoe size and F represents the length of the foot, in inches.
    a. Solve the formula for F.
    b. To the *nearest tenth of an inch,* how long is the foot of a person who wears a size $10\frac{1}{2}$ shoe?

23. Which number below is irrational?

$$\sqrt{\tfrac{4}{9}}, \quad \sqrt{20}, \quad \sqrt{121}$$

Why is the number you chose an irrational number?

24. Simplify: $\dfrac{9x^2 - 15xy}{9x^2 - 25y^2}$

25. Sara's telephone service costs $21 per month plus $0.25 for each local call, and long-distance calls are extra. Last month, Sara's bill was $36.64, and it included $6.14 in long-distance charges. How many local calls did she make?

Part III

Answer all questions in this part. Each correct answer will receive 3 credits. Clearly indicate the necessary steps, including appropriate formula substitutions, diagrams, graphs, charts, etc. For all questions in this part, a correct numerical answer with no work shown will receive only 1 credit. [15]

26. During a 45-minute lunch period, Albert (A) went running and Bill (B) walked for exercise. Their times and distances are shown in the accompanying graph. How much faster was Albert running than Bill was walking, in miles per hour?

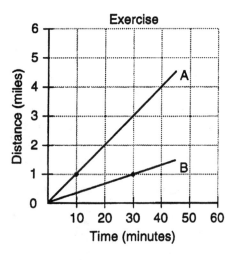

27. The dimensions of a brick, in inches, are 2 by 4 by 8. How many such bricks are needed to have a total volume of exactly 1 cubic foot?

28. A swimmer plans to swim at least 100 laps during a 6-day period. During this period, the swimmer will increase the number of laps completed each day by one lap. What is the *least* number of laps the swimmer must complete on the first day?

29. The mean (average) weight of three dogs is 38 pounds. One of the dogs, Sparky, weighs 46 pounds. The other two dogs, Eddie and Sandy, have the same weight. Find Eddie's weight.

30. In the accompanying diagram, △ABC and △ABD are isosceles triangles with m∠CAB = 50 and m∠BDA = 55. If AB = AC and AB = BD, what is m∠CBD?

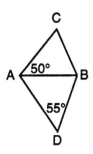

Part IV

Answer all questions in this part. Each correct answer will receive 4 credits. Clearly indicate the necessary steps, including appropriate formula substitutions, diagrams, graphs, charts, etc. For all questions in this part, a correct numerical answer with no work shown will receive only 1 credit. [20]

31. A target shown in the accompanying diagram consists of three circles with the same center. The radii of the circles have lengths of 3 inches, 7 inches, and 9 inches.
    a. What is the area of the shaded region to the *nearest tenth of a square inch*?
    b. To the *nearest percent*, what percent of the target is shaded?

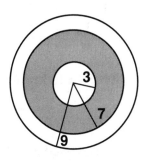

32. A bookshelf contains six mysteries and three biographies. Two books are selected at random without replacement.
    a. What is the probability that both books are mysteries?
    b. What is the probability that one book is a mystery and the other is a biography?

33. The cross section of an attic is in the shape of an isosceles trapezoid, as shown in the accompanying figure. If the height of the attic is 9 feet, BC = 12 feet, and AD = 28 feet, find the length of $\overline{AB}$ to the *nearest foot*.

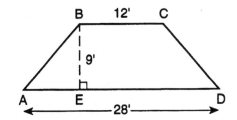

34. Joe is holding his kite string 3 feet above the ground, as shown in the accompanying diagram. The distance between his hand and a point directly under the kite is 95 feet. If the angle of elevation to the kite is 50°, find the height, h, of his kite, to the *nearest foot*.

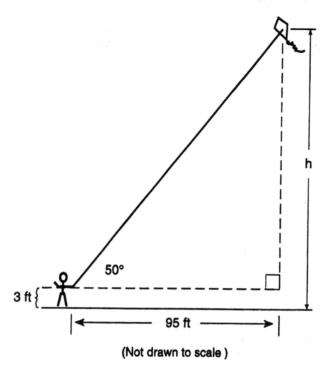

(Not drawn to scale)

35. Solve the following system of equations algebraically *or* graphically for x and y:

$$y = x^2 + 2x - 1$$
$$y = 3x + 5$$

For an algebraic solution, show your work.

For a graphic solution, show your work here.

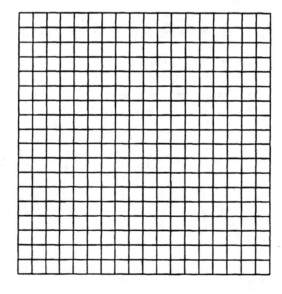

PRACTICE EXAMINATION B

Part I

Answer all questions in this part. Each correct answer will receive 2 credits. No partial credit will be allowed. Record your answers in the spaces provided on the separate answer sheet. [40]

1. A roll of candy is shown in the accompanying diagram.

   The shape of the candy is best described as a
   (1) rectangular solid   (3) cone
   (2) pyramid   (4) cylinder

2. The expression $\sqrt{50}$ can be simplified to
   (1) $5\sqrt{2}$ (2) $5\sqrt{10}$ (3) $2\sqrt{25}$ (4) $25\sqrt{2}$

3. The transformation of $\triangle ABC$ to $\triangle AB'C'$ is shown in the accompanying diagram.

   This transformation is an example of a
   (1) line reflection in line ℓ (3) dilation
   (2) rotation about point A (4) translation

   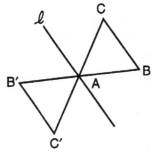

4. Which expression is equivalent to 6.02×10^{23}?
   (1) 0.602×10^{21} (2) 60.2×10^{21} (3) 602×10^{21} (4) 6020×10^{21}

5. The Pentagon building in Washington, D.C., is shaped like a regular pentagon. If the length of one side of the Pentagon is represented by $n + 2$, its perimeter would be represented by
   (1) $5n + 10$ (2) $5n + 2$ (3) $n + 10$ (4) $10n$

6. The product of $4x^2y$ and $2xy^3$ is
   (1) $8x^2y^3$ (2) $8x^3y^3$ (3) $8x^3y^4$ (4) $8x^2y^4$

7. Which equation is an illustration of the additive identity property?
   (1) $x \cdot 1 = x$ (2) $x + 0 = x$ (3) $x - x = 0$ (4) $x \cdot \frac{1}{x} = 1$

8. The formula $C = \frac{5}{9}(F - 32)$ can be used to find the Celsius temperature (C) for a given Fahrenheit temperature (F). What Celsius temperature is equal to a Fahrenheit temperature of 77°?
   (1) 8°   (2) 25°   (3) 45°   (4) 171°

9. In the accompanying diagram of rectangle *ABCD*, m∠*BAC* = 3*x* + 4 and m∠*ACD* = *x* + 28. What is m∠*CAD*?
   (1) 12 (2) 37 (3) 40 (4) 50

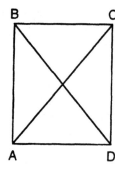

10. On June 17, the temperature in Las Vegas ranged from 90° to 99°, while the temperature in Los Angeles ranged from 60° to 69°. The difference in the temperatures in these two cities must be between
    (1) 20° and 30° (2) 20° and 40° (3) 25° and 35° (4) 30° and 40°

11. Which expression is equivalent to $\frac{a}{x} + \frac{b}{2x}$?
    (1) $\frac{2a+b}{2x}$ (2) $\frac{2a+b}{x}$ (3) $\frac{a+b}{3x}$ (4) $\frac{a+b}{2x}$

12. What is true about the statement "If two angles are right angles, the angles have equal measure" and its converse "If two angles have equal measure then the two angles are right angles"?
    (1) The statement is true but its converse is false.
    (2) The statement is false but its converse is true.
    (3) Both the statement and its converse are false.
    (4) Both the statement and its converse are true.

13. If 6 and *x* have the same mean (average) as 2, 4, and 24, what is the value of *x*?
    (1) 5 (2) 10 (3) 14 (4) 36

14. In a hockey league, 87 players play on seven different teams. Each team has at least 12 players. What is the largest possible number of players on any one team?
    (1) 13 (2) 14 (3) 15 (4) 21

15. In the accompanying diagram of equilateral triangle *ABC*, *DE* = 5 and $\overline{DE} \parallel \overline{AB}$.
    If *AB* is three times as long as *DE*, what is the perimeter of quadrilateral *ABED*?
    (1) 20 (2) 30 (3) 35 (4) 40

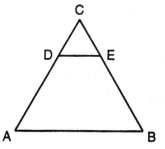

438 Practice Examinations

16. At a concert, $720 was collected for hot dogs, hamburgers, and soft drinks. All three items sold for $1.00 each. Twice as many hot dogs were sold as hamburgers. Three times as many soft drinks were sold as hamburgers. The number of soft drinks sold was
    (1) 120   (2) 240   (3) 360   (4) 480

17. How many different 6-letter arrangements can be formed using the letters in the word "ABSENT," if each letter is used only once?
    (1) 6   (2) 36   (3) 720   (4) 46,656

18. The ratio of the corresponding sides of two similar squares is 1 to 3. What is the ratio of the area of the smaller square to the area of the larger square?
    (1) $1:\sqrt{3}$ (2) 1:3   (3) 1:6   (4) 1:9

19. What is the slope of the line whose equation is $3x - 4y - 16 = 0$?
    (1) $\frac{3}{4}$ (2) $\frac{4}{3}$ (3) 3   (4) -4

20. What is the perimeter of an equilateral triangle whose height is $2\sqrt{3}$?
    (1) 6   (2) 12   (3) $6\sqrt{3}$ (4) $12\sqrt{3}$

Part II

Answer all questions in this part. Each correct answer will receive 2 credits. Clearly indicate the necessary steps, including appropriate formula substitutions, diagrams, graphs, charts, etc. For all questions in this part, a correct numerical answer with no work shown will receive only 1 credit. [10]

21. Solve for x: $2(x - 3) = 1.2 - x$

22. The Grimaldis have three children born in different years.
    a. Draw a tree diagram or list a sample space to show all the possible arrangements of boy and girl children in the Grimaldi family.
    b. Using your information from part *a*, what is the probability that the Grimaldis have three boys?

23. Paloma has 3 jackets, 6 scarves, and 4 hats. Determine the number of different outfits consisting of a jacket, a scarf, and a hat that Paloma can wear.

24. In a recent poll, 600 people were asked whether they liked Chinese food. A circle graph was constructed to show the results. The central angles for two of the three sectors are shown in the accompanying diagram. How many people had no opinion?

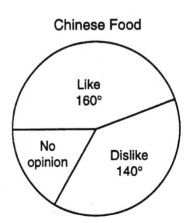

Practice Examination B   439

25. Maria's backyard has two trees that are 40 feet apart, as shown in the accompanying diagram. She wants to place lampposts so that the posts are 30 feet from both of the trees. Draw a sketch to show where the lampposts could be placed in relation to the trees. How many locations for the lampposts are possible?

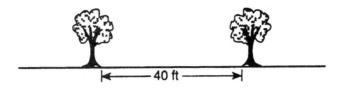

Part III

Answer all questions in this part. Each correct answer will receive 3 credits. Clearly indicate the necessary steps, including appropriate formula substitutions, diagrams, graphs, charts, etc. For all questions in this part, a correct numerical answer with no work shown will receive only 1 credit. [15]

26. Solve for x: $x^2 + 3x - 40 = 0$

27. A person standing on level ground is 2,000 feet away from the foot of a 420-foot-tall building, as shown in the accompanying diagram. To the *nearest degree,* what is the value of x?

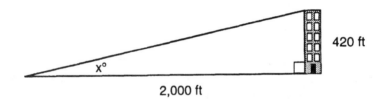

28. Bob and Ray are describing the same number. Bob says, "The number is a positive even integer less than or equal to 20." Ray says, "The number is divisible by 4." If Bob's statement is true and Ray's statement is false, what are all the possible numbers?

29. Line ℓ contains the points (0, 4) and (2, 0). Show that the point (−25, 81) does *or* does not lie on line ℓ.

30. A painting that regularly sells for a price of $55 is on sale for 20% off. The sales tax on the painting is 7%. Will the final total cost of the painting differ depending on whether the salesperson deducts the discount before adding the sales tax or takes the discount after computing the sum of the original price and the sales tax on $55?

Part IV

Answer all questions in this part. Each correct answer will receive 4 credits. Clearly indicate the necessary steps, including appropriate formula substitutions, diagrams, graphs, charts, etc. For all questions in this part, a correct numerical answer with no work shown will receive only 1 credit. [20]

31. The profits in a business are to be shared by the three partners in the ratio of 3 to 2 to 5. The profit for the year was $176,500. Determine the number of dollars each partner is to receive.

32. If asphalt pavement costs $0.78 per square foot, determine, to the *nearest cent*, the cost of paving the shaded circular road with center O, an outside radius of 50 feet, and an inner radius of 36 feet, as shown in the accompanying diagram.

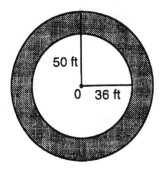

33. An arch is built so that it is 6 feet wide at the base. Its shape can be represented by a parabola with the equation $y = -2x^2 + 12x$, where y is the height of the arch.
    a. Graph the parabola from $x = 0$ to $x = 6$ on the grid below.

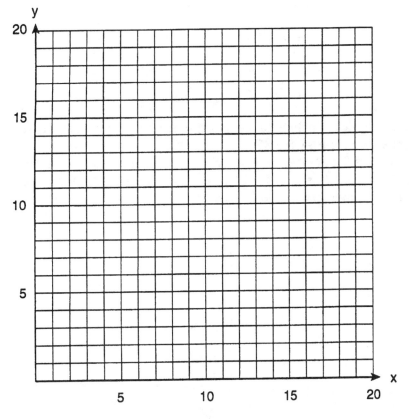

   b. Determine the maximum height, y, of the arch.

34. Mr. Gonzalez owns a triangular plot of land BCD with DB = 25 yards and BC = 16 yards. He wishes to purchase the adjacent plot of land in the shape of right triangle ABD, as shown in the accompanying diagram, with AD = 15 yards. If the purchase is made, what will be the total number of square yards in the area of his plot of land, $\triangle ACD$?

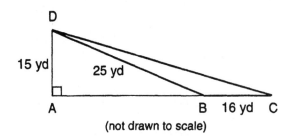
(not drawn to scale)

35. Two health clubs offer different membership plans. The graph below represents the total cost of belonging to Club A and Club B for one year.

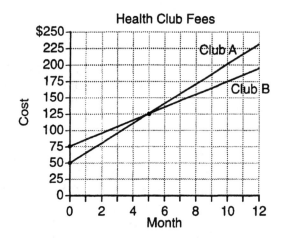

a. If the yearly cost includes a membership fee plus a monthly charge, what is the membership fee for Club A?

b. (1) What is the number of the month when the total cost is the same for both clubs?
(2) What is the total cost for Club A when both plans are the same?

c. What is the monthly charge for Club B?

PRACTICE EXAMINATION C

Part I

Answer all questions in this part. Each correct answer will receive 2 credits. No partial credit will be allowed. Record your answers in the spaces provided on the separate answer sheet. [40]

1. The expression $\sqrt{93}$ is a number between
   (1) 3 and 9   (2) 8 and 9   (3) 9 and 10   (4) 46 and 47

2. Which number has the greatest value?
   (1) $1\frac{2}{3}$ (2) $\sqrt{2}$ (3) $\frac{\pi}{2}$ (4) 1.5

3. Mary says, "The number I am thinking of is divisible by 2 or it is divisible by 3." Mary's statement is false if the number she is thinking of is
   (1) 6   (2) 8   (3) 11   (4) 15

4. Which expression is a factor of $x^2 + 2x - 15$?
   (1) $(x - 3)$ (2) $(x + 3)$ (3) $(x + 15)$ (4) $(x - 5)$

5. What was the median high temperature in Middletown during the 7-day period shown in the table?
   (1) 69   (2) 70   (3) 73   (4) 75

Daily High Temperature in Middletown	
Day	Temperature (°F)
Sunday	68
Monday	73
Tuesday	73
Wednesday	75
Thursday	69
Friday	67
Saturday	63

6. If the number represented by $n - 3$ is an odd integer, which expression represents the next greater odd integer?
   (1) $n - 5$ (2) $n - 2$ (3) $n - 1$ (4) $n + 1$

7. When the point $(2, -5)$ is reflected in the *x*-axis, what are the coordinates of its image?
   (1) $(-5, 2)$ (2) $(-2, 5)$ (3) $(2, 5)$ (4) $(5, 2)$

8. The expression $(x^2z^3)(xy^2z)$ is equivalent to
   (1) $x^2y^2z^3$ (2) $x^3y^2z^4$ (3) $x^3y^3z^4$ (4) $x^4y^2z^5$

9. Twenty-five percent of 88 is the same as what percent of 22?
   (1) $12\frac{1}{2}\%$ (2) 40%   (3) 50%   (4) 100%

10. A plot of land is in the shape of rhombus ABCD as shown.
    Which can *not* be the length of diagonal AC?
    (1) 24 m   (2) 18 m   (3) 11 m   (4) 4 m

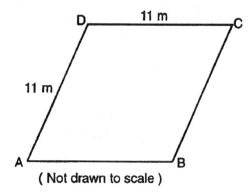
(Not drawn to scale)

11. If $9x + 2a = 3a - 4x$, then x equals
    (1) a (2) $-a$ (3) $\frac{5a}{12}$ (4) $\frac{a}{13}$

12. If the circumference of a circle is 10π inches, what is the area, in square inches, of the circle?
    (1) 10π (2) 25π (3) 50π (4) 100π

13. How many different 4-letter arrangements can be formed using the letters of the word "JUMP," if each letter is used only once?
    (1) 24   (2) 16   (3) 12   (4) 4

14. Sterling silver is made of an alloy of silver and copper in the ratio of $37:3$. If the mass of a sterling silver ingot is 600 grams, how much silver does it contain?
    (1) 48.65 g   (2) 200 g   (3) 450 g   (4) 555 g

15. If $t = -3$, then $3t^2 + 5t + 6$ equals
    (1) -36 (2) -6 (3) 6   (4) 18

16. The expression $\frac{y}{x} - \frac{1}{2}$ is equivalent to
    (1) $\frac{2y - x}{2x}$ (2) $\frac{x - 2y}{2x}$ (3) $\frac{1-y}{2x}$ (4) $\frac{y-1}{x-2}$

17. The party registration of the voters in Jonesville is shown in the table.
    If one of the registered Jonesville voters is selected at random, what is the probability that the person selected is *not* a Democrat?
    (1) 0.333   (3) 0.600
    (2) 0.400   (4) 0.667

| Registered Voters in Jonesville ||
Party Registration	Number of Voters Registered
Democrat	6,000
Republican	5,300
Independent	3,700

18. If the number of molecules in 1 mole of a substance is 6.02×10^{23}, then the number of molecules in 100 moles is
    (1) 6.02×10^{21} (2) 6.02×10^{22} (3) 6.02×10^{24} (4) 6.02×10^{25}

19. When $3a^2 - 2a + 5$ is subtracted from $a^2 + a - 1$, the result is
    (1) $2a^2 - 3a + 6$ (2) $-2a^2 + 3a - 6$ (3) $2a^2 - 3a - 6$ (4) $-2a^2 + 3a + 6$

20. The distance between parallel lines ℓ and m is 12 units. Point A is on line ℓ. How many points are equidistant from lines ℓ and m and 8 units from point A?
    (1) 1   (2) 2   (3) 3   (4) 4

Part II

Answer all questions in this part. Each correct answer will receive 2 credits. Clearly indicate the necessary steps, including appropriate formula substitutions, diagrams, graphs, charts, etc. For all questions in this part, a correct numerical answer with no work shown will receive only 1 credit. [10]

21. The midpoint M of line segment AB has coordinates $(-3, 4)$. If point A is the origin, $(0, 0)$, what are the coordinates of point B? [The use of the accompanying grid is optional.]

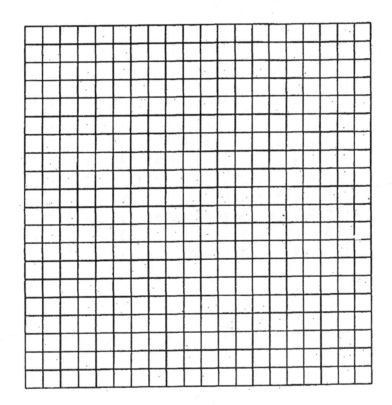

22. Mary and Amy had a total of 20 yards of material from which to make costumes. Mary used three times more material to make her costume than Amy used, and 2 yards of material was not used. How many yards of material did Amy use for her costume?

Practice Examination C **445**

23. A wall is supported by a brace 10 feet long, as shown in the diagram. If one end of the brace is placed 6 feet from the base of the wall, how many feet up the wall does the brace reach?

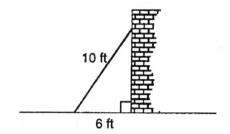

24. A straight line with slope 5 contains the points (1, 2) and (3, K). Find the value of K. [The use of the accompanying grid is optional.]

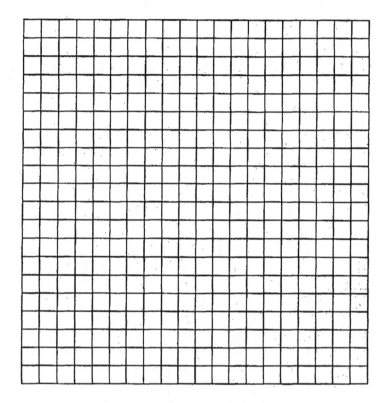

25. Al says, "If ABCD is a parallelogram, then ABCD is a rectangle." Sketch a quadrilateral ABCD that shows that Al's statement is *not* always true. Your sketch must show the length of each side and the measure of each angle for the quadrilateral you draw.

Part III

Answer all questions in this part. Each correct answer will receive 3 credits. Clearly indicate the necessary steps, including appropriate formula substitutions, diagrams, graphs, charts, etc. For all questions in this part, a correct numerical answer with no work shown will receive only 1 credit. [15]

26. Judy needs a mean (average) score of 86 on four tests to earn a midterm grade of B. If the mean of her scores for the first three tests was 83, what is the *lowest* score on a 100-point scale that she can receive on the fourth test to have a midterm grade of B?

27. A truck traveling at a constant rate of 45 miles per hour leaves Albany. One hour later a car traveling at a constant rate of 60 miles per hour also leaves Albany traveling in the same direction on the same highway. How long will it take for the car to catch up to the truck, if both vehicles continue in the same direction on the highway?

28. In the figure below, the large rectangle, *ABCD*, is divided into four smaller rectangles. The area of rectangle $AEHG = 5x$, the area of rectangle $GHFB = 2x^2$, the area of rectangle $HJCF = 6x$, segment $AG = 5$, and segment $AE = x$.

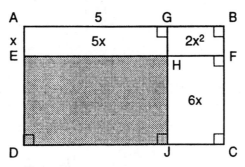

  a. Find the area of the shaded region.
  b. Write an expression for the area of rectangle *ABCD* in terms of *x*.

29. *a.* On the set of axes provided below, sketch a circle with a radius of 3 and a center at (2, 1) and also sketch the graph of the line $2x + y = 8$.

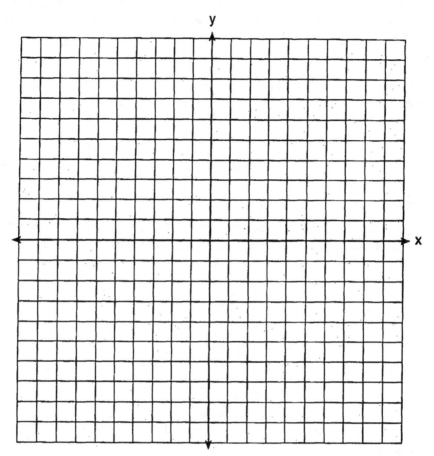

   b. What is the total number of points of intersection of the two graphs?

30. The volume of a rectangular pool is 1,080 cubic meters. Its length, width, and depth are in the ratio 10:4:1. Find the number of meters in each of the three dimensions of the pool.

Part IV

Answer all questions in this part. Each correct answer will receive 4 credits. Clearly indicate the necessary steps, including appropriate formula substitutions, diagrams, graphs, charts, etc. For all questions in this part, a correct numerical answer with no work shown will receive only 1 credit. [20]

31. Amy tossed a ball in the air in such a way that the path of the ball was modeled by the equation $y = -x^2 + 6x$. In the equation, y represents the height of the ball in feet and x is the time in seconds.

    a. Graph $y = -x^2 + 6x$ for $0 \leq x \leq 6$ on the grid provided.

    b. At what time, x, is the ball at its highest point?

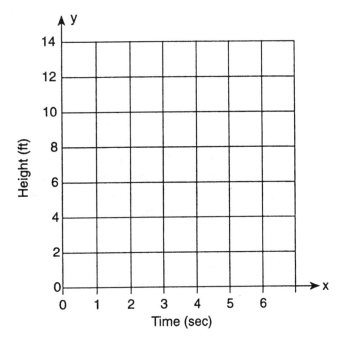

32. In the time trials for the 400-meter run at the state sectionals, the 15 runners recorded the times shown in the table below.

400-Meter Run	
Time (sec)	Frequency
50.0–50.9	
51.0–51.9	II
52.0–52.9	IIII I
53.0–53.9	III
54.0–54.9	IIII

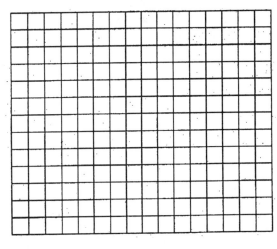

    a. Using the data from the frequency column, draw a frequency histogram on the grid provided at the right.

    b. What percent of the runners completed the time trial between 52.0 and 53.9 seconds?

33. A group of 148 people is spending five days at a summer camp. The cook ordered 12 pounds of food for each adult and 9 pounds of food for each child. A total of 1,410 pounds of food was ordered.
   a. Write an equation *or* a system of equations that describes the above situation and define your variables.
   b. Using your work from part *a,* find:
      (1) the total number of adults in the group
      (2) the total number of children in the group

34. Three roses will be selected for a flower vase. The florist has 1 red rose, 1 white rose, 1 yellow rose, 1 orange rose, and 1 pink rose from which to choose.
   a. How many different 3-rose selections can be formed from the 5 roses?
   b. What is the probability that 3 roses selected at random will contain 1 red rose, 1 white rose, and 1 pink rose?
   c. What is the probability that 3 roses selected at random will *not* contain an orange rose?

35. The Excel Cable Company has a monthly fee of $32.00 and an additional charge of $8.00 for each premium channel. The Best Cable Company has a monthly fee of $26.00 and an additional charge of $10.00 for each premium channel. The Horton family is deciding which of these two cable companies to subscribe to.
   a. For what number of premium channels will the total monthly subscription fee for the Excel and Best Cable companies be the same?
   b. The Horton family decides to subscribe to 2 premium channels for a period of one year.
      (1) Which cable company should they subscribe to in order to spend less money?
      (2) How much money will the Hortons save in one year by using the less expensive company?

PRACTICE EXAMINATION D

Part I

Answer all questions in this part. Each correct answer will receive 2 credits. No partial credit will be allowed. Record your answers in the spaces provided on the separate answer sheet. [40]

1. Which inequality is represented in the graph below?

   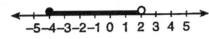

   (1) $-4 < x < 2$ (2) $-4 \leq x < 2$ (3) $-4 < x \leq 2$ (4) $-4 \leq x \leq 2$

2. Which geometric figure has one and only one line of symmetry?

   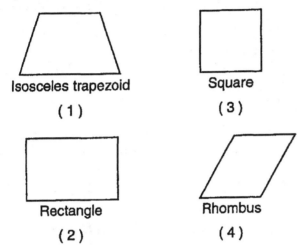

3. Which number is rational?

   (1) π (2) $\frac{5}{4}$ (3) $\sqrt{7}$ (4) $\sqrt{\frac{3}{2}}$

4. Two numbers are in the ratio 2:5. If 6 is subtracted from their sum, the result is 50. What is the larger number?
   (1) 55 (2) 45 (3) 40 (4) 35

5. The quotient of $-\frac{15x^8}{5x^2}$, $x \neq 0$, is

   (1) $-3x^4$ (2) $-10x^4$ (3) $-3x^6$ (4) $-10x^6$

6. What is the inverse of the statement "If it is sunny, I will play baseball"?
   (1) If I play baseball, then it is sunny.
   (2) If it is not sunny, I will not play baseball.
   (3) If I do not play baseball, then it is not sunny.
   (4) I will play baseball if and only if it is sunny.

7. Which ordered pair is the solution of the following system of equations?

$$3x + 2y = 4$$
$$-2x + 2y = 24$$

   (1) (2, −1)   (2) (2, −5)   (3) (−4, 8)   (4) (−4, −8)

8. Which equation represents a circle whose center is (3, −2)?
   (1) $(x + 3)^2 + (y − 2)^2 = 4$
   (2) $(x − 3)^2 + (y + 2)^2 = 4$
   (3) $(x + 2)^2 + (y − 3)^2 = 4$
   (4) $(x − 2)^2 + (y + 3)^2 = 4$

9. The set of integers {3, 4, 5} is a Pythagorean triple. Another such set is
   (1) {6, 7, 8}   (2) {6, 8, 12}   (3) {6, 12, 13}   (4) {8, 15, 17}

10. A truck travels 40 miles from point A to point B in exactly 1 hour. When the truck is halfway between point A and point B, a car starts from point A and travels at 50 miles per hour. How many miles has the car traveled when the truck reaches point B?
    (1) 25   (2) 40   (3) 50   (4) 60

11. If $a \neq 0$ and the sum of x and $\frac{1}{a}$ is 0, then
    (1) $x = a$ (2) $x = -a$ (3) $x = -\frac{1}{a}$ (4) $x = 1 - a$

12. The accompanying figure shows the graph of the equation $x = 5$.
    What is the slope of the line $x = 5$?
    (1) 5   (2) −5   (3) 0   (4) undefined

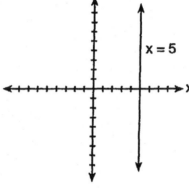

13. Which transformation does *not* always produce an image that is congruent to the original figure?
    (1) translation   (2) dilation   (3) rotation   (4) reflection

14. If rain is falling at the rate of 2 inches per hour, how many inches of rain will fall in x minutes?
    (1) $2x$ (2) $\frac{30}{x}$ (3) $\frac{60}{x}$ (4) $\frac{x}{30}$

15. The expression $(x − 6)^2$ is equivalent to
    (1) $x^2 − 36$ (2) $x^2 + 36$ (3) $x^2 − 12x + 36$ (4) $x^2 + 12x + 36$

16. How many different five-digit numbers can be formed from the digits 1, 2, 3, 4, and 5 if each digit is used only once?
    (1) 120   (2) 60   (3) 24   (4) 20

17. For five algebra examinations, Maria has an average of 88. What must she score on the sixth test to bring her average up to exactly 90?
    (1) 92   (2) 94   (3) 98   (4) 100

18. The graphs of the equations $y = x^2 + 4x - 1$ and $y + 3 = x$ are drawn on the same set of axes. At which point do the graphs intersect?
    (1) (1, 4)   (2) (1, –2)   (3) (–2, 1)   (4) (–2, –5)

19. If $2x^2 - 4x + 6$ is subtracted from $5x^2 + 8x - 2$, the difference is
    (1) $3x^2 + 12x - 8$ (2) $-3x^2 - 12x + 8$ (3) $3x^2 + 4x + 4$ (4) $-3x^2 + 4x + 4$

20. What is the value of 3^{-2}?
    (1) $\frac{1}{9}$ (2) $-\frac{1}{9}$ (3) 9   (4) –9

Part II

Answer all questions in this part. Each correct answer will receive 2 credits. Clearly indicate the necessary steps, including appropriate formula substitutions, diagrams, graphs, charts, etc. For all questions in this part, a correct numerical answer with no work shown will receive only 1 credit. [10]

21. The formula for changing Celsius (C) temperature to Fahrenheit (F) temperature is $F = \frac{9}{5}C + 32$. Calculate, to the *nearest degree,* the Fahrenheit temperature when the Celsius temperature is –8.

22. Using only a ruler and compass, construct the bisector of angle *BAC* in the accompanying diagram.

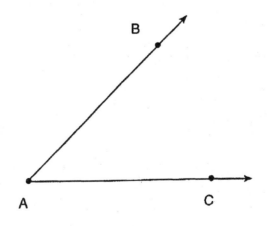

23. All seven-digit telephone numbers in a town begin with 245. How many telephone numbers may be assigned in the town if the last four digits do *not* begin or end in a zero?

24. The Rivera family bought a new tent for camping. Their old tent had equal sides of 10 feet and a floor width of 15 feet, as shown in the accompanying diagram.

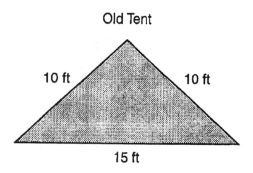

Old Tent

If the new tent is similar in shape to the old tent and has equal sides of 16 feet, how wide is the floor of the new tent?

25. The accompanying graph represents the yearly cost of playing 0 to 5 games of golf at the Shadybrook Golf Course. What is the total cost of joining the club and playing 10 games during the year?

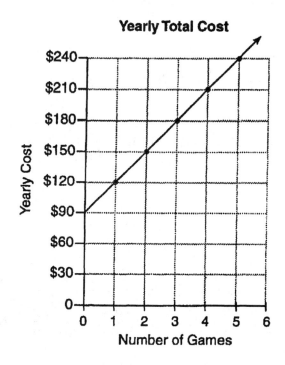

Part III

Answer all questions in this part. Each correct answer will receive 3 credits. Clearly indicate the necessary steps, including appropriate formula substitutions, diagrams, graphs, charts, etc. For all questions in this part, a correct numerical answer with no work shown will receive only 1 credit. [15]

26. The accompanying Venn diagram shows the number of students who take various courses. All students in circle A take mathematics. All in circle B take science. All in circle C take technology. What percentage of the students take mathematics or technology?

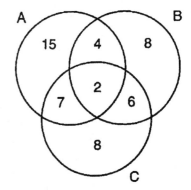

27. Hersch says if a triangle is an obtuse triangle, then it cannot also be an isosceles triangle. Using a diagram, show that Hersch is incorrect, and indicate the measures of all the angles and sides to justify your answer.

28. Tamika has a hard rubber ball whose circumference measures 13 inches. She wants to box it for a gift but can only find cube-shaped boxes of sides 3 inches, 4 inches, 5 inches, or 6 inches. What is the *smallest* box that the ball will fit into with the top on?

29. The distance from Earth to the imaginary planet Med is 1.7×10^7 miles. If a spaceship is capable of traveling 1,420 miles per hour, how many days will it take the spaceship to reach the planet Med? Round your answer to the *nearest day*.

30. A surveyor needs to determine the distance across the pond shown in the accompanying diagram. She determines that the distance from her position to point P on the south shore of the pond is 175 meters and the angle from her position to point X on the north shore is 32°. Determine the distance, PX, across the pond, rounded to the *nearest meter*.

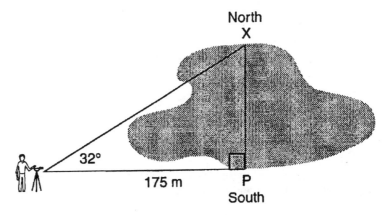

Part IV

Answer all questions in this part. Each correct answer will receive 4 credits. Clearly indicate the necessary steps, including appropriate formula substitutions, diagrams, graphs, charts, etc. For all questions in this part, a correct numerical answer with no work shown will receive only 1 credit. [20]

31. The owner of a movie theater was counting the money from 1 day's ticket sales. He knew that a total of 150 tickets were sold. Adult tickets cost $7.50 each and children's tickets cost $4.75 each. If the total receipts for the day were $891.25, how many of *each* kind of ticket were sold?

32. A treasure map shows a treasure hidden in a park near a tree and a statue. The map indicates that the tree and the statue are 10 feet apart. The treasure is buried 7 feet from the base of the tree and also 5 feet from the base of the statue. How many places are possible locations for the treasure to be buried? Draw a diagram of the treasure map, and indicate with an **X** *each* possible location of the treasure.

33. The scores on a mathematics test were 70, 55, 61, 80, 85, 72, 65, 40, 74, 68, and 84. Complete the accompanying table, and use the table to construct a frequency histogram for these scores.

Score	Tally	Frequency
40–49		
50–59		
60–69		
70–79		
80–89		

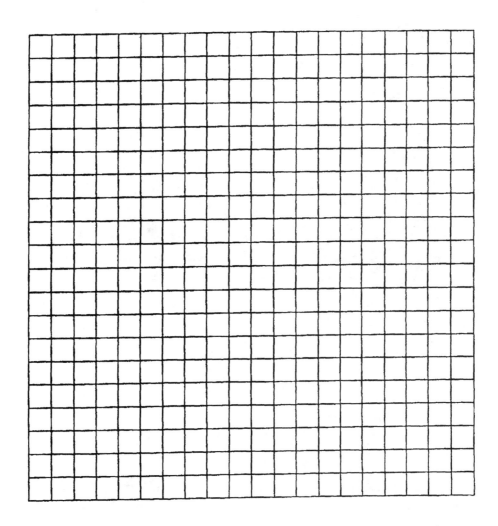

34. Paul orders a pizza. Chef Carl randomly chooses two different toppings to put on the pizza from the following: pepperoni, onion, sausage, mushrooms, and anchovies. If Paul will not eat pizza with mushrooms, determine the probability that Paul will *not* eat the pizza Chef Carl has made.

35. The area of the rectangular playground enclosure at South School is 500 square meters. The length of the playground is 5 meters longer than the width. Find the dimensions of the playground, in meters. [*Only an algebraic solution will be accepted.*]

Index

Index

A

AA similar right triangle, 233
AAS congruent triangles, 229
Abscissa (x-coordinate), 146
Absolute value, 35–36, 299
Acute angle, 199
Acute angle-acute angle similar right triangles, 233
Acute triangle, 215
Addition, 29
   of algebraic expressions, 74–76
   associative property of, 29
   commutative property of, 29
   of decimals, 47
   of fractions, 40–43
   of rational expressions, 83–85
   of signed numbers, 36–37
   of square roots, 54
   of terms with exponents, 50
Addition method of solving quadratic-linear pairs, 137, 139–140
Addition property of inequality, 103
Addition-subtraction method of solving systems of equations, 107–108
Additive identity, 30
Additive inverse (opposite), 30, 35
Adjacent angles, 201
Algebraic equations, 68
   literal equations and formulas, 85–88, 100–102
Algebraic expressions, 68
   adding and subtracting, 74–76
   dividing, by monomial, 81–83
   evaluating, 72–73
   multiplication of, 76–81
Algebraic inequalities, 68
Alternate exterior angles, 202
Alternate interior angles, 202
Altitude, *see* Height
Angle(s), 198–200
   adjacent, 201
   alternate interior and exterior, 202
   angle measures preserved under transformations, 337
   base, 215, 220
   central, 206
   classification of, 199
   classification of triangles by, 215
   complementary, 201
   congruent, 201, 228
   consecutive, 219
   consecutive interior, 203
   corresponding, 202, 228, 232, 233
   of depression, 294
   of elevation, 294
   inscribed, 206
   line and angle relationships, 201–205
   naming, 198–199
   opposite, in quadrilateral, 219
   in right triangle, finding, 295–297
   sides of, 198
   supplementary, 201
   vertex, of isosceles triangle, 215
   vertex of, 198
   vertical, 201
Angle-angle (AA) similar triangles, 233
Angle-angle-side (AAS) congruent triangles, 229
Angle bisector, 213, 351–353
Angle-side-angle (ASA) congruent triangles, 229
Antecedent of conditional, 246
Apothem, 268
Arc, 206
   degrees of, 206, 257
Arc length, 206

B

Bar graphs, 360–362
   double, 361
   frequency histogram as vertical, 359–360
Base(s), 25, 49
   negative, 51
   of polygon, 264
   of prism, 224
   of trapezoid, 220, 264
   of triangle, 215, 264
Base angles, 215, 220
Betweenness preserved under transformation, 337
Bias, 355
Biconditionals, 228, 251–253
Bimodal set of numbers, 368
Binary operations, 29
Binomials, 74
   multiplying, 78
   special products of, 79–81
Bisector, 201
   angle, 213, 351–353
   perpendicular, 320, 344–347
Boundary line, 178
Box-and-whisker plot, 371, 372
Braces { }, 14, 32
Broken-line graphs, 362–363

C

Calculator, using your, 398–405
   graphing calculator, 403–405
   scientific calculator, 398–403
Capacity, 255, 256, 257
Cartesian coordinate system, 146
   see also Coordinate plane
Census, 355
Center
   of circle, 206, 340
   of sphere, 225
Central angle, 206
Central tendency, measures of, 367–369
Certain event, probability of, 381
Checking solution, 89
Chord, 206
Circle(s), 206–207, 340–344
   area of, 266
   center of, 206, 340
   circumference of, 206, 262–264
   circumscribed about polygon, 210
   concentric, 206, 342
   congruent, 228
   construction of, 340–344
   in coordinate plane, finding, 340, 341, 342
   inscribed in polygon, 210
   radius of, 206, 340
Circle graphs, 364–367
Circular cone, right, 271–272
Circular cylinder, right, 271
Circumference, 206, 262–264
Closed sentence or statement, 240

Closure, 32–33
Coefficient, 52, 74
   fractional, 90, 93
Coin problems, 70, 109
Collinearity preserved under transformations, 337
Collinear points, 197
Column method
   to find product of sum and difference, 79
   to find square of binomial, 80
   for multiplying polynomials, 78
Combinations, 389, 391–396
   defined, 389, 391
   guidelines for determining problem involving, 393
   probability problems involving, 394–396
   on scientific calculator, 400
Common external tangent, 207
Common internal tangent, 207
Common monomial factors, 112–114
Commutative property of addition or multiplication, 29
Comparison property of numbers, 103
Compass, 315
Complement, 382
Complementary angles, 201
Composite number, 17
Compound inequality, 102
Compound locus, 340
Compound statements, 243–246
   biconditional, 228, 251–253
   conjunctions, 243–244, 245
   disjunctions, 244–246
   see also Conditionals
Concentric circles, 206, 342
Conclusion of conditional, 246
Concurrency, point of, 347
Conditional equation, 86
Conditionals, 246–248
   biconditionals, 228, 251–253
   contrapositive of, 249
   converse of, 249
   inverse of, 249
Cone, 225
   right circular, surface area of, 271–272
   volume of, 273
Congruence statement, 228
Congruent angles, 199, 201, 228
Congruent faces, 224
Congruent figures, 228–232
   constructing, 318–319
   properties of, 228
Congruent lines, 228
Congruent line segments, 197
Conjunctions, 243–244, 245
Consecutive angles, 219
Consecutive interior angles, 203
Consistent system of equations, 106, 176
Constant, 74
   quadratic equations with missing, solving, 131, 133–134
Constant of variation, 157–158, 276
Construction
   of box-and-whisker plot, 371
   of broken-line graph, 362–363
   of circle, 340–344
   of circle graph, 364
   of duplicate of congruent figures, 318–319
   of frequency histogram, 359–360
   geometric, 315
   of parallel line, 349–350
   of perpendicular bisector, 346–347
   of perpendicular line, 327–329
   of scatter plot, 363–364
   of similar figures, 331–334
Continued ratio, 56

Contradiction, 86
Contrapositives, 249–251
Converses, 249–251
Conversion factors, 256, 257
Conversion problems, 90
Convex polygons, 210
Coordinate geometry, 298–313
    distance formula, 298–302, 308, 340
    finding areas in, 305–307
    midpoint formula, 302–305, 308
    proofs, 308–313
Coordinate plane, 146
    angle bisector in, finding, 351
    circle in, finding, 340, 341, 342
    dilations in, 329–331
    half-planes, 178
    line parallel to given line in, finding, 347–348
    lines on, 150–168
    perpendicular bisector in, finding, 344–345
    quadratic equations on, 168–174
    third parallel between two lines in, finding, 349
Coordinates, 146
Coplanar lines, 197
Coplanar points, 197
Correlation, 364
Corresponding angles, 202, 228
    in similar figures, 232, 233
Corresponding sides, 228
    proportional pairs of, in similar figures, 232, 233
Corresponding vertices, 228
Cosine, 291–293
Counting method, 322
Counting numbers (natural numbers), 14
Counting principle, 386–387
Cross multiplication, 59, 135
Cross products, 59
Cube
    surface area of, 270
    volume of, 273
Cubic equation, 127
Cubic units, 255
Cumulative frequency histograms, 373–376
Curved line, 196
Customary system (English system), 256, 257
Cylinder, 225
    right circular, surface area of, 271
    volume of, 273

D

Data
    graphing, on graphing calculator, 405
    statistical, 355
Data analysis, 367–376
Data collection, 355–356
Data organization, 356–367
Day (d), 257
Decagon, 209
Deciles, 374
Decimal method of computing with percent, 62, 63
Decimal numbers
    changing to or from a fraction, 22–23
    changing to or from percent, 22
    comparing, 19–20
    operations with, 47–49
    repeating, 15, 23
    rounding, 23
    terminating, 15, 22
Degree(s), 199
    of arc, 206, 257
    of polynomial, 88
Delta (Δ), 154
Dependent axis, 188
Dependent events, 386
Dependent system of equations, 106, 176
Depreciation, 11

Depression, angle of, 294
Diagonal(s)
    of isosceles trapezoid, 220
    of polygon, 209
    of quadrilateral, 219
    of rectangle, 220
    of rhombus, 220
    of square, 220
Diagrams, problem solving by drawing, 3–5
Diameter, 206
Difference of two squares, factoring, 114–115
Dilations, 232, 329–334
    constructing similar figures, 331–334
    in coordinate plane, 329–331
Dimensional analysis, 255, 260–282
    area, 255, 264–269
    circumference, 206, 262–264
    error in measurement, 279–282
    perimeter, 260–264
    similar figures, measures of, 275–278
    surface area, 255, 269–272
    volume, 255, 273–275
Direct isometry, 338
Direct variation, 157–161, 276
Discount, 65
Disjunctions, 244–246
Distance formula, 298–302, 308, 340
Distance preserved under transformations, 337, 338
Distributions, probability, 382–385
Distributive method
    to find product of sum and difference, 79
    to find square of binomial, 80
Distributive property, 29, 77, 114
Division
    of decimals, 48–49
    of fractions, 44–46
    indicators of, 29
    of rational expressions, 83–85
    of signed numbers, 39–40
    of square roots, 54–55
    of terms with exponents, 50
Division property of inequality, 103
Domain
    of algebraic expression, 72
    graphing parabola with/without, 168–170
    of inequality, 102
    of relation, 147
Double bar graph, 361
Duplicating an image, 315–319
    constructing duplicate or congruent figures, 318–319
    translations, 315–317, 338

E

Edges of polyhedron, 224
Elevation, angle of, 294
Elimination method of solving quadratic-linear pairs, 137, 139–140
Empirical probability, 380, 382
Empty (null) set, 14
Endpoint, 197
English system (customary system), 256
Equality, sign of, 69
Equation(s)
    algebraic, 68
    conditional, 86
    cubic, 127
    graphing, on graphing calculator, 404
    literal, 85–88, 100–102
    quadratic, see Quadratic equations
    systems of linear, 106–110, 175–178
Equiangular triangle, 215
Equilateral polygon, regular, 261
Equilateral triangle, 215, 265, 288
Error in measurement, 279–282
Estimating, 23
Evaluating algebraic expressions, 72–73
Evaluating mathematical expression, 33
Events in probability, 381
    dependent and independent, 386

Experimental (empirical) probability, 380, 382
Exponential expressions, 25
Exponents, 25–27
    defined, 49
    in like terms, 74
    operations with, 49–52
    positive and negative, 26, 50
Exterior angle(s)
    alternate, 202
    of polygon, 210
    of triangle, 213
Extremes, 59, 371

F

Face (surface), 224
Factor(s), 17–19
    common monomial, 112–114
    greatest common monomial (GCF), 17, 112–114
Factorial, 389–390, 400
Factoring, 112–122
    common monomial factors, 112–114
    completely, 119–120
    defined, 112
    difference of two squares, 114–115
    factoring out -1, 121–122
    solving factorable quadratic equations, 123–137
    trinomials, 116–119
    types of, 112
Factorization, prime, 17
Factor tree, 17
Field properties, 29–31, 32, 117
    see also Closure; Operations
Figures, 197
    congruent, 228–232, 318–319
    plane, 205–223
    similar, 232–237
    solid, 224–227
Finite set, 14
FOIL method, 114
    to find product of sum and difference, 79
    to find square of binomial, 80
    for multiplying binomials, 78
Formula(s), 85–88
    for area, 265–266
    for circumference of circle, 262
    distance, 298–302, 308, 340
    midpoint, 302–305, 308
    for perimeter, 261
    for permutations, 390
    probability, 381, 382
    quadratic, 127, 402
    slope, 154, 308
Fraction(s)
    adding and subtracting, 40–43
    changing repeating decimal to, 23
    changing terminating decimal to, 22
    changing to or from percent, 22
    comparing, 20
    improper, 40
    multiplying and dividing, 43–46
    operations with, 40–46
    raising, to power, 50
    reciprocal of, 44, 83
    on scientific calculator, 401
    slope expressed as, 154
Fractional coefficient, 90, 93
Fractional quadratic equations, 134–137
Fractional radicand, rationalizing, 53
Fraction method of computing with percent, 62, 63
Frequency histograms, 359–360
    cumulative, 373–376
Frequency tables, 358–359
Function(s), 148–150
    notation, 148
    parabola as, 170
    trigonometric, on scientific calculator, 401
    vertical line test for, 148

G

Geometric construction, 315
 see also Construction
Geometry, 196–239
 angles, 198–200
 circles, 206–207
 congruent figures, 228–232
 coordinate, 298–313
 defined, 196
 line and angle relationships, 201–205
 plane figures, 205–223
 points, lines, and planes, 196–198
 polygons, 208–212
 quadrilaterals, 219–223
 similar figures, 232–237
 solid figures, 224–227
 symbols used in drawings, 209
 triangles, 213–218
Gram (g), 256
Graph(s)
 bar, 360–362
 broken-line, 362–363
 circle, 364–367
 reading, 189
Graphing
 graphic solutions of real situations, 188–192
 linear equations, 150–152
 linear inequalities, 178–181
 parabola, 168–172
 points, 147
 quadratic-linear pairs, 184–187
 systems of linear equations, 175–178
 systems of linear inequalities, 181–184
Graphing calculator, using, 403–405
Greater than symbol, 19
Greatest common monomial factor (GCF), 17, 112–114
Grouping symbols, 33
Guessing and checking, 8–9
 wild guesses, caution about, 2

H

Half-planes, 178
Height
 of cone, 225
 of parallelogram, 264
 of pyramid, 224
 of quadrilateral, 219
 slant, 224, 225, 271, 273
 of trapezoid, 264
 of triangle, 213, 264
Hemisphere, 275
Hexagon, 209, 210
Histograms, frequency, 359–360, 373–376
HL congruent right triangle, 229
Horizontal lines, 152–153, 155
Hour (hr), 257
Hypotenuse, 215
Hypotenuse-leg (HL) congruent right triangle, 229
Hypothesis of conditional, 246

I

Identity, 86
 additive, 30
 multiplicative, 30
Image under tranformation, 315
Impossible event, probability of, 381
Improper fractions, 40
Incomplete quadratic equations, 130–134
Inconsistent system of equations, 106, 176
Independent axis, 188
Independent events, 386
Inequality(ies), 102
 algebraic, 68
 compound, 102
 linear, 102–106, 178–181
 properties of, 103
 signs of, 69, 102

Infinite set, 14
Initial conditions, 189
Initial error in subsequent calculations, 280
Inscribed angle, 206
Integers, 14
Intercepted arc, 206
Intercepts, 151
 slope-intercept form of line, 161–162
Interior angle(s)
 alternate, 202
 consecutive, 203
 of polygon, 210
 of triangle, 213
Interquartile range, 370
Intersecting lines, 198, 351–353
Intervals, 358–359
Inverse(s), 249–251
 additive (opposite), 30, 35
 multiplicative (reciprocal), 30
 on scientific calculator, 400
Irrational numbers, 15, 23
Isometry, 338
Isosceles trapezoid, 220, 309
Isosceles triangle, 215
 right triangle, 287–288

L

Lateral area, 269
 of regular right pyramid, 271
 of right circular cone, 272
 of right circular cylinder, 271
 of right prism, 270
Lateral faces, 224
Lateral surface of cylinder, 225
Least common multiple (LCM), 17–18, 93
Legs
 of trapezoid, 220
 of triangle, 215
Length, 255
 arc, 206
 in customary system, 257
 in metric system, 256
 of rectangle, 261, 264
Less than symbol, 19
Like terms, 74
 linear equations with, 94–96
Line(s), 196–198
 boundary, 178
 congruent, 228
 coplanar, 197
 curved, 196
 intersecting, 198, 351–353
 line and angle relationships, 201–205
 parallel, 166, 202, 203, 308, 347–350
 perpendicular, 166, 202, 308, 327–329, 351
 skew, 197
 straight, 196
Linear equations
 graphing, 150–152
 point-slope form of, 163–165
 slope-intercept form of, 161–162
 standard form of, 150
 systems of, 106–110, 175–178
Linear equations, solving, 88–102
 of form $ax = b$, 90–92
 of form $ax + b = c$, 92–94
 of form $x + a = b$, 89–90
 with like terms, 94–96
 literal equations, 100–102
 with parentheses and rational expressions, 96–100
Linear inequalities
 graphing, 178–181
 in one variable, 102–106
 systems of, 181–184
Linear units, 255
Line graph, 362–363
Line of best fit, 405
Line reflection, 320–322, 338
Line segment, 197
Line symmetry, 322–324

Lines on coordinate plane, 150–168
 direct variation, 157–161
 graphing linear equations, 150–152
 horizontal and vertical, 152–153, 155, 156
 from partial information, 166–168
 point-slope form of, 163–165
 slope-intercept form of, 161–162
 slope of, 154–157
Liquid measure, 255, 256, 257
Lists, using, 5–6
Literal equations and formulas, 85–88, 100–102
Liter (L), 256
Locus, 315, 340–354
 compound, 340
 distance from line (two parallel lines), 347–348
 distance from point (circle), 340–344
 equidistant from two points (perpendicular bisector), 344–347
 points equidistant from intersecting lines (angle bisectors), 351–353
 points equidistant from two parallel lines, 348–350
Logically equivalent statements, 249

M

Magic square, 9
Mapping, 147, 315
 dilation mapping rule, 329
 line reflection mapping rules, 320
 point reflection mapping rule, 325
 rotation mapping rules, 335
 translation mapping rule, 316
Margin of error, 279
Markup, 65
Mass, measure of, 256
Mathematical sentence, 240–242
Maximum point, 170
Mean proportional, 283
Means, 59, 367–369
Measurement, 255–260
 conversions in, 256–260
 error in, 279–282
 of similar figures, 275–278
 systems of, 255–260
 triangles and, 282–297
 see also Dimensional analysis
Measures of central tendency, 367–369
 on a graphing calculator, 405
Median, 368–369
 of triangle, 213
Memory on scientific calculator, 401
Meter (m), 256
Metric conversion factors, 256
Metric system, 256
Midpoint formula, 302–305, 308
Midpoints, 213, 234
Minimum point, 170
Minute (min), 257
Minutes ('), 257
Mixed number
 changing to or from percent, 22
 operations with, 40
Mode, 368–369
Monomial(s), 74
 dividing monomial by another, 81
 dividing polynomial by, 82–83
 finding product of, 76
 finding product of polynomial and, 76–77
 with like terms, adding or subtracting, 74
Monomial factors, common, 112–114
Multiples, 17–19
 least common (LCM), 17–18, 93
Multiplication
 of algebraic expressions, 76–81
 associative property of, 29
 commutative property of, 29
 cross, 59, 135
 of decimals, 47
 of fractions, 43–46

indicators of, 29
raising product to power, 50
of rational expressions, 83–85
of signed numbers, 38–39
of square roots, 54
of terms with exponents, 50
Multiplication property of inequality, 103
Multiplicative identity, 30
Multiplicative inverse, 30
  see also Reciprocal
Mutually exclusive events, 382

N

Negations, 242–243
Negative correlation, 364
Negative reciprocals, 166
n-gon, 209
Notation
  function, 148
  radical, 52–56
  scientific, 25–27, 400
Null set, 14
Number line, 14
  signed numbers on, 35
  solution set of inequality on, 102
Number problems, 140–142
Numbers
  composite, 17
  counting (natural numbers), 14
  irrational, 15, 23
  prime, 17
  rational, 14
  real, see Real numbers
  signed, operations with, 35–40
  whole, 14
Numeration, 14–28

O

Obtuse angle, 199
Obtuse triangle, 215
Octagon, 209, 210
Odometer, 49
Open sentence, 240
Operations, 29–67
  binary, 29
  closure under, 32–33
  with decimals, 47–49
  with exponents, 49–52
  with fractions, 40–46
  order of, 33–34, 398
  percent, applications of, 62–66
  properties (field properties) of, 29–31, 32, 117
  proportions, 59–61
  radical notation and, 52–56
  with rational expressions, 83–85
  ratios and rates, 56–58, 255
  on scientific calculator, 398
  with signed numbers, 35–40
  signs of, 68
  see also Addition; Division; Multiplication; Subtraction
Opposite (additive inverse), 30, 35
Opposite angles, 219
Opposite isometry, 338
Opposite sides, 219
Opposite vertices, 219
Ordered pairs, 146
  sets of, 147, 388–389
Ordered triples, sets of, 388
Order of operations, 33–34
  on scientific calculator, 398
Order of real numbers, 19–21
Ordinate (y-coordinate), 146
Orientation preserved under transformations, 337, 338
Origin, 146
Outcome, in probability, 381
Outliers, 371

P

Parabola
  axis of symmetry, 169
  as graphic solution of real situation, 189
  graphing, 168–174
  intersecting x-axis at exactly one point, 173
  intersecting x-axis at two distinct points, 172–173
  not intersecting x-axis, 174
  turning point of, 169–170, 173
Parallelism preserved under transformations, 337
Parallel lines, 166, 202, 308
  construction of, 349–350
  cut by transversal, 203
  line parallel to given line in coordinate plane, finding, 347–348
  locus involving, 348–350
Parallelogram, 219
  area of, 265
  height of, 264
  proof of, 308
Parentheses
  on scientific calculator, 399
  solving linear equations with, 96–100
Pentagon, 208, 210
Percent(s), 22, 62
  applications of, 62–66
  computing with, 62–64
  converting to or from, 22
  of decrease, 64–66
  of increase, 64–66
  on scientific calculator, 401
Percentage error, 279–280
Percentiles, 373–376
Perfect square, 52
Perfect-square trinomial, 124
  factoring, 116
Perimeter, 260–264
Permutations, 389–391, 394–396
  formulas for, 390
  guidelines for determining problem involving, 393
  probability problems involving, 394–396
  on scientific calculator, 400
  of selected members of set, 389–390
  of set taken as whole, 389
  special case of, 390
Perpendicular bisector, 320, 344–347
Perpendicular lines, 166, 202, 308, 327–329, 351
Pi (π), 262, 400
Plane figures, 205–223
  circles, 206–207, 340–344
  polygons, 208–212
  quadrilaterals, 208, 219–223
  triangles, 213–218
Planes, 196–198
Point(s), 196–198
  collinear, 197
  of concurrency, 347
  in coordinate plane, finding distance between two, 298–302
  coplanar, 197
  graph of, 147
  maximum and minimum, 170
  of rotation, 334
  of symmetry, 325
Point reflection, 325–327
Point-slope form, 163–165
Point symmetry, 325–327, 334
Polygon(s), 208–212
  angle properties of, 210
  base of, 264
  circumscribed around circle, 210
  congruent, 228
  convex, 210
  in coordinate plane, finding area of, 305–307
  diagonal of, 209
  equilateral, 261
  general properties of, 208–212
  inscribed in circle, 210
  interior and exterior angles of, 210
  perimeter of, 260–261
  quadrilaterals, 219–223
  regular, 209, 261, 264
  sides of, 208
  types of, 208–209
  vertices of, 208
Polyhedron, 224, 270, 276
Polynomial(s), 74
  adding, 74–76
  degree of, 88
  dividing, by monomial, 82–83
  multiplying, 76–77
  prime, 112
  second-degree, in one variable, 123
  simplifying, 74
  special products of, 112
  standard form of, 74
  subtracting, 75–76
  see also Factoring
Positive correlation, 364
Power(s), 25, 29
  defined, 49
  raising number to, 50
  raising number to, on scientific calculator, 399
Power expressions, 25
Preimage points, 315
Premise of conditional, 246
Prime factorization, 17
Prime number, 17
Prime polynomial, 112
Principal square root, 52
Prism, 224
  surface area of, 270
  volume of, 273
Probability, 380–397
  combinations, 389, 391–396
  counting principle, 386–387
  defined, 380
  experimental (empirical), 380, 382
  formulas for, 381, 382
  permutations, 389–391, 394–396
  principles of, 380–385
  random variables and probability distributions, 382–385
  rules of, 380–382
  sample spaces, calculating, 385–396
  sets of ordered pairs, 388–389
  theoretical, 380
  tree diagrams, 386
Probability distributions, 382–385
Problem solving applications
  age problems, 69
  coin problems, 70, 109
  compound ratio problems, 95
  consecutive integer problems, 99, 140–142
  distance-rate-time problems, 87, 97
  graphing real situations, 188–192
  indirect measure problems, 293–297
  interest problems, 86
  money problems, 12, 95, 105, 159
  number problems, 140–142
  percent problems, 62–66
  sequence problems, 9–10
Problem solving strategies, 3–13
  by drawing diagram, 3–5
  by finding pattern, 9–11
  by guessing and checking, 8–9
  by solving simpler problem, 11–13
  steps in, 3
  strategies for, 3–13
  by using tables and lists, 5–6
  by working backward, 6–7
Products
  cross, 59
  of slopes, 308

see also Multiplication
Proportional pairs of corresponding sides, 232, 233
Proportion problems, 90
Proportions, 59–61
    direct variation problems solved by, 158
    mean proportional, 283
    of right triangle, 282–285
    true proportion, 283
Pyramid, 224
    regular (right), 224, 271
Pythagorean theorem, 215, 216, 285–287, 298
Pythagorean triple, 216

Q

Quadrants, 146
Quadratic equations, 123–137
    on coordinate plane, 168–174
    fractional, 134–137
    incomplete, 130–134
    quadratic trinomial equations, 124–130
    roots of, 172–174
    solving graphically, 172–174
    standard form of, 123
    verbal problems involving, 140–145
Quadratic formula, 127, 402
Quadratic-linear pairs
    graphing solution set to, 184–187
    solving algebraically, 137–140
Quadratic trinomial, 117, 123
Quadrilaterals, 208, 219–223
    see also Parallelogram; Rectangle; Rhombus; Square (figure); Trapezoid
Quartiles, 370, 371–372, 374

R

Radical notation, operations with, 52–56
Radical sign, 52
Radicand, 52
    fractional, 53
Radius, 206, 340
Random sampling, 355
Random variables, 382–385
Range, 370, 371–372
    interquartile, 370
    of relation, 147
Rates, 56–58, 255
Ratio(s), 56–58
    continued, 56
    percent of increase/decrease as, 64–66
    proportional, 59
    of similitude, *see* Scale factor
    trigonometric, 290–297
Rational expressions, 83
    operations with, 83–85
    solving linear equations with, 96–100
Rationalizing fractional radicand, 53
Rational numbers, 14
Ray, 197
Real numbers
    operations with, *see* Operations
    order of, 19–21
    sets of, 14–16
Real situations, graphic solutions of, 188–192
Real-world problems, quadratics applied to, 142–145
Reasoning, mathematical, 240–254
Reciprocal, 30
    of fraction, 44, 83
    negative, 166
Rectangle, 220
    area of, 265
    length and width of, 261, 264
    perimeter of, 261
    proof of, 308
Rectangular prism (rectangular solid), 224, 270
    volume of, 273
Reflections, 320–329
    line, 320–322, 338
    point, 325–327
Reflex angle, 199
Regular polygon, 209, 261, 264
Regular polyhedron, 276
Regular prism, 224
Regular (right) pyramid, 224
    surface area of, 271
    volume of, 273
Relations, 147–150
    see also Function(s)
Repeating decimals, 15, 23
Replacement set, 72
    see also Domain
Rhombus, 220
    area of, 265
    proof of, 308
Right angle, 199
Right circular cone, 225
    surface area of, 271–272
    volume of, 273
Right circular cylinder, 225
    surface area of, 271
    volume of, 273
Right prism, 224
    surface area of, 270
    volume of, 273
Right pyramid, *see* Regular (right) pyramid
Right trapezoid, 220, 309
Right triangle, 215, 216, 287–290
    angles in, finding, 295–297
    congruent, 229
    45°–45°–90° (isosceles), 287–288
    proportions of, 282–285
    Pythagorean theorem and, 215, 216, 274–275, 285–287, 298
    similar, 233
    30°–60°–90°, 288–289
    trigonometric ratio, 290–293
Root(s), 29, 52
    of quadratic equation, 172–174
    square, 52–56, 399
Rotation(s), 334–337, 338
    point of, 334
Rotational symmetry, 334
Rounding decimals, 23

S

Sample spaces, calculating, 381, 385–396
Sampling, 355–356
SAS congruent triangles, 228
SAS similar triangles, 234
Scale factor, 232, 233, 275, 276
Scalene triangle, 215
Scatter plots, 363–364
Scientific calculator, using, 398–403
Scientific notation, 25–27, 400
Second-degree polynomial in one variable, 123
Seconds ("), 257
Second (sec), 257
Semicircle, 206
Sentences, open and closed, 240–242
Set(s), 14
    braces to indicate, 32
    closed, 32
    empty (null), 14
    finite, 14
    infinite, 14
    of ordered pairs, 147, 388–389
    of ordered triples, 388
    permutations of selected members of, 389–390
    permutations of set taken as whole, 389
    of real numbers, 14–16
    replacement (domain), 72
    solution, 85
Side-angle-side (SAS) congruent triangles, 228
Side-angle-side (SAS) similar triangles, 234
Sides
    of angle, 198
    corresponding, 228, 232, 233
    opposite, of quadrilateral, 219
    of polygon, 208
    of regular polygon, 264
    of triangles, 213, 214, 215
Side-side-side (SSS) congruent triangles, 229
Side-side-side (SSS) similar triangles, 233
Sign(s)
    of equality, 69
    of inequality, 69, 102
    of operations, 68
Sign change on scientific calculator, 399
Signed numbers, operations with, 35–40
    adding and subtracting, 36–38
    multiplying and dividing, 38–40
Similar figures, 232–237
    construction of, 331–334
    measures of, 275–278
    properties of, 232
Simplifying mathematical expressions, 33
Simplifying polynomials, 74
Simplifying square roots, 52–53
Simultaneous equations, 106–110
    system of, 175
Sine, 291–293
Skew lines, 197
Slant height, 271, 273
    of cone, 225
    of pyramid, 224, 271
Slope(s), 154–157
    of parallel line, 166
    of perpendicular lines, 166
    point-slope form, 163–165
    product of, 308
    undefined, 156
Slope formula, 308
Slope-intercept form, 161–162
Solid figures, 224–227
Solution
    checking, 89
    to system of linear equations, 106
Solution set, 85
    of inequality, 102
Special products of polynomials, 112
Sphere, 225
    surface area of, 272
    volume of, 274
Square(s)
    of binomial, 80
    factoring difference of two, 114–115
    factoring perfect trinomial, 116
    perfect, 52
    squaring number on scientific calculator, 399
Square (figure), 220
    area of, 265
    magic, 9
    proof of, 308
Square root, 52
    operations with, 54–56
    principal, 52
    on scientific calculator, 399
    simplifying, 52–53
Square units, 255
SSS congruent triangles, 229
SSS similar triangles, 233
Standard form
    of linear equation, 160
    of polynomial, 74
    of quadratic equation, 123
Statement, 240–242
    compound, 243–246
    logically equivalent, 249
    of similarity, 233
Statistics, 355–379
    data analysis, 367–376
    organizing data, 356–367
    sampling and collecting data, 355–356

Index   **465**

Stem-and-leaf plots, 356–357
  back-to-back, 357
Straight angle, 199
Straightedge, 315
Straight line, 196
Substitution method
  of solving quadratic-linear pairs, 137, 138–139
  of solving systems of equations, 108–110
Subtraction, 29
  addition-subtraction method of solving systems of equations, 107–108
  of algebraic expressions, 74–76
  of decimals, 47
  of fractions, 40–43
  of rational expressions, 83–85
  of signed numbers, 37–38
  of square roots, 54
  of terms with exponents, 50
Subtraction property of inequality, 103
Supplementary angles, 201
Surface area, 255, 269–272
  of cube, 270
  of rectangular prism, 270
  of regular right pyramid, 271
  of right circular cone, 271–272
  of right circular cylinder, 271
  of sphere, 272
Surface (face), 224
Symbols
  in geometric drawings, 209
  greater and less than, 19
  grouping, 33
  for similarity and negation, 242
Symmetry, 320–329
  axis of, 169, 322
  line, 322–324
  point, 325–327, 334
  point of, 325
  rotational, 334
System of simultaneous linear equations, 175
Systems of linear equations, solving, 106–110
  with addition-subtraction method, 107–108
  graphing and, 175–178
  with substitution method, 108–110
Systems of linear inequalities, graphing solution set of, 181–184

T

Table(s)
  frequency, 358–359
  using, 5–6
  of values, 147
Tally, 358–359
Tangent, 207, 291–293
Terminating decimals, 15, 22
Terms, 9, 74
  like, 74, 94–96
  in ratio, 56

Test-taking strategies, 1–2
Tetrahedron, 224
Theoretical probability, 380
Time, units of, 257
Transformations, 315–339
  dilations, 232, 329–334
  duplicating an image, 315–319
  properties under, 337–339
  reflections and symmetry, 320–329
  rotations, 334–337, 338
Translations, 315–317, 338
Transversal, 202, 203
Trapezoid, 220
  area of, 266
  bases of, 220, 264
  height of, 264
  isosceles, 220, 309
  proof of, 309
  right, 309
Tree diagrams, 386
Triangle(s), 213–218
  angle properties of, 214
  area of, 265
  base of, 215, 264
  classification of, by angles and sides, 215
  congruent, proving, 228–232
  equilateral, 215, 265, 288
  height of, 213, 264
  interior and exterior angles of, 213
  legs of, 215
  parts of, 213
  as polygon, 208
  right, see Right triangle
  side properties of, 214
  similar, proving, 233–235
Triangles, measurement of, 282–297
  indirect measure of, 293–297
  Pythagorean theorem and, 215, 216, 274–275, 285–287, 298
  right triangle, proportions of, 282–285
  special right triangles, 287–290
  trigonometric ratios, 290–297
Trigonometric functions on scientific calculator, 401
Trigonometric ratios, 290–293
  solving for parts of right triangle using, 293–297
Trinomial equations, quadratic, 124–130
Trinomials, 74
  factoring, 116–119
  perfect-square, 116, 124
  quadratic, 123
Truth value, 240–242
  for biconditional, 251
  for conditionals, 247
  for conjunction, 244, 245
  for contrapositive, 249
  for converse, 249
  for disjunctions, 245
  for inverse, 249
  for negation, 242
Turning point of parabola, 169–170, 173

U

Unit conversion, 256–260
Unit price, 56
Unit rate, 56

V

Variable(s), 14
  in like terms, 74
  from literal equation or formula, finding value of, 86
  random, 382–385
Variation
  constant of, 157–158, 276
  direct, 157–161, 276
Venn diagram, 4, 381
Vertex, vertices
  of angle, 198
  corresponding, 228
  opposite, in quadrilateral, 219
  of polygon, 208
  of polyhedron, 224
  of pyramid, 224
  of triangle, 213
Vertex angle of isosceles triangle, 215
Vertical angles, 201
Vertical lines, 152–153, 156
Vertical line test for functions, 148
Volume, 255, 273–275
  of cube, 273
  in customary system, 257
  of right prism or right cylinder, 273
  of right pyramid or cone, 273
  of sphere, 274

W

Weight, in customary system, 257
Whole numbers, 14
Width of rectangle, 261, 264
WINDOW of a graphing calculator, 404, 405
Working backward, 6–7

X

x-axis, 146, 188
x-coordinate (abscissa), 146
x-intercept of vertical line, 153

Y

y-axis, 146, 188
y-coordinate (ordinate), 146
Year (yr), 257
y-intercept of horizontal line, 152

Z

Zero, 14
  absolute value of, 35
  as exponent, 50
  lack of reciprocal, 30
  opposite of, 35
Zero product property, 30